International Conference on Advanced Materials (ICAM 2015)

IOP Conference Series: Materials Science and Engineering
Volume 92

Irbid, Jordan
27 – 29 April 2015

Editors:

Mohammed Y. El-Khateeb   Nathir A.F. Al-Rawashdeh
Borhan A. Albiss

ISBN: 978-1-5108-1463-9
ISSN: 1757-8981

Printed from e-media with permission by:

Curran Associates, Inc.
57 Morehouse Lane
Red Hook, NY 12571

proceedings.com

Some format issues inherent in the e-media version may also appear in this print version.

Printed by Curran Associates, Inc. (2016)

For permission requests, please contact the Institute of Physics
at the address below.

Institute of Physics
Dirac House, Temple Back
Bristol BS1 6BE UK

Phone: 44 1 17 929 7481
Fax: 44 1 17 920 0979

techtracking@iop.org

Additional copies of this publication are available from:

Curran Associates, Inc.
57 Morehouse Lane
Red Hook, NY 12571 USA
Phone: 845-758-0400
Fax: 845-758-2634
Email: curran@proceedings.com
Web: www.proceedings.com

Table of contents

Volume 92

International Conference on Advanced Materials (ICAM 2015)
27–29 April 2015, Irbid, Jordan

Accepted papers received: 13 August 2015
Published online: 12 October 2015

IOP Conf. Series: Materials Science and Engineering **92** (2015) 011001 doi:10.1088/1757-899X/92/1/011001

PREFACE/ ICAM2015

It is with great pleasure to welcome you to the "International Conference of Advanced Materials ICAM 2015" that will take place at Jordan University of Science and Technology (JUST), Irbid, Jordan. This year, the conference coincides with the coming of spring in Jordan; we hope the participants will enjoy the colors and fragrance of April in Jordan.

The call for papers attracted submissions of over a hundred abstracts from twenty one different countries. These papers are going to be classified under four plenary lectures, fifteen invited papers, thirty five oral presentations and more than sixty posters covering the different research areas of the conference.

The ICAM conference focuses on new advances in research in the field of materials covering chemical, physical and biological aspects. ICAM includes representatives from academia, industry, governmental and private sectors. The plenary and invited speakers will present, discuss, promote and disseminate research in all fields of advanced materials. Topics range from synthesis, applications, and solid state to nano-materials. In addition, talented junior investigators will present their best ongoing research at a poster session.

We have also organized several workshops contiguous to the main conference, such as the one-day workshop on "Particle Surface Modification for Improved Applications". The purpose of this short course was to introduce interested materials technologists to several methodologies that have been developed to modify the surfaces of particulate matter. Moreover, a pre-conference workshop on "Communication in Science" was conducted for young scientists. The main goal of this workshop was to train young scientists in matters of interdisciplinary scientific communications.

In addition to the scientific program, the attendees will have a chance to discover the beauty of Jordan, a land of rich history and varied culture. Numerous social events that will provide opportunities to renew old contacts and establish new ones were scheduled to accompany and follow the scientific activities. They also aim to acquaint the participants with the capital, Amman, and other famous historic and touristic sites in Jordan, such as Petra, one of the new seven wonders of the world, Jerash, desert palaces, Jedara, Wadi Rum and the Dead Sea. The social program will be extraordinary for the participants and their companions, reflecting the Jordanian sense of hospitality and generosity combined with fun.

A special issue of IOP Conference Series: Material Science and Engineering is dedicated to publishing selected papers presented at the conference. Papers submitted for publication will be peer reviewed. The journal offers open access making the proceedings available to a wider scope of readership.

I would like to express my gratitude to all participants for their contributions to the conference program. I am also thankful to the members of the Organizing Committee for their endless efforts and diligence. Special thanks go to the conference Scientific Committee members whose inputs were important and who were always supportive to the organizers.

It is the Committee's honor to acknowledge the financial sponsorship of the conference from the following organizations: COMESTIC, Stevin Rock JJC, Al-Safadi Housing Company, Al-Haiek Housing Company, Cairo Amman Bank, Ramtha Chamber of Commerce, JUST Library, Jordan Engineers Association, and El-Far Importing. Sincere thanks and appreciation go to the keynote speakers, session organizers, session chairs, and the JUST President, staff and students for their support and efforts.

Finally, we wish to thank you for your participation in the conference and contributing to its success by sharing your scholarly achievements with other colleagues and scholars.

Best wishes for an enjoyable and memorable experience.

Prof. Mohammad Y. El-Khateeb
Conference Chair

IOP Conf. Series: Materials Science and Engineering **92** (2015) 011001 doi:10.1088/1757-899X/92/1/011001

ICAM2015 Participants

IOP Conf. Series: Materials Science and Engineering **92** (2015) 011001 doi:10.1088/1757-899X/92/1/011001

ICAM2015
International Conference on Advanced Materials
Irbid, Jordan 27-29 April 2015

GUEST EDITORS:

1- Prof. Mohammad Y. El-Khateeb
Jordan University of Science and Technology, JORDAN.
Email: kateeb@just.edu.jo

2- Prof. Borhan A. Albiss
Jordan University of Science and Technology, JORDAN.
Email: baalbiss@just.edu.jo

3- Prof. Nathir A. F. Al- Rawashdeh
Jordan University of Science and Technology, JORDAN.
Email : nathir@just.edu.jo

ORGANIZING COMMITTEE:

SCIENTIFIC COMMITTEE:

Prof. Brian S. Sorg, National Cancer Institute, USA

Prof. Mohammad Y. El-Khateeb, JUST, Jordan

Prof. Syed Ismat Shah, University of Delaware, USA

Prof. Isabel C. Escobar, The University of Toledo, USA

Prof. Juergen Popp, Leibniz Institute of Photonic Technology, Germany

Prof. Ahmed Alomary, Yarmouk University, Jordan

Prof. Marwan Mousa, Mutah University and Chairman of JPS, Jordan

Prof. C. N. R. Rao, Jawaharlal Nehru Centre for Advanced Scientific Research, India

Prof. Nabil Y. Ayoub, German-Jordan University, Jordan

Prof. M. Samy El-Shall, Virginia Commonwealth University, USA

Prof. Ihab M. Obaidat, University of UAE

Prof . Martin Kordesch, Ohio University, USA

Prof. Mark Hernandez, University of Colorado, USA

Prof. Susannah Scott, University of California, USA

Prof. Eqab M. Rabei, AL- al Bait University, Jordan

Prof. Yaseen Al-Soud, AL- al Bait University, Jordan

Prof. Khaled H. Abu-Elteen, Hashemite University, Jordan

Prof. Abeer Al-Bawab, University of Jordan, Jordan

Prof. Sami Mahmoud, University of Jordan, Amman, Jordan

Dr. Vinka Oyanedel-Craver, University of Rhode Island, USA

Dr. Abdelraouf Al-Dairy, Yarmouk University, Jordan

Dr. Yahya Al Khatatbeh, Princess Sumaya University for Technology, Jordan

Dr. Mahmoud Al-Hussein, Jordan University, Jordan

Dr. Tequila A. Harris, Georgia Tech, USA

Dr. Renat Sabirianov, University of Nebraska-Omaha, USA

Dr. Pance Naumov, NYU Abu Dhabi, UAE

Dr.Edreese Alsharaeh, Al Fisal University, Saudi Arabia

Dr. Jebreel Khoshman, Al-Hussein Bin Talal University, Jordan

IOP Conf. Series: Materials Science and Engineering **92** (2015) 011002 doi:10.1088/1757-899X/92/1/011002

Peer review statement

All papers published in this volume of *IOP Conference Series: Materials Science and Engineering* have been peer reviewed through processes administered by the proceedings Editors. Reviews were conducted by expert referees to the professional and scientific standards expected of a proceedings journal published by IOP Publishing.

IOP Conf. Series: Materials Science and Engineering **92** (2015) 012001 doi:10.1088/1757-899X/92/1/012001

Nanotechnology – A path forward for developing nations

S Ismat Shah[1,2] and Thomas M. Powers[3, 4]

[1]Physics and Astronomy, University of Delaware, Newark, DE 19716, USA.
[2]Materials Science and Engineering, University of Delaware, Newark, DE 19716, USA.
[3]Philosophy, University of Delaware, Newark, DE 19716, US
[4] Public Policy and Administration, University of Delaware, Newark, DE 19716, USA.

E-mail: Ismat@udel.edu

Abstract. One of the major issues with technology in general, and nanotechnology in particular, is that it could exacerbate the divide between developed and developing nations. If the benefits of the research do not flow beyond the national and geographical borders of the traditional major bastions of R&D, these benefits will not be equally and globally available. The consequence is that the technological divide becomes wider at the expense of mutual reliance. As much as developed nations need to rethink the strategy and the policy to bring nanotechnology products to market with the goal of global prosperity, developing nations cannot afford to simply wait for the lead from the developed nations. In the spirit of collaboration and collegiality, we describe issues with the current practices in nanotechnology R&D in the developing world and suggest a path for nanotechnology research in energy, water and the environment that developing nations could follow in order to become contributors rather than simply consumers.

Several decades have passed since nanotechnology was introduced to the world of scientific inquiry. What is now being called a new industrial revolution has changed the paradigm in scientific research and industrial production. Perhaps the transition from science to production was easy for nanotechnology due to the fact that examples of it existed for a long time before being designated by that term. No matter what the reason, we now see over 1,600 nanotechnology-based products in the market [1]. Predictions of a 1 trillion $USD market by 2015 for all nanomaterials-dependent products appear to be on track [2].

This revolution, nevertheless, is yet to make its impact on the majority of the population of the world that lives in the developing nations. The scientific organizations and the governments in these countries, unable to put in the mammoth amount of investment required for industrial-scale nanotechnology research, development and production, are still standing by to benefit from the trickle-down effects of this revolution. There are small efforts, with minimal investment, to get the research started at the least, and there are some success stories. However, there is a reasonable fear that, just as with the IT bubble, most of these developing nations will be unable to get a piece of the pie and the nanotechnology revolution will also pass them by. These nations will eventually become consumers rather than producers of the nanotechnology products and processes, making them even more dependent on technologies developed by the industrialized nations.

The good news is that as much as the nanotechnology revolution seems to be advanced in some regions of the world, there is still "plenty of room at the bottom" for everyone to participate—to engage in collaborative research, aid with user-centric design of products, and make other significant contributions. If there is a serious effort invested in becoming a part of this development, no matter how small their economies are, these countries can still become participants rather than mere consumers of the nanotechnology. This paper describes a path which could allow a healthy participation of these countries as contributors to the developments in nanotechnology. The path prescribed does not merely sketch one of many possible ways to achieve technological relevance, it also takes into account the sociological and ethical challenges that nanotechnology brings with it to help select the best path. As such, this methodology might be more sustainable than what is currently being followed by the industrialized nations with the old "producer-consumer" model. Inclusion of developing nations in the growth of nanotechnology will be necessary to complete a genuine revolution, since a large portion of humanity resides in these countries. In the top 11 most populous countries, 7 countries have GDP per capita of $5000 or less with the only exception being China at $9800. Together, these 7 countries have a combined population of close to 3.5 billion, almost half of the world's population.

Table 1: Richard Smalley's list of global concerns and the modified list for developing nations.

Richard Smalley's List	List for the Developing Nations
Energy	Poverty
Water	Education
Food	Water
Environment	Health (Disease)
Poverty	Energy
Terrorism and War	Food
Disease	Political Stability
Education	Environment
Democracy	Shelter
Population	Population

Table 2: List of UN Millennium Development Goals

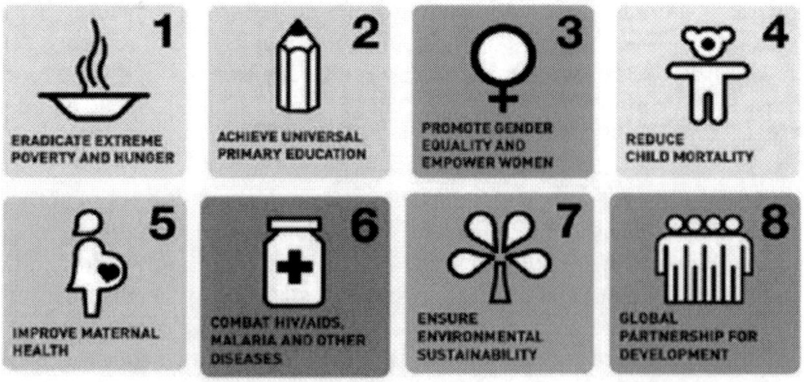

The big question for developing-world participation is that of the prioritization of the several areas in which work is still to be done. Professor Smalley's list of top ten global concerns is a good guideline but it may not completely fit the needs and requirements of developing nations [3]. A slight realignment of the priorities is warranted. Table 1 shows Professor Smalley's list and the modified list for the developing nations. The modified list aligns well with the UN millennium development goals displayed in table 2. Since this article is focused on the research and technological development in nanotechnology, we will skip past the poverty and education concerns and selectively concentrate on Water, Energy and the Environment. It is to be noted that many of these issues need to be further aligned at the regional level. A region may have enough potable water but few energy resources and vice-versa, so the area-relevant identification of issues and appropriate solutions should be sought.

Water: Perhaps the most important area of interest is the availability of potable water. The earth's hydrological cycle works with a fixed amount of water that is constantly re-circulated. With the rapid increase in world population comes the consequential decrease in the amount of water per capita. Additionally, there are industrial and agricultural practices that emanate from the ever-increasing demand for materials and food that further reduce the availability of and access to water due to inevitable pollution.

Nanotechnology presents several pathways to address the availability, purification, and treatment of water. Most of these methodologies and materials are relatively cheap to develop and implement and range from materials and devices that adsorb water from the atmosphere, to reverse osmosis, ceramic based nanofilters, and visible light photocatalysis. The research mostly involves materials and processes that are relatively inexpensive and could easily be undertaken by research communities in the developing nations and be applied in those same nations with close attention to resources, needs, and existing infrastructure.

Energy: One of the main reasons that a country is counted as a developing nation is that it does not have many internal resources, particularly energy resources. Since all major economic sectors—especially industry, transportation, and agriculture—rely heavily on energy, the energy needs of developing nations often dominate their foreign exchange expenditure, using up capital that could be more fruitful for education and research. There are several possibilities related to nanotechnology-based research and development in energy, but most are dependent on solar energy, including photovoltaics, biomass, and wind. Although there are high returns and many investment options for R&D in this field, the cheaper alternatives are also plentifully available. Polymer based solar cells, for example, are a much cheaper alternative to the currently dominating Si based solar cells but suffer from a lack of durability. This area of research is completely open. Developing nations especially near the equator could take advantage of new solar energy technologies.

Environment: Due to their general lack of environmental regulations, most developing nations suffer from water and air pollution that exacerbate public health problems. There are numerous opportunities for R&D in the area of environmental remediation, and nanotechnology could provide cheap and scalable alternatives. An example is photocatalysis in which solar energy is used by nanoparticles of titania to mineralize some of the environmental pollutant.

In all these discussions, we have only considered technological solutions to alleviate some of the concerns relevant to people living in developing countries. This approach has historically been proven inadequate Here are some of the reasons why.

1. *Complete dependency on technological solutions to the problems*: It has historically been the case that we have sought answers to the energy, water and environmental problems from the stewards of science and technology. In such endeavors only natural scientists and engineers are involved, which limits our capacity to solve the problem. Thinkers such as E.F. Schumacher have drawn attention to the folly of seeking all solutions from technology, when it has also been the source of the problems one is trying to solve [4]. Since the magnitude of many problems has

become immense, it will be shortsighted if we continue to go to the same well of science and technology that provided us with stop-gap solutions in the past. The magnitude of problems calls for a multifaceted cross-disciplinary approach that includes humanities and social sciences in addition to the natural science. Some of the problems have roots in social structures that the natural scientists are incapable of factoring into their formulae. This necessitates involvement of other scholarly fields, practitioners, and policy experts.

2. *Inappropriate fixes that do not take into account developmental realities*: Purely scientific solutions do not consider geographical and socioeconomic constraints. A good example is solar cells in off-grid applications—often, villages and far-flung communities, typically poor and of low economic standing. Most of these places have limited energy availability and little access to water as well. Solar cell panels in these communities, say in Sub-Saharan and North Africa, and in Central Asia, have to be kept clean for efficient power production. A 10 micron dust coating on the solar cells reduces the efficiency of the cells by 1% [5]. These are the places where rain is scarce. Therefore, a simple solar cell might not be a viable solution for these areas. Also, purely scientific solutions are unrestricted by economic realities. Nanosilver-based materials for antibacterial applications are abundantly available elsewhere but totally economically unfeasible in most developing countries.

3. *Diverting resources from bigger problems*: Unless the social and economic priorities in development are fully analyzed and recognized, there is a danger of wastefully using limited resources. The socioeconomic structure of a region determines what it needs. Science and technology are generally unencumbered by these realities. Inappropriately large investments are made in areas of scientific research that do not yield solutions to more mundane issues that affect the majority of the population. When widespread problems are neglected, while niche problems are addressed with expensive technology, social divisions can grow."

4. *The excitement of new technology overshadows other issues:* This is perhaps the most troublesome aspect of rapid development of technology, particularly nanotechnology, when the science and technology, for reasons of competition, ownership, profits, etc., leave behind other more important concerns. Environmental and ethical issues become secondary, and sometime not even a consideration at all. Safety is intentionally or unintentionally sacrificed. Progress becomes synonymous with taking chances.

Under these considerations and concerns, what path should a developing nation follow so that it is not left behind in the "scientific revolution" but still does not sacrifice the health and safety of its population and its economy? There are pathways that could make the journey successful once a commitment is made to the general betterment of all its people. A nation so committed may find some of the following practices useful for effective and conscientious progress.

1. *Shared Resources* There are not enough resources for each developing nation to work in isolation, therefore it is absolutely necessary that these resources are efficiently utilized and shared. One way is to form small consortia and scientific partnerships, perhaps at the small provincial level up to a larger multi-national level. Instead of individual research teams, multi-user regional centers can be formed where interdisciplinary teams can utilize a single facility to view problems from various angles simultaneously. Sharing not only reduces redundancy and increases efficiency, it also provides added opportunities for collaborations and cross-disciplinary approaches to problem solving.

2. *Research for the Community* As much as "science for science's sake" is a valid motto, its applicability under constrained economic capability is restricted. There has to be proportionality between the efforts of science and education and technology and application. Academic institutions that train students with undergraduate and post-graduate degrees have an obligation to both the academy and society in general. While preparing experts for the next generation, institutions ought to focus on technological solutions and the needs of the next generation. A "best fit for the future" is achieved not just by involving technocrats but also through an alliance of the

STEM workforce with those humanists and social scientists who understand and help to shape society. In many developing areas, religious values are of utmost importance to people and that should be a consideration in the design and manufacturing of materials. Nanotechnology particularly brings into sight applications that tread near religious, social and cultural boundaries and at times seem to cross them. A sensitivity of the local customs and traditions, religious or social, might reduce alienation of a group of potential contributors and participants.

3. *Collaboration between Developed and Developing Worlds* There is an undeniable interdependence between these worlds but the interdependence has to grow away from the old "producer-consumer" model. There have to be shared values for doing science and technology for a better world for all humanity. The developed world enjoys a wonderful academic system that is available for use by the developing world. This starting point for expansion should be utilized by the current academics in developing regions who have appropriate knowledge of local needs.

4. *Academia – Government – Industry Collaboration* Academia, government, and industry have to collaborate in deciding the path forward and identifying areas of research and development to pursue. This is particularly important for nanotechnology due to its potential to challenge policy and regulatory guidelines. Since new policies will probably be required in dealing with these new materials, pre-emptive measures could reduce or remove tragic outcomes, like asbestosis. Governments and regulatory bodies are vitally important in keeping the consequences of nanotechnology in check as it spreads from industry to household. Industry, in its own way, holds the key for recognition of the potential of the technology and to take action for responsible development of products from new research. Academia is the link between the government and industry that not only provides for scientific discovery and innovation, it also provides proper training for the workforce in terms of ethics, social responsibility, health and safety. Academia also interacts with government to develop policies based on the social, economic and scientific expertise that is uniquely and simultaneously available at the academic institutions.

In as much as all the arguments listed in this essay are equally applicable to any new technology that comes into the picture, there are specific issues related to nanotechnology that make the exercise of going through all of these points very important, particularly for developing economies. The small size of nanoparticles requires visualization tools that are very specific and expensive, including transmission electron microscopes and atomic force microscopes. Therefore, availability of these instruments for researchers in the developing world becomes difficult. A regional facility could solve some of that problem. The small size of applications may eventually bring into question ethical issues related to surveillance and therefore to privacy. Thus, close regulation by governments concerning the appropriate uses of these technologies is imperative. The small size of nanomaterials warrants attention to health and exposure, since general protections, natural and artificial, fail to perform at this size level. Attention requires awareness, education, and research in areas that have so far not been needed for safe development of technologies based on bulk materials. There are some lessons learned already; the developing nations do not have to commit the same mistakes. The risk of failure in these developing nations in much larger due to very limited resources, but the potential for benefit from conscientious participation in appropriate nanotechnology development is almost unlimited.

References

[1] Project for Emerging Nanotechnologies, Woodrow Wilson International Centre for Scholars (http://www.nanotechproject.org/cpi/ accessed on May 14, 2015)
[2] Roco M 2011 The long view of nanotechnology development: the National Nanotechnology Initiative at 10 years, *J Nanopart Res* **13** 427 - 445
[3] Smalley, R 2005 Future global energy prosperity: The terawatt challenge MRS Bulletin **30** 412 - 417
[4] Schumacher E.F 1998 *Small is Beautiful* (London, Harper Perennial)

[5] Sulaimana S, Singh S, Mokhtara M and Bou-Rabeec, M 2014 Effect of dust on the performance of PV panels, Energy Procedia **50** 50 - 56

IOP Conf. Series: Materials Science and Engineering **92** (2015) 012002 doi:10.1088/1757-899X/92/1/012002

Carbon Nanotubes Reinforced Al-11 wt% Si Alloy via Plasma Spray

Ahmed A Moosa[1,*], Mohamed I Mohamed[2] and Mustafa K Ismael[3]

[1]Department of Materials Technology , Technical College, Middle Technical University , 54 Al Zafaraniya Street, Baghdad, Iraq.
[2] Department of Chemical Engineering, University of Technology, Al Sina'a Street, Baghdad, Iraq.
[3]Department of Mechanical Techniques, Institute of Technology, 54 Al Zafaraniya Street, Baghdad, Iraq
[*] E-mail ahmeda1752@yahoo.com

Abstract. In this work, multi-walled carbon nanotubes (MWCNTs) with different portions (0.5, 1, 2, 4) wt% were added to a gas atomized Al-11 wt% Si powder. The Al-11 wt% /MWCNTS nanocomposite powder was examined by FESEM, Raman spectroscopy, X-ray diffraction (XRD). Air plasma spraying (APS) was used to spray Al-11 wt% Si/MWCNTs nanocomposite powder on aluminum alloy AA6082-T6 substrates. Al-11 wt% Si/MWCNTs nanocomposite coating layer was examined using FESEM/EDS, Raman spectroscopy, XRD and HRTEM. SEM/EDS showed that Al_4C_3 is formed at the interface e between the coating layer and the substrate in Al-11 wt% Si/4 wt% MWCNTs plasma spray coating. The adhesion test showed good adhesion in the ranges 5-15 MPa between the coating layer and the substrate. Microhardness test of the air plasma sprayed (APS) Al-11 wt% Si/MWNTs nanocomposite layer is increased with the MWCNTs wt%.

1. Introduction

Composite materials combined distinct physical and mechanical properties like thermal conductivity, electrical conductivity, strength, toughness and stiffness properties [1, 2]. Composite materials with at least one of their constituent phases being less than 100 nm are commonly termed nanocomposites. Nanocomposites will make remarkable improvements in the mechanical and physical properties of these nanocomposites [3]. Aluminum silicon alloys are widely used in internal combustion engine, owing to their good characteristics such as, high corrosion resistance and low density [4]. Wear resistance of aluminum silicon alloys strongly depends on the alloy composition [5].

Aluminum-Silicon composite powder coating material such as (Al -7Si -26 Graphite) are used to provide consistent quality clearance control in applications where rotating components may come into contact with the coating as a result of design or operational limitation in the low temperature sections of turbine engines. These powder coatings are used to minimize wear to the rotating components while maximizing gas path efficiency in seal areas [6,7].

Carbon nanotubes (CNTs) have excellent properties such as: High aspect ratio, low mechanical, electrical and thermal properties [8-10]. Carbon fibers were used to reinforce Al-Si alloys in order to combine high strength and high modulus of carbon fiber with the good ductility and high toughness of aluminum alloys [11]. Actually, CNTs are the strongest and stiffest materials yet discovered in terms of tensile strength and elastic modulus respectively. High Van der Waals forces between individual tubes cause them to agglomerate into CNT 'ropes' or randomly ordered free-floating particles. Therefore, the major challenges is to disperse CNTs in metal matrix alloy with less damaging of CNTs [12,1].

The aim of this work is to use Al-11 wt% Si /MWNTs (MMCs) nanocomposite powder to coat aluminum alloy substrates (AA6082-T6) using plasma spraying process. The effect of MWNTs as reinforcement in Al-11 wt% Si/MWNTs nanocomposite coating will be studied together with the microstructure, microhardness and bond strength of the Al-11 wt% Si/MWNTs nanocomposite coating .

2. Materials

In this research a 3 mm of flat sheet of aluminium alloy AA6082-T6 (AALCO Metals Limited, UK) was used as a substrate to be coated with Al-11 wt% Si /MWNTs using plasma spray process . The chemical composition of the aluminium alloy substrate AA6082-T6 is shown in Table 1.

Table 1. Chemical composition of wrought aluminium alloy AA6082-T6.

Chemical composition, wt%									
Element	**Cr**	**Cu**	**Fe**	**Mg**	**Mn**	**Si**	**Ti**	**Zn**	**Al**
Nominal composition	0.25	0.1	0.5	0.6-1.2	0.4-1	0.7-1.3	0.1	0.2	Bal.
Measured Composition	0.19	0.07	0.35	1.08	0.76	1.14	0.06	0.14	Bal.

The specifications of MWNTs (ARRY International Group Limited-Germany) are shown in Table 2. A gas atomized Al-11 wt% Si alloy powder (Phoenix Scientific Industry Ltd., UK (PSI)) was used with an average grain size 15-75 µm (-200 mesh/+15 µm). The chemical composition of gas atomized powder of Al-11 wt% Si illustrated in Table 3. The chemical composition analysis was carried out by X-Ray fluorescence (XRF) ,Spectro Xepos, Germany at the Applied Physics Department , Ministry of Science and Technology, Iraq.

Table 2. Specifications of MWNTs [ARRY® Company, Germany].

Specifications	
wt%	>80
Diameter (nm)	30-50
Length (µm)	0.5-2
SSA (m^2/g)	220
Amorphous carbon	<2
Ni/Fe	<2

Air plasma spray system consist of plasma spray gun (Sulzer Metco, model F4MB-XL, Switzerland) which is mounted on flexible 6-axis robotic manipulator (ABB, model IRB -6400, Switzerland). The process parameters are tabulated in Table 4. The tensile adhesion test had been done according to ASTM standard C633-01.

Table 3. Chemical composition of Al-11 wt% Si powder [PSI* Company Ltd., UK].

Element	Si	Cu	Fe	Mg	Mn	Pb	Sn	Zn	Al
Al-11Si Standard	11-13	0.3	0.25	0.1	0.1	0.02	0.02	0.2	Bal.
Al-11Si Measured	11.47	0.12	0.20	0.08	0.01	0.04	-	0.1	Bal

The Tensile adhesion tests of Al-11 wt% Si /MWNTs coating were performed using Hounsfield Test Equipment (UK, model 50H-K-S) with 50 kN capacity load cell and 0.05 mm/sec crosshead speed . The microhardness of the Al-11 wt% Si /MWNTs coating was measured using Buehler 60044, UK for well-polished cross-section samples using applying load of 50 gram at 20 second. The XRD was measured using diffractometer Shimadzu , Japan with Cu- Kα radiation at 40 kV and 30 mA , wave length (λ) 1.5418 Å, using nickel filters. The diffraction angle (2θ) is 10°-90° at scanning speed was 6° deg/min. The XRD analysis was carried out at S.C. Geological Survey and Mining, Baghdad. High Resolution Transmission Electron Microscope (HRTEM), 200kV, A FEI, Model: Tecnai G2 F30, USA, high sensitivity combining diverse applications like bright and dark field with STEM capability equipped with X-ray energy dispersive spectroscopy (EDS). Field emission scanning electron microscope (FESEM) model (Carl Zeiss Supra 40VP, Germany) and FEI Quanta 250 ESEM-US used to characterize the coating layer.

Table 4. Process sheet of plasma spray parameters.

Spraying parameters		
Gun type: F4MB-XL	**Pressure**	**Flow rate**
Argon	3bar	SLPM
Hydrogen	3bar	SLPM
Carrier/Argon	3bar	SLPM
Current	500 Amp	
Power	33 kW	
Nozzle diameter	6 mm	
Powder feed rate	40 ± 5 (g/min.)	
Spray distance	140 mm	

3. Experimental Work

To obtain a good dispersion of CNT with Al-11 wt% Si powder, two steps mixing processes were used. First, different weigh percent (0.5, 1, 2 and 4 wt %) of Carbon nanotubes were put in beaker then methanol (HPLC grade) was added (CNTs: Methanol ratio is 1:250 ml). The solution was then sonicated for 5 minutes using a sonicator, followed by magnetic stirring at 250 rpm for 5 minutes followed by sonication for 5 minutes. Then, a 250 g of Al-11 wt% Si powder were then added to CNT / methanol solution in paraffin oil bath at 70°C by using mechanical stirrer at 250 rpm. Argon gas was purged to the Al-11 wt% Si /MWNTs /methanol solution to make sure no oxidation takes place during methanol evaporation. After entire methanol was evaporated, the Al-11 wt% Si /MWNTs mixture was transferred to a drying oven at 70°C for 24 hours and then was saved in desiccators under vacuum.

In the second mixing step, Al-11 wt% Si /MWNTs mixture was ball milled at ambient temperature using a ball mill (Pascall Engineering, UK) with a 2.4 litter porcelain jar and 1/2 and 9/16 inches stainless steel balls. The ball to powder ratio (BPR) was taken as 10:1 and ball mill speed and milling time are 100 rpm and for 5 hours respectively. Methanol (HPLC) was added as a lubricant to prevent powder sticking on the balls and for each 30 grams of Al-11 wt% Si powder; a 400 microliter (µl) of methanol was added. The Al-11 wt% Si /MWNTs powder was then collected and dried in an oven for 48 hours and then stored in desiccators under vacuum.

Plasma spraying process was done by few passes. After the first plasma spray pass, the Al-11 wt% Si /MWNTs coating layer was cleaned with compressed air and the coating thickness was measured by digital micrometer. Four passes were made until the coating thickness measured more than 100 micrometer. The gauge radius means the distance from the powder injector to plasma nozzle centre. After thermal spraying process, the samples were cut with low speed diamond saw (Buehler, Isomat, UK), followed with cold mounted with Acrylic resin. Each sample was then ground using automatic grinding machine with silicon carbide, washed with distilled water, rinsed with Acetone and then dried. The samples were then polished with 0.1µm diamond paste on PoliCloth, washed with distilled water, and then followed by polishing with aluminum oxide 0.05 µm , washed again with distilled water rinsed with Acetone and dried.

4. Results and Discussion

Figure 1 , shows High Resolution Transmission Electron Microscopy (HRTEM) image of purified CNTs produced by CVD using Ni/Fe catalyst supplied by Arry.Co. The MWNTs image of high magnification HRTEM contains some defects that were introduced during purification process.

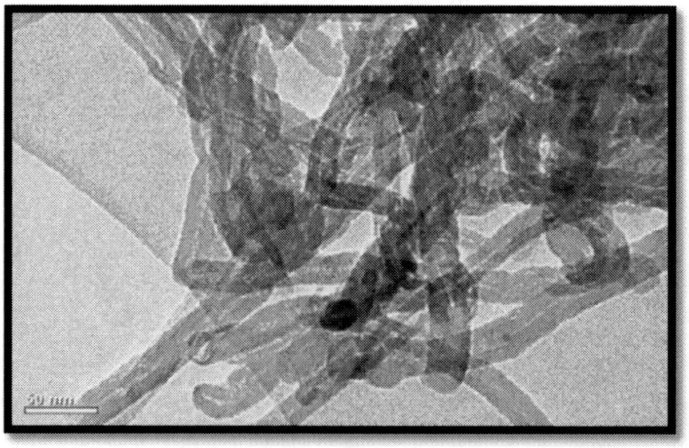

Figure 1. HRTEM image shows purified CNTs produced by CVD using Ni/Fe catalyst supplied by Arry.Co.

Figure 2 shows HRTEM image of MWCNT produced by CVD with Fast Fourier Transformation (FFT) to analyses the interlayer spacing and to measure the number of walls. There are about 23-22 uneven walls with uneven outer and inner diameter. These defects would affect carbon nanotubes properties. The outer diameters were found 20-40 nm, the inner diameter were 3-5 nm, with 0.341 nm interlayer spacing which is typically similar of interlayer found in standard XRD card 0.3434 nm.

FESEM micrograph for Al-11 wt% Si/0.5MWCNTs powder after 5 hours of ball milling showed the absence of any clusters on the stubs as shown in Figures (3a,b). The dark area is the individual MWNTs that are embedded into the Al-Si particles. Many other dispersion techniques has been studied in order to get uniform spherical particles, including mechanical alloying [13], nanoparticle dispersion [14], molecular-level mixing [15] and spray drying [16]. It was found that ball milling is a good technique to obtain good dispersion of CNTs with Al-Si powder as shown by Kang et al. [17].

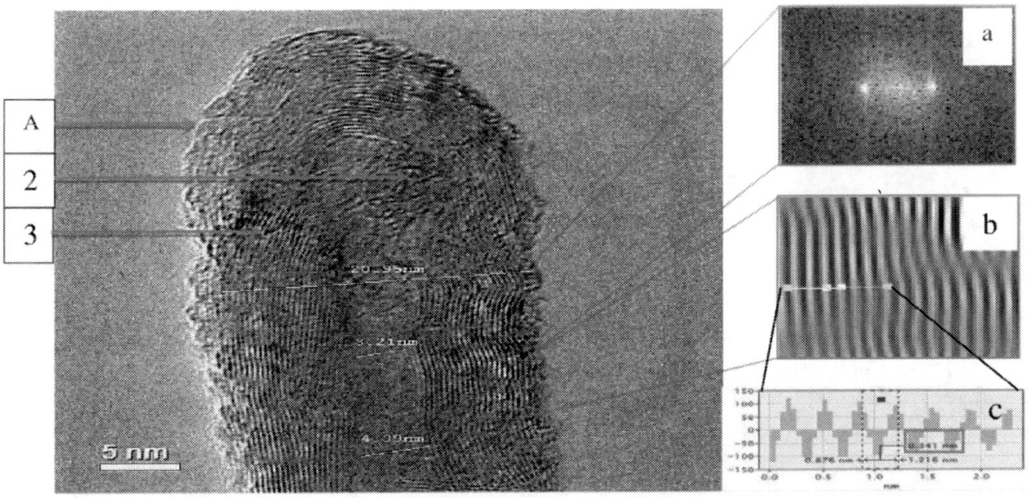

Figure 2. HRTEM image of MWCNT produced by CVD using Ni/Fe catalysts (Arry Ltd.), (a) a Fast Fourier Transform (FFT), (b) Magnified filtered image shows the interlayer of MWNTs, (c) the MWNTs interlayer scale measurement.

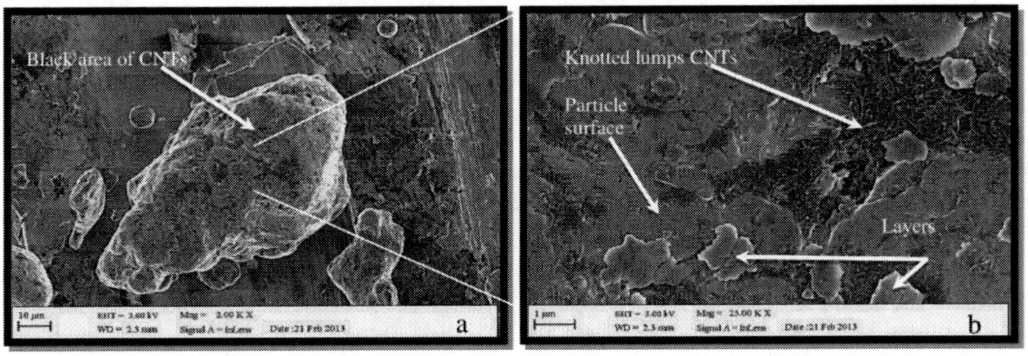

Figure 3. FESEM micrograph of Al-11 wt% Si/0.5 wt% MWNTs powder, solution dispersion followed by ball mill mixing for 5 hours, (a, b) LM and HM.

At higher magnification (Figure 3b) the CNTs are embedded into powder particles due to layering of particles formation during ball milling. These layers cover the new forming particle and only knotted-lumped agglom-erated CNTs settles in pores of particle surface. However, in some areas a knotted-lump of CNTs appears with larger area due to increasing the amount of CNTs. In ball milling, the layering formation was explained by Suryanarayana [13] where the particles are repeatedly flattened, cold welded, fractured and re-welded to form large particles. The deformation of particles in ball milling led to embedding of MWNTs in Al-Si particles [17]. The small particles sizes of Al-Si have been reduced after the ball milling. This explains the small layering formation through which CNTs were embedded. Agglomerates of CNTs (knotted lumps) were found in some areas after ball mill mixing due to entanglement of carbon nanotubes with Al-Si powders.

X-ray diffraction (XRD) patterns of Al-11 wt% Si/ MWNTs powder mixture is shown in Figure 4. The peaks for aluminum and silicon are shown without appearance of CNTs peaks only very small peaks in case of 2 wt% and 4 wt% of CNTs this is due to the small amount of CNTs which is below the detection limit of the X-ray counter and also the dispersion of CNTs in Al-Si particles. The XRD patterns of Al-11 wt% Si/MWNTs coatings layer is identical to X-ray diffraction (XRD) patterns of Al-11 wt% Si/ MWNTs powder which indicate no significant phase trans-formation occur during plasma spraying. SEM/EDS examination of the coating layer cross-section was performed in order to identify the elements after plasma spraying.

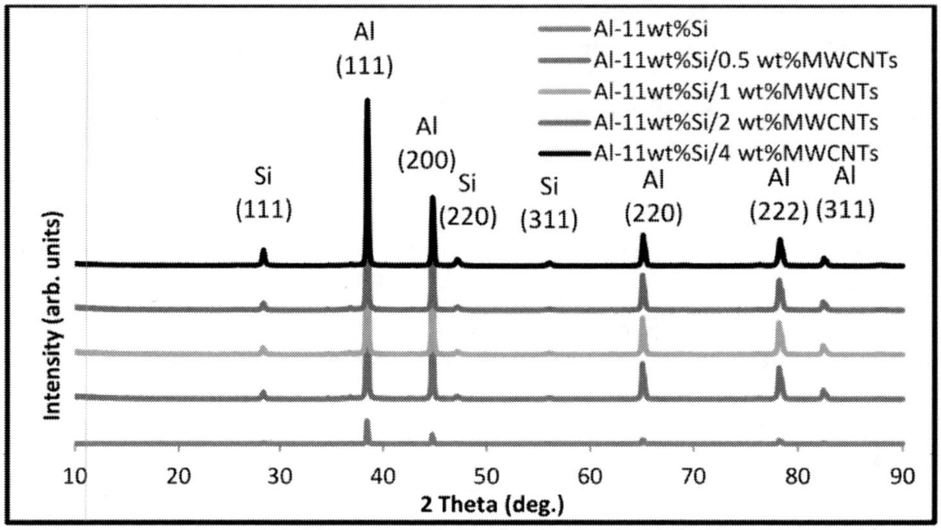

Figure 4. XRD patterns of Al-11 wt% Si/MWNTs powder mixture with various amounts of MWNTs.

Figure 5 shows SEM/EDS micrograph of Al-11 wt% Si coating layer for as received powder. The optical microstructure of the coating layer is shown in Figure 5a. The selected area spectrum for the coating layer is highlighted in Figure 5b. The EDS spectrum of the coating layer shows Al, Si and Oxygen peaks , Figure 5c.

SEM/EDS micrograph of Al-11 wt% Si/0.5 wt% MWNTs coating layer is shown in Figure 6. The optical microstructure of the coating layer is shown in Figure 6a which reveals aluminum oxide, needles shape silicon embedded in aluminum matrix. The selected area spectrum for the coating layer is highlighted in Figure 6b. The EDS spectrum is shown in Figure 6c which contains carbon, aluminum, silicon. Voids were also seen clearly in the optical micrograph at high magnification, as well as un-melted particles formed during APS.

The cross section of the Al-11 wt% Si/1 wt% MWNTs nanocomposite coating layer contains aluminum oxide, needle shape silicon embedded in aluminum matrix. The EDS spectrum shows some amount of iron and this may cause some deleterious effects on the composite coating. SEM/EDS micrograph of coating layer cross-section shows more voids and there are no needles like silicon. The selected area spectrum for the coating layer shows the silicon wt% decreased.

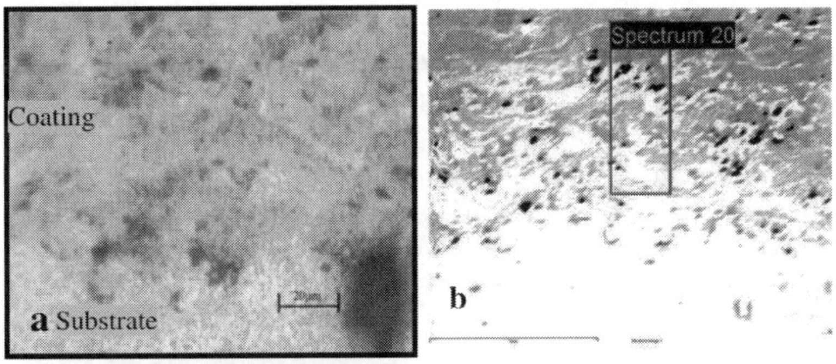

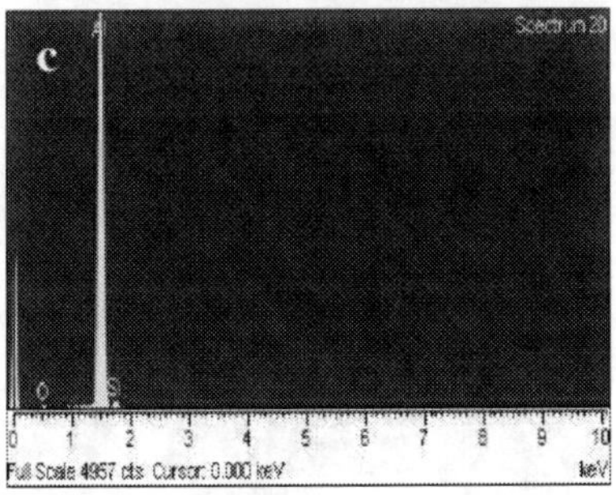

Spectrum processing :
Peak possibly omitted

Element	Weight
O K	4.62
Al K	89.79
Si K	5.59
Totals	100.00

Figure 5. SEM/EDS analysis micrograph and OM for Al-11 wt% Si coating layer cross section; (a) OM, (b) SEM area selected spectrum, (c) EDS spectrum peaks.

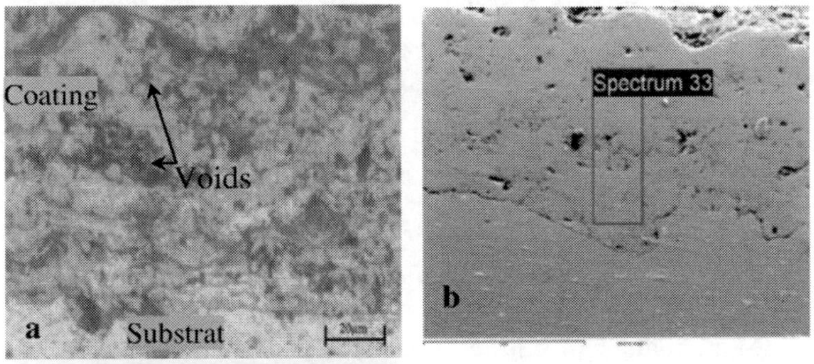

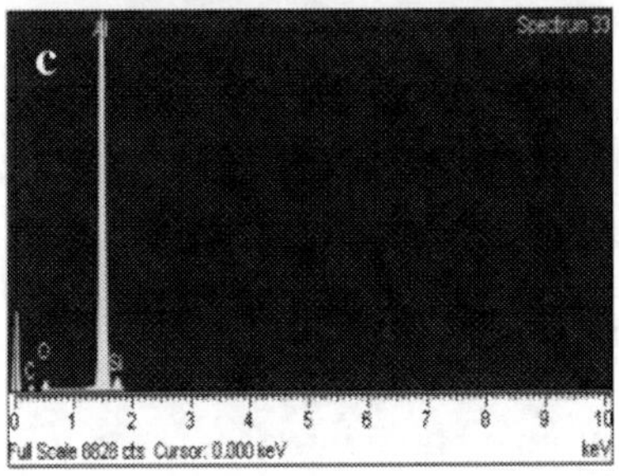

Spectrum processing :
Peak possibly omitted

Element	Weight%
O K	6.82
Al K	84.52
Si K	8.03
C K	0.63

Figure 6. SEM/EDS micrograph and OM for Al-11 wt% Si/ 0.5 wt% MWNTs coating layer cross section; (a) OM, (b) SEM area selective spectrum, (c) EDS spectrum peaks.

SEM/EDS micrograph of Al-11 wt% Si/4 wt% MWNTs coating layer cross-section is shown in Figure 7. SEM micrograph of the coating layer Figure 7a shows a significant black clusters and a black interlayer between the coating and the substrate. The selected area spectrum for the coating layer is highlighted in Figure 7b. EDS examination shows relatively high peak of carbon in the interlayer (Spec. 2). This can be explained by three possibilities as: First the carbon nanotubes segregate form the lamellae during cooling and create mat-network on the surface. Secondly, the particles were covered with MWNTs during mixing process, so during impact the carbon nanotubes become the first impacting layer. Thirdly, the probability of graphitization of some CNTs occurs during high temperature of APS. The higher concentration of CNTs leads to clustering. This phenomenon occurs due to the surface tension forces of CNTs in the molten droplet [18]. Aluminum carbide (Al_4C_3) interface also can be formed during thermo-mechanical processing when the temperature is higher than 800°C [19-22]

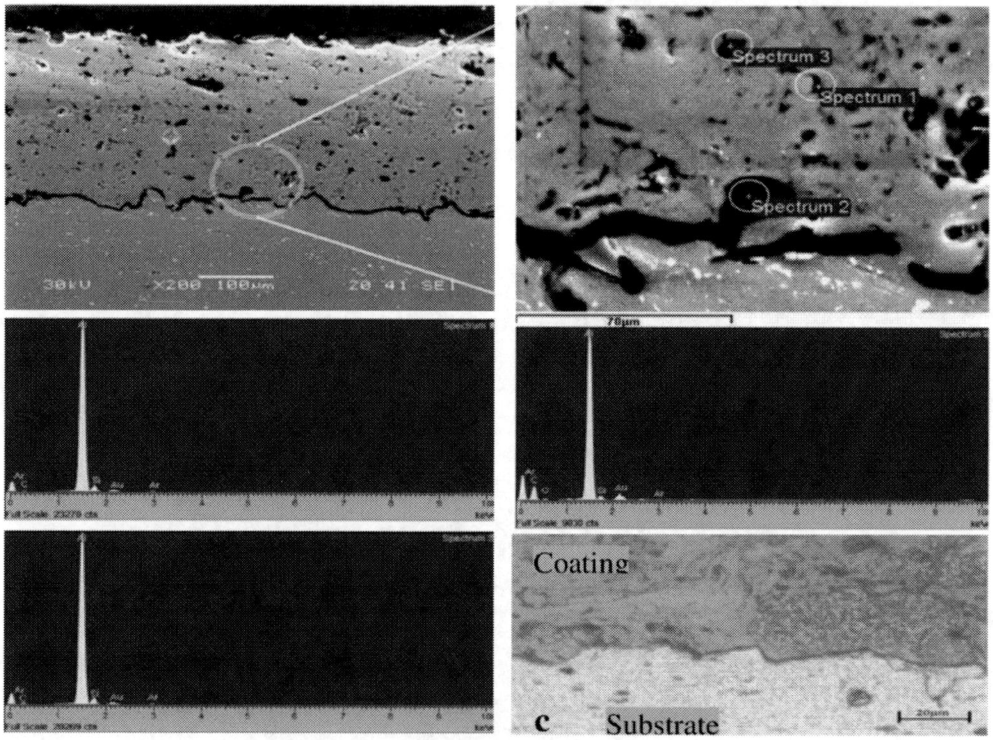

Figure 7. SEM/EDS analysis micrograph and OM for Al-11 wt% Si/ 4 wt% MWNTs coating layer, (a) SEM micrograph of selected area, (b) HM of the selected area shows the point EDS, (c) OM microstructure of the coating cross-section.

The stress vs. strain curves for the coating layer of Al-11 wt% Si and Al-11 wt% Si/MWNTs are shown in Figure 8 which gives the axial force vs. displacement curve. The maximum load at failure is 7.93 kN for Al-11 wt% Si/2 wt% MWNT, while the minimum load at failure is 2.86 kN for Al-11 wt% Si. The load range for the samples is (2.86 – 7.93 kN). The adhesion strength can be calculated according to ASTM by (F_{max}/A) where F_{max} is the maximum load and A is cross-section area. The adhesion strength increases with MWNTs loading. But less bonding strength obtained at increasing the proportion of MWNTs to 4% due to formation of aluminium carbide interface.

The microhardness of the substrate Al alloy AA6082-T6 has an average of 116.78 ± 2.9 HV. Figure 9 shows the microhardness curve for the coating layer with different wt% MWNTs. In general the microhardness increases with increasing MWNTs weight percent reinforcement and the maximum hardness found at 4 wt% MWNT reinforcement. The microhardness test of 1 wt% and 2 wt% MWNTs the nanocoating composite of Al-11 wt% Si takes downward. Thus due to the increasing of the close pores inside the coating layer (as shown in Figure (15b) and effect on the hardness values. Increasing the amount of MWNTs more than 2 wt% MWNTs increases the hardness value because CNTs resist the flow of material and impede dislocation motion.

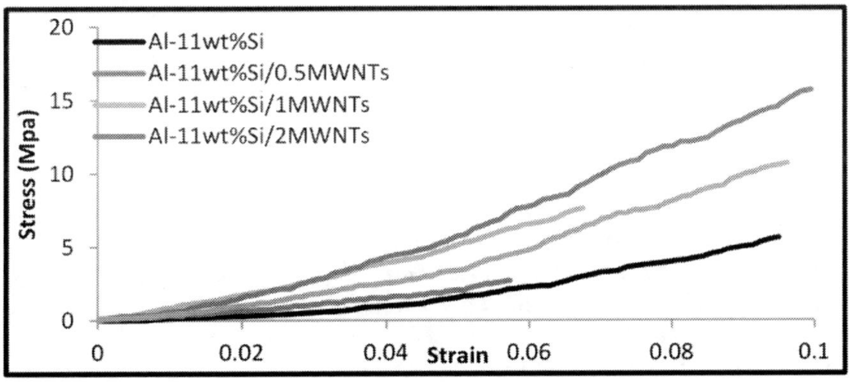

Figure 8. Stress vs. strain curve for the coating layer of Al-11 wt% Si and Al-11 wt% Si/MWNTs.

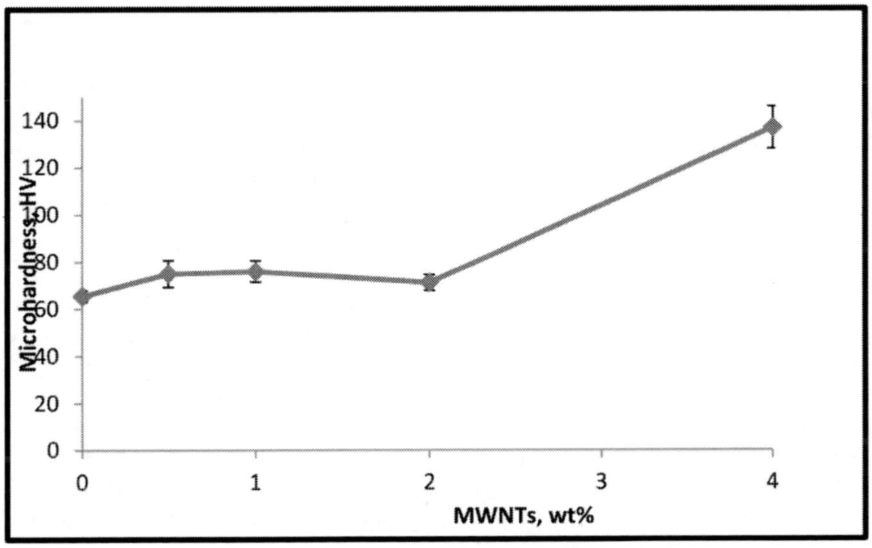

Figure 9. Microhardness curve for the coating layer of plasma spared Al-11 wt% Si and Al-11 wt% Si/MWNTs.

5. Conclusions

The result of the current study reached to the following conclusions;
From the present work it can be concluded that:

1. The MWNTs produced by CVD (Arry Ltd.) have about 23 uneven walls with uneven outer and inner diameter.

2. The mixing process using methanol solution and mechanical stirring followed by ball milling for Al-11 wt%Si with MWNTs are a good dispersion technique, very convenient and suitable for industrial processes to achieve flowability of powder in thermal spray process.
3. Good adhesion of nanocomposite coating layer on aluminium substrate can be obtained by plasma spraying of Al-11 wt% Si/MWNTs. Where the adhesion ranges was from 5-15 MPa.
4. The microhardness is increased in general due to reinforcement effect of MWNTs.

Strite S and Morkoc H 1992 *J. Vac. Sci. Technol.* B **10** 1237

Kurata M 1982 *Numerical Analysis for Semiconductor Devices* (Lexington, MA: Heath)

Selberherr S 1984 *Analysis and Simulation of Semiconductor Devices* (Berlin: Springer)

References

[1] Agarwal A, Bakshi S R, Lahiri D, 2010 *Carbon nanotubes: reinforced metal matrix composites* (US: Taylor and Francis Group)

[2] Ramesh B T 2014 Characterizat-ion of al based nano composites using powder metallurgy technique *Inte. J. of Res. in Aero. and Mech. Eng.* **2** 131

[3] Tjong S C 2009 *Carbon nanotube reinforced composites, metal and ceramic matrices* (Weinheim: Wiley-VCH Verlag GmbH and Co. KGaA)

[4] Hemanth J 2005 Tribological behavior of cryogenically treated B_4Cp/Al–12% Si composites *Wear* **258** 1732

[5] Altnkok N, Özsert I and Findik F 2013 Dry sliding wear behavior of Al_2O_3/SiC particle reinforced aluminium based MMCs fabricated by stir casting method *Acta Phys. Polonica A* **124** 1

[6] Metco S 2006 *Thermal spray materials guide* (New York: Sulzer Metco Inc.)

[7] Metco S *2012 Aluminum silicon graphite abradable powders* (New York: Sulzer Metco)

[8] Tjong S C 2006 Structural and mechanical properties of polymer nanocomposites, *Mat. Sci. and Eng. R* **53** 73.

[9] Tjong S C, Liang G.D and Bao S.P 2007 Electrical behavior of polypropylene/multi-walled carbon nanotube nanocomposites with low percolation threshold, *Scripta Materialia*, **57** 461

[10] Tjong S C, Liang G.D and Bao S.P 2008 Effects of crystallization on dispersion of carbon nanofibers and electrical properties of polymer nanocomposites', *Polymer Eng. and Sci.*, **48** 177

[11] Qiurong Y, Jinxu L, Shukui L, Fuchi W and Tengteng W 2014 Fabrication and mechanical properties of Cu-coated woven carbon fibers reinforced aluminum alloy composite, *Materials and Design*, **14** 442

[12] Girifalco L A, Hodak M and Lee R S 2000 Carbon nanotubes, buckyballs, ropes and a universal graphitic potential *Phys. Rev.* **B 62** 12104

[13] Suryanarayana C 2001 Mechanical alloying and milling, Prog. *In Mat. Sci.* **46** 1

[14] Noguchi T, Magario A, Fukazawa S , Shimizu S , Beppu J and Seki M Carbon nanotube/aluminum compo-sites with uniform dispersion, *Mat. Trans.*, 45(2004)602-604.

[15] S.I. Cha, K.T. Kim, S.N. Arshad, C.B. Mo, S.H. Hong, Extraordinary strengthening effect of carbon nanotubes in metal-matrix nanocomposites processed by molecular-level mixing', Adv. Mat., 17(2005)1377-1381.

[16] S.R. Bakshi, V. Singh, S. Seal, A. Agarwal, Aluminum composite reinforced with multiwalled carbon nanotubes from plasma spraying of spray dried powders, Sur. and Coat. Tech., 203(2009)1544-1554.

[17] K. Kang, G. Bae, B. Kim, C. Lee, Electrical and mechanical properties of multi-walled carbon nanotube reinforced Al composite coatings fabricated by high velocity oxygen fuel spraying', Surf. and Coat. Tech., 206(2012)4060-4067.

[18] Biro L, Horváth Z, Márk G, Osváth Z, Koós A, Benito A, Maser W and Lambin P 2004, Carbon nanotube Y-junctions: growth and properties, *Diamond and related materials*, **13** 241-249.
[19] Bakshi S R, Singh V, Balani K , McCartney D G , Seal S and Agarwal A 2008 Carbon nanotube reinforced aluminum composite coating via cold spraying *Surf. and Coat. Tech.* **202** 5162
[20] Landry K , Kalogeropoulou S and Eustathopoulos N 1998 Wettability of carbon by aluminum and aluminum alloys, *Mat. Scie. and Eng. A* **254** 99.
[21] Favre A, Fuzellier H and Suptil J 2003 An original way to investigate the siliconizing of carbon materials *Ceramics Inter.* **29** 235
[22] Bakshi SR, Keshri A K and Agarwal A 2011 A comparison of mechanical and wear properties of plasma sprayed carbon nanotube reinforced aluminum composites at nano and macro scale *Mat. Sci.. and Eng.*: A **528** 3375

IOP Conf. Series: Materials Science and Engineering **92** (2015) 012003 doi:10.1088/1757-899X/92/1/012003

Continuous and Delayed Photohemolysis Sensitized With Methylene Blue and Iron Oxide Nanoparticles (Fe_3O_4)

M-Ali AL-Akhras[*,1], Khaled Aljarrah[1,2], Borhan Albiss[1], Abba Alhaji Bala[3]

[1]Bio-Medical physics Laboratory, Department of Physics, Jordan University of Science & Technology (JUST), P.O. Box 3030, Irbid 22110, Jordan

[2]College of Applied Medical Science, King Saud Bin Abdulaziz University for Health Science, Al-Ahsa , KSA

[3]Kano State Collage of Arts, Science and Remedial Studies, P.M.B. 3145, Kano-Nigeria

E-mail: alakmoh@just.edu.jo

Abstract. This research present the sensitization of methylene blue (MB), as a potential photodynamic therapy photosensitizer which showed phototoxicity for many tumor cells *in vitro* incorporated with iron oxide nanoparticles (Fe_3O_4, IO-NP), which offer magnificent interaction both inside and outside the surface of biomolecules together with red blood cells (RBC's) with significant change in hemolysis process. The study investigated the sensitization of continuous photohemolysis (CPH) for MB and MB with IO-NP, delayed photohemolysis (DPH) at different irradiation temperature (T_{irr}). The photohemolysis rate for CPH at room temperature has a power dependence of 0.39 ± 0.05 with relative of steepness of 1.25 ± 0.02 and for different concentration of MB and power dependent of 0.15 ± 0.03 with relative steepness of 1.34 ± 0.01 for different MB and IO-NP. Logistic and Gompertz functions were applied as appropriate mathematical models to fit the collected experimental data for CPH and DPH respectively, and to calculate fractional photohemolysis rate with minimum errors. The Logistic function parameter; $\boldsymbol{\alpha}$, the hemolysis rate, increases with increasing concentrations of MB and decreases with increasing IO-NP concentrations in the presence of 6 μg/ml of MB. The parameter $\boldsymbol{\beta}$; the time required to reduce the maximum number of RBCs to one half of its value, decreases with increasing MB concentration and increases with increasing IO-NP concentrations in the presence of 6 μg/ml of MB. In DPH at different $T_{irr,}$ the Gompertz parameter; $\boldsymbol{a}$, fractional hemolysis ratio, is independent of temperature in both case MB and MB plus IO-NP, while the parameter; $\boldsymbol{b}$, rate of fractional hemolysis change, increases with increasing $T_{irr,}$ in both case MB and MB plus IO-NP. The apparent activation energy of colloid-osmotic hemolysis is 9.47±0.01 Kcal/mol with relative steepness of 1.31 ± 0.05 for different MB and 6.06±0.03 Kcal/mol with relative steepness of 1.41 ± 0.09 for MB with iron oxide. Our results suggest that Logistic equation is the best fit for the CPH and Gompertz function for the DPH. Both models predict also that the relative steepness is independent of the light dose, sensitizer and IO-NP concentrations.

1. Introduction

Cancer is a complex deadly disease mainly caused by factors in the environment that mutate genes and encrypting decisive cell-regulatory proteins. The agents that cause cancer can be present in Food, chemicals, air and sunlight which people use in their daily activities [1, 2]. It has been predicted that, in

2020 it will be about 20 million new cases of cancer and almost twelve million deaths. Different methods of treatment and therapies were employed such as radiation therapy, cytostatic chemotherapy, surgical operation, and photodynamic therapy (PDT).

PDT appeared to be a promising mode of treatment for many diseases most especially localized cancer with minimal side effects [3, 4]. In PDT, light photosensitizing drug (photosensitizer), oxygen (O_2), and light are combined to have therapeutic effect. PDT is a kind of treatment that affects only the designated tissue. The activation of the photosensitizer by exposure to a carefully regulated dose of light of appropriate wavelength for a specified length of time elicits a cytotoxic action resulting in cell death [2]. PDT uses selective irradiation in therapeutic treatment which does not damage tissues that are healthy. It develops no resistance in case of repeated treatment. It can also be used to treat cancers that are resistance to other medication and heal very fast as the collagen contain in the tissue is not affected by photodynamic damage [5].

Nanoparticles provide new interaction with biomolecules on cutaneous and sub-cutaneous tissues which may raise cancer treatment and diagnostic. Nanoparticles are been recognized as a drug transporting system because of their size and extreme surface to volume ratio results to their usage in PDT [6], and their ability to concentrate the photosensitizer (PS) on the cells wall or diffuse inside the membrane. Researchers have shown that, magnetic carriers are widely applicable in drug targeting, isolation and sorting cell [7, 8]. Magnetic nanoparticles might be used as carriers for methylene blue (MB) photosensitizer because of their low toxicity, superparamagnetic properties and high saturation magnetization [9].

Red blood cells (RBC), human erythrocytes shown to be ideally targeted in PDT since they are somewhat simple structured design to study the ability of compounds to produce photo-oxidation process [10] in cells *in vitro* and amenable for hemotasis in the vascular bed of tumor. On the other hand methylene blue (MB) as a photosensitizer has low toxicity and high quantum production of O_2 generation. It has fascinating photochemical and photo-physical properties that give magnificent results in *in vivo* and *in vitro* [9, 11].

This study investigated the RBC`s sensitization and compare the efficiency of continuous photohemolysis (CPH) for free MB and MB with IO-NP, delayed photohemolysis (DPH) at room temperature, at different irradiation temperature (T_{irr}) and at different irradiation time (t_{irr}).

logistic function (Eq. 1) and Gompertz function (Eq. 2) were applied as appropriate mathematical models to calculate the fractional photohemolysis rate [12,13] and analyzes the resulting sigmoid curves to fit the collected experimental data for CPH and DPH respectively, with insignificant errors [14].

$$N = \frac{A}{1+exp(-\alpha(t-\beta))} \quad \text{(Eq. 1)}$$

$$H = H_0 e^{-ae^{-bt}} \quad \text{(Eq.2)}$$

Whereas: ***N*** and ***H*** is the percentage of hemolysis during the lysis time ***t*** (i.e. the time measured from the beginning of rupturing the RBCs), the number of RBCs, **A** and ***Ho*** is the maximum number of RBCs (always normalized to one), ***α*** and ***a*** is the fractional hemolysis rate, and ***β*** is the time required to reduce the maximum number of RBCs to one half of its value, and ***b*** is the rate of fractional hemolysis change.

The dependence of continuous photohemolysis on the photosensitizer concentration (MB) can be empirically determined by the power law [11] as:

$$1/t_{50} = kC^{P} \quad \text{..................(Eq. 3)}$$

Where: ***k*** is constant of time- intensity, ***C*** is the sensitizer concentration and ***p*** is the power dependent parameter.

The objective of this work was to investigate photosensitization of methylene blue as a drug for PDT and to present the IO-NP influence in the therapy. The MB has been selected in this work due to interested

IOP Conf. Series: Materials Science and Engineering **92** (2015) 012003 doi:10.1088/1757-899X/92/1/012003

optical properties and to its high sensitivity to temperature and its high ability to light absorption in UV-Vis regions.

2. Materials and Methods

Red blood cells were obtained from healthy donor and isolated by repeated centrifugation at 1000g for 8 min each time. The cells were then suspended in 7.4 pH Phosphate buffer saline (PBS). The light scattering absorption (optical density) of the isolated *RBC* at 680nm was approximately 2.0 which correspond to 7.86×10^6 cells/mm^3 erythrocytes concentration as counted using haemocytometer. The light source used for irradiating the samples was a 200W Hg-Xe arc lamp (Power supply model: 68907) housed in an Oriel Research Arc Lamp focused to the sample spot size using a concave lens. Optical density of the samples was measured using Shimadzu 2450 UV-VIS spectrophotometer/Japan. Methylene blue (MB), an available commercial photosensitizing agent, was acquired in the form of powder from Aldrich (Milwaukee, WI). The iron (II, III) oxide nanopowder, 98% certified of 50nm size was purchased from Sigma-Aldrich (St Louis, MO, USA). Stock suspension was prepared by suspending 7.5 mg of iron oxide nanopowder in 10 ml of PBS to obtain a suspension of 0.75 mg/ml which is then diluted to 0.15 mg/ml and ultrasonic wave generator was used to completely suspend the particles. A concentration of 0.3 mg/ml of MB was prepared by dissolving 0.3 mg of MB soluble powder in 10 ml of PBS with all the measurement taking in darkness. The samples were a mixture of methylene blue (MB), iron oxide nanoparticles and Red blood cells (RBCs) suspended in PBS of 7.4 pH. Samples were incubated at 37° C for 30 min with gently shaking every 10 min to have a homogeneous solution and to allow the photosensitizer and IO-NPs to penetrate through RBCs membrane or to be bounded to the cell walls. The samples were centrifuged again at 1000g for 8 min to remove the aggregated and unbounded IO-NPs. Figure 1 represents these samples with two UV absorption peaks and four VIS peaks.

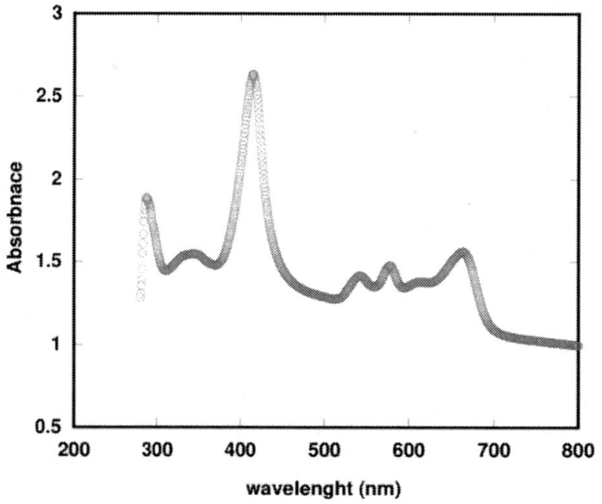

Fig. 1: Absorption spectrum of suspension containing RBCs, MB and IO-NP.

3. Results and Discussions

In the past decades, metallic inorganic nanoparticles have taken the attention of many nano-biotechnologist due to the effect they shown on biological imaging and bioconjugation in delivering drug and diagnostic work. This metallic particle assists in the interaction within the range of drug molecules due to their wide surface area relative to the volumes [15]. One of the ideal reasons set to improve the novel photosensitizers was the expansion and their improvement in cancer treatment using photodynamic therapy (PDT). It involves selective transfer of photosensitizer such as methylene blue to tumors, and when irradiated with a suitable wavelength of light, the MB produces a singlet- oxygen ($^1O_2$) and reactive oxygen species (ROS) that cause damage to tumor cell with minimal effect to normal tissue. Sensitization of *RBC* for continuous and delayed photohemolysis *in vitro* has been used as a medium of $^1O_2$ and ROS generation. Therefore, this work demonstrates the photohemolysis behavior of RBCs and methylene blue incorporated with iron oxide nanoparticles.

There are many researches that show MB mediated phototherapy against fungal, tumor and bacterial infections [9, 11, 16, 21, 22, 25]. It was discovered that, small amount of MB concentration inhibits high percentage of bacteria on irradiation, which is consistence with the results of this study; whereas MB incorporated with iron oxide nanoparticles regulate the inhibition of normal cell rupture [16].

3.1. Continuous Irradiation of MB and MB with IO-NP:

Five sets of samples were prepared with different MB concentration of 2 - 10 µg/ml. Another six sets of samples were prepared with fixed concentration of 6 µg/ml MB and different concentration of IO-NP (0 - 40 µg/ml). The prepared samples are covered with aluminum foil and kept in refrigerator to prevent it from exposure to light and temperature then irradiated continuously sample by sample at room temperature to the total rupture of the RBCs. The typical sigmoidal CPH curves are shown in Fig. 2. The curves represent the effect of different concentrations of methylene blue (2 – 10 µg/ml) on RBCs under irradiation at room temperature.

For all curves, the points in the figures represent the experimental data while the lines show their best fit with logistic function (Eq. 1), which was applied as best fit equation among several equations [17-20] for CPH curves. The power law, (Eq. 3) gives the relationship between the drug concentration and the time to rapture 50% of RBCs (***t_{50}***). Therefore; the dependence of ***$1/t_{50}$*** versus methylene blue (**MB**) concentration ***(C)*** that corresponds to Fig. 2 is presented in Fig. 4. Table 1A, shows the experimental and the theoretical times required for 50% lysis of red blood cells sensitized with different concentrations of *MB*, Logistic function fitted parameters corresponding to *CPH* measurements, average values of power dependent parameter ***(P)***, and relative steepness for lysis curves ***(S)***which is defined as; ***$S = t_{80}/t_{30}$*** , where ***t_{80}*** and ***t_{30}*** are the time to lysis 80% and 30% of the RBC`s respectively.

In Fig. 2, the increasing concentration of MB leads to decrease in time of generating ROS which in turn increases the cell death time. The relationship shown in Fig. 4 indicates that, ***t_{50}*** increases with increasing MB concentration. The high concentration of MB is relatively associated with high time percentage of cell rupture under irradiation and vice versa. Similar results were found by Fahmida et al [21]. The power dependence found to be 0.39±0.05 and a typical hemolysis curves with steepness above one was observed at different MB concentrations. The hemolysis rate, ***α*** increases with increasing concentration of MB while the time required to reduce the maximum number of RBCs to one half of its value, ***β***, decreases with increasing MB concentration. Resemblance of such result was obtained by *Ankita at el.* [22] that concludes, β depend on MB concentration, were β decreases with increasing MB concentration (See Table

Table1. (A); RBCs with different concentrations of MB, (2, 4, 6, 8, and 10 μg/ml)**. (B)** RBCs with different concentration of IO-NPs (0, 10, 20, 30, and 40 μg/ml) and fixed concentration of 6 μg/ml MB. The results are presented as mean ± S.D

Group	**$(t_{50})_{exp}$**	**C**	**Logistic function parameters**		
	(min)	**(μg/ml)**	***A***	**α (min)$^{-1}$**	**β (min)**
A	98.167	2		0.089±0.006	98.07±0.82
	82.833	4	1.014	0.104±0.007	82.56±0.67
	79.500	6	±	0.104±0.006	79.19±0.64
	68.000	8	0.029	0.121±0.007	67.83±0.49
	62.833	10		0.113±0.008	62.58±0.66
	Power dependent ***P*** = 0.39±0.05, and Steepness ***S*** = 1.25 ± 0.02				
B	51.50	0		0.145±0.009	51.20±0.34
	68.17	10	1.032	0.104±0.005	69.33±0.62
	81.17	20	±	0.075±0.003	82.44±0.76
	85.17	30	0.008	0.067±0.003	87.44±1.06
	92.17	40		0.085±0.003	93.18±0.47
	Power dependent ***P*** = 0.15±0.03, and Steepness ***S*** = 1.34 ± 0.01				

1A).

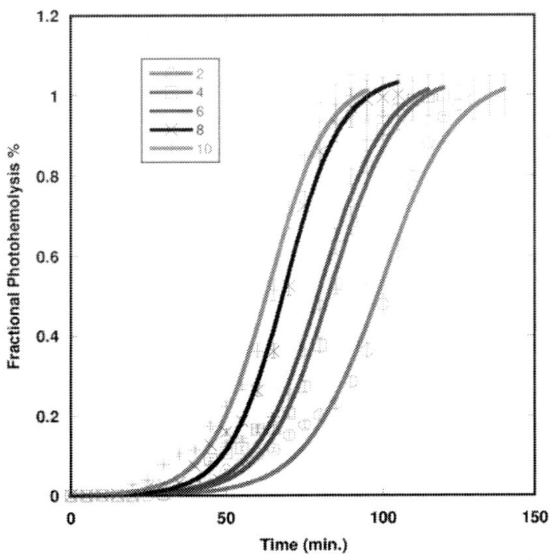

Fig. 2; Photosensitization of CPH for RBCs with different concentrations of MB irradiated at room temperature. The solid lines are theoretical fitted with Logistic function and the points are the experimental data.

The effect of different concentration of IO-NP (0 - 40 μg/ml) and fixed concentration of 6 μg/ml MB incubated at 37° C and irradiated at room temperature are presented in Fig. 3. The dependence of $1/t_{50}$ versus IO-NP concentrations are shown in Fig. 4. Table 1B shows the experimental and theoretical times required for 50% lysis of Red blood cells sensitized with different concentrations of IO-NP, Logistic function parameters corresponding to CPH measurements, average values of power dependant parameter *(P)*, and relative steepness for lysis curves *(S)*.

IOP Conf. Series: Materials Science and Engineering **92** (2015) 012003 doi:10.1088/1757-899X/92/1/012003

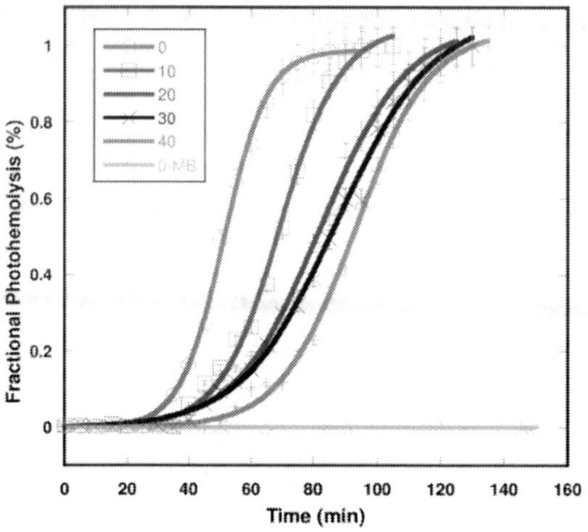

Fig. 3; Photosensitization of CPH for RBCs with different concentration of IO-NPs and fixed concentration of 6 μg/ml MB irradiated at room temperature. The solid lines are theoretical fitted with Logistic function and the points are the experimental data.

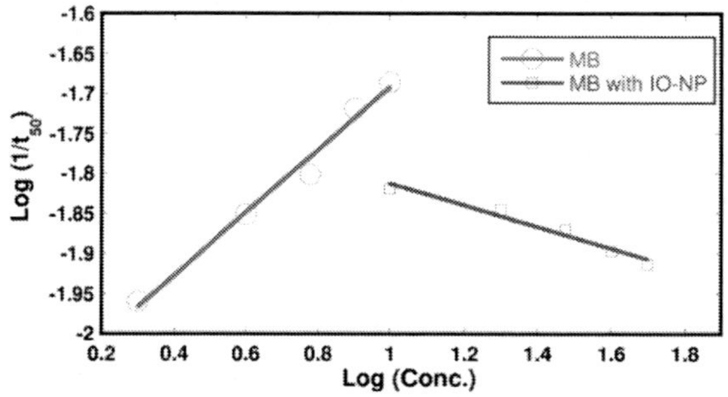

Fig. 4 Relation between $\boldsymbol{1/t_{50}}$ of continuous irradiation of MB and the concentrations results to power dependence of 0.393±0.002 and relation between $\boldsymbol{1/t_{50}}$ of continuous irradiation of IO-NP with fixed 6μg/ml MB and the concentration led to power dependence of 0.159±0.003.

The size of the spherical iron oxide nanoparticles has an average diameter of 50nm. In agreement with Aravind et al. [23], a mixture of MB with iron oxide indicated the succeeding of $^1O_2$ production and cell death efficiency *in vitro*, whereas the *MB* loaded nanoparticles are targeted directly to the cell membrane [11, 24]. Fig. 3 gives the typical photohemolysis curves of RBCs sensitized with MB and IO-NP [17, 18, 20]. Fig. 5 demonstrates high lysis time of RBCs with increasing IO-NP concentration and low percentage of cell rupturing time at low iron oxide concentration. This indicates that, the interaction that occur between MB and iron oxide influence the lysis of cell rupture as also represented by Sonal et al. [25]. This interaction enhances the activities of protein peroxides as well as reducing the intensity of light entering into the cells that reduces normal cell membrane damage.

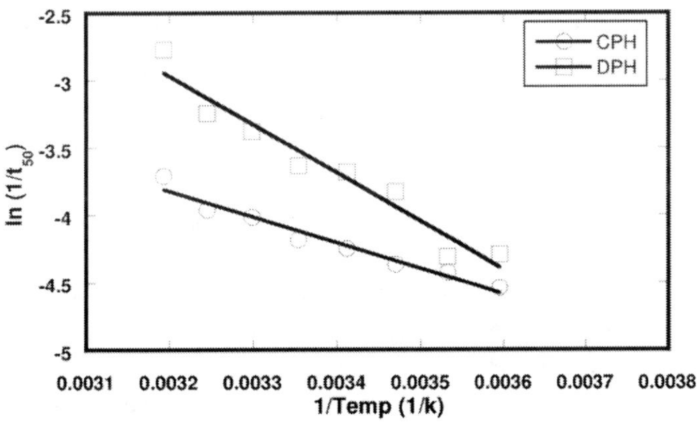

Fig. 5 Arrhenius plot for CPH and DPH dependence rate on 20 min irradiation temperatures of 5 °C, 10 °C, 15 °C, 20 °C, 25 °C, 30 °C, 35 °C and 40 °C.

Similar results using Chinese chlorella were reported by Al-Akhras et al. [26] and Fahmida et al. [21] using MB. The curves have power dependence of 0.15 ± 0.03 and relative steepness of 1.34± 0.01 with the presence IO-NP concentration. Values of the Logistic Function parameter; α decreases with increasing iron oxide concentration and the fitting parameter β; increases with increasing iron concentration. That is, β is dependent on iron oxide concentration. (Table 1B) Al-Akhras et al. [27] obtained similar using Cichorium pumilum.

3.2. Delayed Irradiation of MB and MB with IO-NP:

Eight samples were prepared with fixed concentration of 6 μg/ml MB and another set of eight samples prepared with fixed 6 μg/ml MB and fixed 20 μg/ml IO-NP. Samples were irradiated at different temperatures (5 °C - 40 °C) for 20 min at each temperature.

IOP Conf. Series: Materials Science and Engineering **92** (2015) 012003 doi:10.1088/1757-899X/92/1/012003

The Gompertz curves for delayed photohemolysis (DPH) of fixed 6 μg/ml methylene blue are shown in Fig. 6. The behavior of the RBC lysis for fixed 6 μg/ml methylene blue with 20 μg/ml IO-NP incubated at 37°C and irradiated at different temperatures varying from 5 - 40°C (Fig. 7). The dependence of $\boldsymbol{1/t_{50}}$ for the concentrations versus different temperature of irradiation obtained from the curves (Fig.6 and 7) is represented in Fig. 5. Gompertz function fitted parameters corresponding to DPH measurements (Fig. 6 & 7), average values of power dependant parameter (**P**) and relative steepness for lysis curves (**S**) are shown in Table 2. The DPH rate was known to rely on concentration of the photosensitizer, incident fluence, $\boldsymbol{T_{inc}}$ and $\boldsymbol{T_{irr}}$. In the measurements, the accelerating outcome at high $\boldsymbol{T_{inc}}$ is as a result of thermal activation of colloid-osmotic hemolysis. Higher $\boldsymbol{T_{irr}}$ is attributed to achieving practical $\boldsymbol{t_{50}}$ but photobleaching may occur at low $\boldsymbol{T_{irr}}$. This result came abreast with that of Al-Akhras [13] using protoporphyrin IX. In DPH curves, (Fig. 6 and 7) the curves represent the photohemolysis of MB and MB with IO-NP both in which at low $\boldsymbol{T_{irr}}$, higher hemolysis rate is experienced. It also reveals that, by increasing the $\boldsymbol{T_{irr}}$, the ROS generation also decrease which decrease the cell damage. The values of the Gompertz function parameter, ***a*** decreases with increasing MB and IO-NP concentration. The parameter, ***a*** and ***b*** are almost independent of $\boldsymbol{T_{irr}}$. For further analysis of the temperature dependence of intoxicating cells and to calculate the activation energy, the Arrhenius equation has been applied and described as:

$$\boldsymbol{(1/t_{50}) = Ae^{-E/RT}} \qquad \text{(Eq. 4)}$$

Where $\boldsymbol{(1/t_{50})}$ is the activation rate which is described by steepness of the killing curve and ***A*** is constant; ***E*** is the activation energy; ***R*** is the gas constant, and ***T*** is the absolute irradiation temperature.

The linear plot of $(1/t_{50})$ *vs* the reciprocal of irradiation temperature ($1/T$irr) leads to an apparent activation energy equal to 9.47±0.01 Kcal/mol for the MB with (R=0.98) and 6.06±0.03 kcal/mol for the MB+IO-NP with (R=0.97) (Fig.5). In this Figure, the $\boldsymbol{t_{50}}$ decreases with increasing irradiation temperature, $\boldsymbol{T_{irr}}$. The typical sigmoidal curves of DPH are shown in Fig. 6 and Fig. 7 fitted with Gompertz function. the fitting parameters from the curves shows that, the parameter ***a*** increase with increasing MB and iron concentration while the parameter, ***b*** is dependent on MB and MB with IO-NP concentrations (Table 2).

The slight decrease in this energy for MB with and without IO-NP is due to temperature and the light effect can be seen clearly in which higher irradiation time accelerates hemolysis much faster compared to incubation. This came alongside with the results obtained by AL-Akhras [13, 28]. From both curves, the relative average steepness are ***S***= 1.31±0.05 and 1.41±0.09.

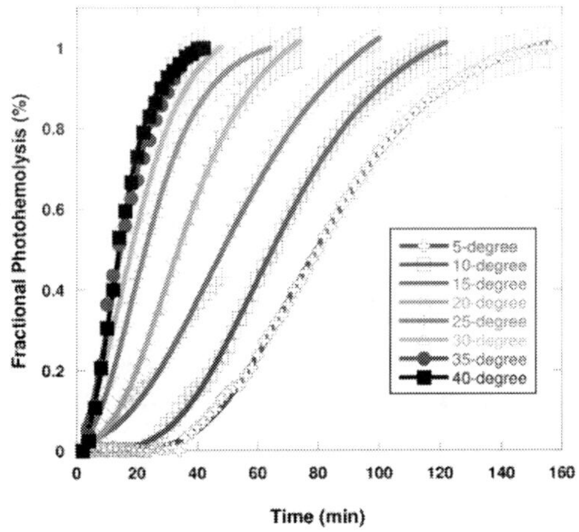

Fig. 6 Photosensitization of delayed photohemolysis (DPH) of fixed 6 μg/ml MB irradiated with arc lamp for 20 min. The solid lines are theoretical fitted with Gompertz function and the points are the experimental data.

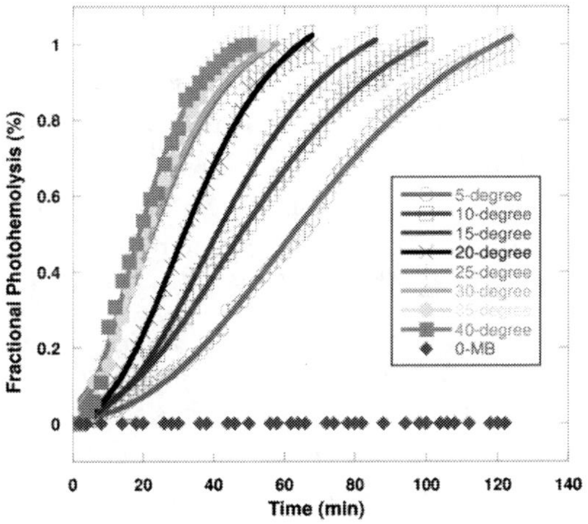

Fig. 7 Photosensitization of delayed photohemolysis (DPH) of fixed 6 μg/ml MB and fixed 20 μg/ml IO-NPs irradiated at different temperatures for 20 min each. The solid lines are theoretical fitted with Gompertz function and the points are the experimental data.

Table 2: **(A)** RBCs with fixed 6 μg/ml MB incubated at 37°C for 30min and irradiated with different temperatures, **(B)** RBCs with fixed 6 μg/ml MB and fixed 20 μg/ml IO-NPs incubated at 37°C for 30min and irradiated at different temperatures. The results are presented as mean ± S.D

Group	$(t_{50})_{exp}$ (min)	T_{irr} (°C)	Gompertz function parameters		
			Ho	*a*	*b* (min)$^{-1}$
A	80	5	1.01±0.01	12.33±0.25	0.0352±0.0003
	65	10		9.49±0.21	0.0383±0.0004
	49	15		4.58±0.16	0.0345±0.0012
	33	20		5.18±0.11	0.0558±0.0009
	23	25		4.27±0.08	0.0784±0.0011
	18	30		3.41±0.13	0.0825±0.0032
	15	35		3.85±0.31	0.1156±0.0074
	14	40		4.96±0.25	0.1399±0.0043
	Activation energy ***E*** = 9.47±0.01 Kcal/mol, and ***S*** = 1.31 ± 0.05				
B	62	5	1.03±0.01	4.95±0.12	0.0279±0.0007
	48	10		4.28±0.06	0.0343±0.0006
	41	15		4.89±0.19	0.0438±0.0015
	32	20		4.79±0.15	0.0557±0.0016
	23	25		3.32±0.10	0.0644±0.0022
	22	30		3.50±0.12	0.6882±0.0023
	21	35		4.35±0.14	0.0849±0.0021
	19	40		3.96±0.26	0.0888±0.0480
	Activation energy ***E*** = 6.06±0.03 Kcal/mol, and ***S*** = 1.41 ± 0.09				

4. Conclusion:

In this work we reported the photosensitivity of the methylene blue associated with Iron Oxide Nanoparticles (Fe_3O_4) on red blood cells. Both delayed and photohymolysis were assessed under different concentration and different irradiation temperature. Logistic and Gompertz functions were valid to fit the experimental results and to predict the cell lysis parameters. For both CPH fitted with Logistic equation and DPH fitted with Gompertz function, the combination of different concentration of MB and Iron-Oxide nanoparticles inversely affect the RBC`s $t_{50}$ for different $T_{irr}$. Moreover the Photoactivation of MB associated with iron oxide nanoparticles causes a delay in cells membrane rupturing. Iron-Oxide nanoparticles in combination with MB have no significant effect on the RBC`s rupturing rate. The power dependent factor for CPH has less effect in MB with and without iron oxide concentrations but has significant effect at T_{irr} for DPH. Furthermore, the T_{irr} has more impact on the photohemolysis rate in the presence of MB as compared to that of MB plus IO-NP.

Acknowledgements: This work was supported by a grant from the Deanship of Scientific Research at Jordan University of Science and Technology (JUST) No. 185/2013. The Authors would like to thank Miss Duaa Al-Khalili for her help and useful discussions.

References:

[1] Darrell RS, Carucci JA 2000 Malignant Melanoma: prevention, early detection, and treatment in the 21st century, *CA: A Cancer Journal for Clinicians.* **50** 215.

[2] His RA, Rosenthal DI, Glatstein E 1999 Photodynamic therapy in the treatment of cancer. *Drugs* **57** 725.

[3] Mills E, Ernst E, Singh R, Ross C, Wilson K, 2003 Health food store recommendations: implications for breast cancer patients. *Breast Cancer Res.* **5** R170.

[4] Nyman ES, Hynninen PH. 2004 Research advances in the use of tetrapyrrolic photosensitizers for photodynamic therapy. *Journal of Photochemistry and Photobiology B: Biology.* **73** (1-2) 1.

[5] Andre SR, Pelegrino AC, Tedesco AC, Jorge RA. 2008 Photodynamic activity of chloro (5, 10, 15, 20-tetraphenylporphyrinato) indium (III). *Journal of the Brazilian Chemical Society.* **19** 491.

[6] Cai w, Gao T, Hong H, Sun J. 2008 Applications of gold nanoparticles in cancer nanotechnology. *Nanotechnology, Science and Applications.* **1** 17

[7] Aljarrah K, Mhaidat NM, Al-Akhras MA, Aldaher AN, Albiss BA, Aledealat K, et al. 2012 Magnetic nanoparticles sensitize MCF-7 breast cancer cells to doxorubicin-induced apoptosis. *World J Surg Oncol.* **10** 62.

[8] Dong HD, Guang LC, Hai L. 2014 Observation of the dynamics of magnetic nanoparticles induced by a focused laser beam by using dark-field microscopy. *Journal of Magnetism and Magnetic Materials.* **263**103.

[9] Tada DB, Vono LL, Duarte EL, Itri R, Kiyohara PK, Baptista MS, et al. 2007 Methylene blue-containing silica-coated magnetic particles: a potential magnetic carrier for photodynamic therapy. *Langmuir.* **23** 8194.

[10] Sonoda M, Krishna CM, Riesz P. 1987 The role of singlet oxygen in the photohemolysis of red blood cells sensitized by phthalocyanine sulfonates. *Photochemistry and photobiology.* **46** 625.

[11] Wei T, Xu H, Kopelman R, Philbert MA. 2005 Photodynamic Characterization and In Vitro Application of Methylene Blue - containing Nanoparticle Platforms. *Photochemistry and photobiology.* **81** 242.

[12] Fujikawa H, Kai A, Morozumi S. 2003 A new logistic model for bacterial growth, Shokuhin eiseigaku zasshi. *Journal of the Food Hygienic Society of Japan.* **44** 155.

[13] Akhras M-Ali. 2006 A new application of Gompertz function in photohemolysis: the effect of temperature on red blood cells hemolysis photosensitized by protoporphyrin IX. *Med. and Bio Eng & computing.* **44** 703.

[14] Xinyou Y, Goudriaan JA, Lantinga EA, Vos JA, Spiertz HJ. 2003 A flexible sigmoid function of determinate growth. Annals of Botany. **91** 361.

[15] West JL, Halas NJ. 2000 Applications of nanotechnology to biotechnology: Commentary. Current opinion in Biotechnology. **11** 215.

[16] Khan SS, Alam F, Azam A, Khan AU. 2012 Gold nanoparticles enhance methylene blue–induced photodynamic therapy: a novel therapeutic approach to inhibit Candida albicans biofilm. International journal of nanomedicine. **7** 3245.

[17] Chetan TG. 2012 Analyzing the dynamics of cell growth and protein production in mammalian cell fed-batch systems using logistic equations. *Journal of industrial microbiology & biotechnology.* **39** 1061.

[18] Li MY, Sun XM, Zhao GM, Huang XQ, Zhang JW, Tian W. et al. 2013 Comparison of Mathematical Models of Lactic Acid Bacteria Growth in Vacuum Packaged Raw Beef Stored at Different Temperatures. *Journal of food science.* **78** 600.

[19] Budimulyati S, Noor RR, Saefuddin A, Talib C. 2012 Comparison on accuracy of logistic, Gompertz and von Bertalanffy models in predicting growth of new born calf until first mating of Holstein Friesian heifers. *Journal of the Indonesian Tropical Animal Agriculture.* **37** 151.

[20] William K. H. 2012 Logistic Function Profile Fit: A least-squares program for fitting interface profiles to an extended logistic function. *Journal of Vacuum Science & Technology A.* **30** 051101.

[21] Fahmida G, Yanagida S, Kameshima Y, Isobe T, Nakajima A, Okada K. 2010 Various factors affecting photodecomposition of methylene blue by iron-oxides in an oxalate solution. *Water Research.* **44** 2876.

[22] Ankita A, Rakshit A, Mamta A. Photocatalytic 2013 Degradation of Methylene Blue over Ferric Tungstate. *Science Revs. Chem. Commun.* **3** 172.

[23] Aravind A, Nair R, Raveendran S, Veeranarayanan S, Nagaoka Y, Fukuda T, et al. 2013 Aptamer conjugated paclitaxel and magnetic fluid loaded fluorescently tagged PLGA nanoparticles for targeted cancer therapy. *Journal of Magnetism and Magnetic Materials.* **344** 116.

[24] Bechet D, Couleaud P, Frochot C, Viriot ML, Guillemin F, Barberi-Heyob M. 2008 Nanoparticles as vehicles for delivery of photodynamic therapy agents. *Trends in Biotechnology.* **26** 612.

[25] Sonal S, Sharma R, Singh C, Bansal S. 2013 Enhanced Photocatalytic Degradation of Methylene Blue Using/MWCNT Composite Synthesized by Hydrothermal Method. *Indian Journal of Materials Science.* **10** 356025.

[26] Al-Akhras MA, Aljarrah K, Makhadmeh GN, Shorman A. 2013 Introducing the Effect of Chinese Chlorella as a Photosensitizing Drug at Different Temperatures. *Journal of Molecular Pharmaceutics & Organic Process Research.* **1** 1000.

[27] Al-Akhras MA, Amin A, Mohammad K, AlHaddad F, Hamza A. 2007 Sensitization of Photohemolysis by a New Extraction from Flowers and Aerial Parts of Cichorium Pumilum Jacq: Effects of Inulin and Hydrogen Peroxide. *American Journal of Pharmacology and Toxicology.* **2** 75.

[28] Al-Akhras MA, and Al-Khalili D. 2015 The Influence of Iron Oxide Nanoparticles on the Red Blood Cells Photohemolysis Sensitized with Photofrin: Temperature Effect. *Jordan Journal of Biological Sciences* **8**(1) 55.

Structural Studies of Silver Nanoparticles Obtained Through Single-Step Green Synthesis

Siva Prasad Peddi[1]*, Bilal Abdallah Sadeh[1]

Department of Physics, College of Arts and Sciences, Aljouf University,

Al Qurayyat, KSA

***e-mail:** prof.siva@ju.edu.sa

Abstract: Green synthesis of silver Nanoparticles (AGNP's) has been the most prominent among the metallic nanoparticles for research for over a decade and half now due to both the simplicity of preparation and the applicability of biological species with extensive applications in medicine and biotechnology to reduce and trap the particles. The current article uses Eclipta Prostrata leaf extract as the biological species to cap the AGNP's through a single step process. The characterization data obtained was used for the analysis of the sample structure. The article emphasizes the disquisition of their shape and size of the lattice parameters and proposes a general scheme and a mathematical model for the analysis of their dependence. The data of the synthesized AGNP's has been used to advantage through the introduction of a structural shape factor for the crystalline nanoparticles. The properties of the structure of the AGNP's proposed and evaluated through a theoretical model was undeviating with the experimental consequences. This modus operandi gives scope for the structural studies of ultrafine particles prepared using biological methods.

1. Introduction

Bio-convivial green synthesis of nanoparticles (NPs) in general and Silver Nanoparticles (AGNPs) in particular have been in vogue for a long time now with pronounced applications in Physics, Chemistry, and Medical research. Ethno-botanical species for mitigating and surmounting the NPs are being pursued as a large scale substitute to overcome many complex problems that humans have been facing while organizing themselves in a non-chaotic way, withal due to the achievability of desired dimensions and distribution without loss of morphology[1-7].

The species of plant in the family Asteraceae viz., Eclipta prostrata (syn. Eclipta prostrata) has been used for the synthesis of AGNPs[21-22, 28].

Due to the sui generis electrical, optical, conductivity, and thermal properties of AGNPs they possess unbound appositeness in different territories of sciences [8-19].The extrapolation of the complex structural characteristics needs further insights and studies on its distinction from the bulk characteristics, though extensively in the literature[1-16].

The current study emphasizes the dovetailing of the size, shape, surface, and aggregation state of the AGNPs mathematically adopting basic principles of the properties of matter to study of the relative changes of the lattice parameters of the NPs relating the elastic energy to the surface energy of the NPs per unit area[20].

2. Experimental Method

Aqueous solution of silver nitrate (1mM) and the extract of Eclipta Prostrata used as the reaction medium have been mixed in the proportion of 1:9 and the contents were maintained at room temperature for about two hours until the reduction and capping of the AGNPs was distinctly visible through the change in the color of the solution to brown, which was the initial confirmation test of the formation of AGNPs. Later the sample was repeatedly centrifuged and a colloid of AGNPs has been separated and stored for characterization.[28]

The process of characterization of the sample has been carried out using SEM, XRD, FTIR, UV-Visible spectrophotometer, which have been reported in detail elsewhere by the same authors[28]. The results have been compendiously summed up to sustain continuity of the analysis of the structural studies of the AGNPs.

The structural features of the synthesized silver nanoparticles using Eclipta prostrata leaf extract were characterized using SEM analysis (PHILIPS-XL-30SEM) machine. It depicted high density nanoparticles synthesized by the Eclipta prostrata leaf extract and are relatively spherical in shape. X-ray diffraction has been a convenient method for determining the mean size of single-crystal nanoparticles.

After repeated centrifugation followed by re-dispersion of the pellet of silver nanoparticles into 10 ml of sterile distilled water and freeze drying the purified silver nanoparticles, the structure of the synthesized silver nanoparticles have been investigated with an XRD (RIGAKU-D Machine). Indexing process has been done and *Miller Indices* (h k l) to each peak have been assigned during the first step. The XRD results therefore suggest that the crystallization of the bioorganic phase occurs on the surface of the silver nanoparticles [29]. The particles have size ranging between 11 to 60 nm. From this study, considering the peak at degrees, average particle size has been estimated by using Debye-Scherrer formula [40,41]. And the average size of AGNPs thus synthesized was 34 nm

3. Results and Discussion

The Chemical interaction between the phytochemicals present in Eclipta Prostrata leaf extract and AgNPs and how it has led to stabilization of AgNPs has been briefly presented here and has been reported by the same authors elsewhere[28].

The FTIR spectrum of Ag nanoparticle is shown in Fig 6. The spectrum of Ag nanoparticles show stretching at (units: wavenumber cm-1) 3106 , 2962, 2885, 2791 , 2601 corresponds to O-H stretching for Carboxylic acids stretching, 2274 corresponds to N-H bond stretching of amines, 2169 corresponds to C-N stretching of the aromatic amino group and C-O stretching of alcohols and ethers respectively, 1126 corresponds to C–O stretch alcohols, 1392, 1436, 1497 corresponds to CH2 bend alkenes, 609, 685 and 890 corresponds to C–H bend alkenes. These stretching values indicate that the carbonyl group formed amino acid residues and that these residues ‘‘capped’’ the silver nanoparticles to prevent agglomeration, thereby stabilizing the medium. When the metal nanoparticles form in solution, they must be stabilized against the van der Waals forces of attraction which may otherwise cause coagulation. Physisorbed surfactant and polymers may cause steric or electrostatic barriers or purely electrostatic barriers around the particle surface and may thereby provide stabilization. FTIR peaks that were corresponding to aromatic rings, germinal methyls, and ether linkages indicate the presence of flavones and terpenoids responsible for the stabilization of the AgNPs synthesized by the Sesuvium portulacastrum leaf extract. The FTIR spectrum of Ag nanoparticles suggested that Ag nanoparticles were surrounded by different organic molecules such as terpenoids, alcohols, ketones, aldehydes and carboxylic acids.

3.1 Structural studies using UV-visible Spectra

The brown colour in the aqueous solution has been due to the surface Plasmon vibrations[29]. The polyol compounds, proteins, metabolites present in the plant system, the electrons released during glycolysis and the water soluble heterocyclic compounds were mainly responsible for the reduction of the silver ions and the stabilization of the nanoparticles respectively[30-33]. The UV-Vis spectral studies used to examine size and shape controlled nanoparticles in aqueous suspensions revealed an absorption peak at 416 nm and the broadening of the peak indicated that the particles are poly-dispersed[33, 38-39].

The much heavier ionic core of these particles induce polarization of the electrons from an incoming wave of the electric field[39]that would result in a restoring force which creates an in phase dipolar oscillation.

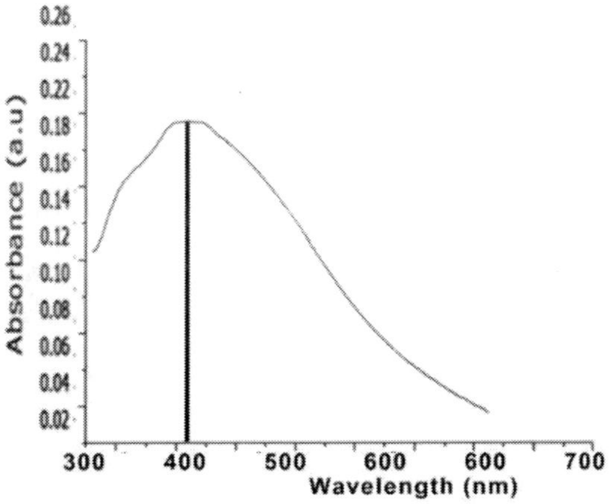

Figure 1. UV-Visible spectra of Ag nanoparticles

And the origin of the observed colour has been due to a strong absorption owing to the resonance between the frequency of the EM field and coherent electron motion. The difference in the energies of the conduction and valency bands in the case of AG is very low permitting free movement of electrons thus, leading to the oscillation of free electrons of AGNPs in resonance with the light wave ultimately resulting in SPR [37, 40-42]. Furthermore , UV-Vis a paramount tool of analysis of the metastable solution of single particles [43, 45]. Beyond this metastable state the dominant Van der Waal's forces sequel the cluster formation [45-46]. The flattened spectrum indicated the saturated state of the reaction and reduction of AGNPs for this concentration of the Eclipta Prostrata leaf extract. [47-48]. The framework of Mie scattering theory to analyze optical spectra has been applied for the determination of the particle size of the AGNPs present in the stable suspension [34-52]. The *full width at half maximum*(FWHM) of the peak ω has been obtained from the spectral graph Fig.1 as follows.

$$\omega = \frac{(\epsilon_0 + 2n^2)c\ m\ u_F}{2\,N_c\,e^2\,D}$$

where, and , the frequency independent part of complex form of the particle are, ε_0 – dielectric constant = .49, n - refractive index = 0.15016, c - velocity of light, m - mass of the electron, u_F - electron velocity at the Fermi Energy, N_c - Number of electrons per unit volume, e - charge of the electron, and D - Diameter of the particle. The particle diameter has been determined. It was found to be ~ 27 nm. The UV-vis spectra are fitted using log-normal function $P(D)dD = \frac{1}{\sqrt{2\pi}\sigma D} e^{\left(-\frac{ln^2\left(\frac{D}{D_0}\right)}{2\sigma^2}\right)}$ to obtain the standard deviation (σ) as the system of the poly dispersed nanoparticles obeys log-normal size distribution function. Using σ and the mean particle size obtained particle size distribution curve has been generated as in Fig.3.

4. Model to study the variation of the lattice parameters

The synthesized AGNPs have been observed to be of a varying size and shape, the direct repercussion of which would be on the lattice parameters(LP). There would be an abstraction of the LPs as an inverse function with the particle size.

Recently a general mathematical model has been developed and employed [20, 55-59] to account for size effect on the LPs of the AGNPs. According to this method initially a particle from the bulk has been isolated assuming no loss of the structural properties. The surface tension of such an isolated particle would contract its size elastically, and finally aid its formation in an equilibrium state tending to attain a shape that possesses minimum energy.

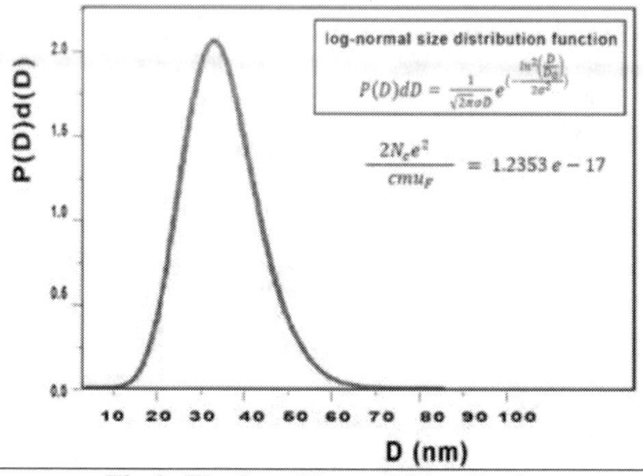

Fig.2. The Log-Normal Distribution

A shape factor α [55] has been introduced to include non-spherical NPs. By minimizing the sum of the increased surface energy and the elastic energy, a formula has been obtained in the generation of the lattice parameters. These lattice parameters have been compared with the corresponding experimental values. For an ideal crystal lattice, the lattice parameter contraction is proportional to the diameter of nanoparticle

$$\frac{\Delta a}{a} = \frac{(a_p - a)}{a} = \frac{(1-\epsilon)R - R}{R} \quad (1)$$

where a_p and a are the lattice parameters of the nanoparticle and the corresponding bulk material. Inserting Equation $\epsilon = \frac{1}{1+\left(\frac{2\sigma}{\gamma}\right).R\,.\alpha^{\frac{1}{2}}}$ From eqn. (10) in reference [20] the study of the variation of the LPs of the AGNPs have been possible.

σ Shear module (298 K) [60]	**3.03 X 10^{10} (N/m^2)**
γ_0 Surface energy (0 K) [61]	**1.250 (J/m^2)**
dγ/dT coefficient of surface energy to Temp. [61]	**- 15 X 10^{-5} ($J/(K\ m^2)$)**
γ Surface energy (298 K) (From Eqn.(3))	**1.205 (J/m^2)**

Table 1. The input physical parameters for the Silver Nanoparticles

Value of Alpha	Value of D in (nm) 27	28	48	55.2
1	-0.170	-0.1639	-0.095	-0.0832
1.15	-0.158	-0.152	-0.089	-0.0776
1.18	-0.156	-0.1509	-0.0881	-0.0766
1.24	-0.1527	-0.1472	-0.0859	-0.0747
1.49	-0.139	-0.134	-0.0784	-0.0682
1.5	-0.138	-0.133	-0.0781	-0.0679

Table 2: Lattice parameter variation

$$\frac{\Delta a}{a} = \frac{1}{1+K.D} \qquad (2)$$

Equation (2) is the basic relation for the size and shape dependent lattice parameters of metallic nanoparticles,

where D (= 2R) is the diameter of the nanoparticle, and $\boldsymbol{K} = \frac{\alpha^{\frac{1}{2}}\sigma}{\gamma}$.

Generally, both of the shear modulus and the surface energy are positive; therefore, the lattice parameter of the metallic nanoparticles will decrease with decrease in particle size. The shape dependent lattice parameters of nanoparticles in specific size have been simulated using MATLAB program and the results have been presented below.

There was a decrease in the variation of the lattice parameter with increased shape factor, which in turn depended on the particle dimensions and has been stronger in the case of smaller particle sizes contributing about 10-15% to the lattice parameter variation as depicted in Fig.3. As the model has been successful in bringing out a definitive relation to the experimental results to that of the theory with which many new characteristics of the nanostructures can be understood, it stands out as a 'continuous media model' and can be used to predict the lattice parameter variation of metallic nanoparticles.

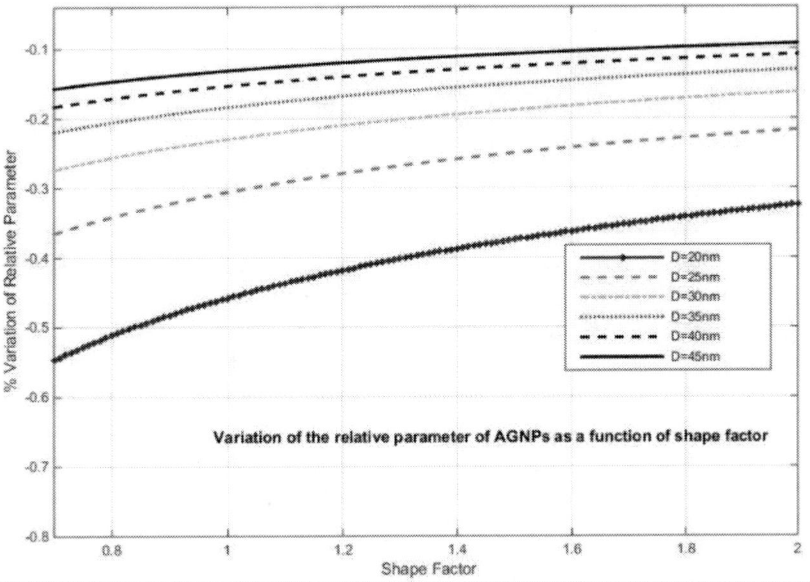

Figure 3 Variation of the relative parameter of AGNPs as a function of shape factor.

5. Conclusions:

The UV- Visible spectra confirmed the reduction of silver ions at 416 nm. The dispersions of silver nanoparticles displayed intense colours due to the plasmon resonance absorption inveigled by the size, shape, inter-particle interactions, free electron density and surrounding medium. The dependence of bandwidth upon particle size $\omega \sim 1/D$, results from the changing mean free path of electron[38] and shows that the bandwidth increases with the decrease of the cluster size. The normalized absorption coefficient which matches with the experiment within the experimental error has been presented in Fig.4. The mathematical model employed was successful in accounting for the size and shape dependent lattice parameters of metallic nanoparticles. And the results were consistent with the experiment. It has also been observed that about 10-15% variation of the LPs has been due to the particle shape. The model uses physical constants of the bulk material and is a continuous media model applicable for particles size up to about 90nm, beyond which the results deter.

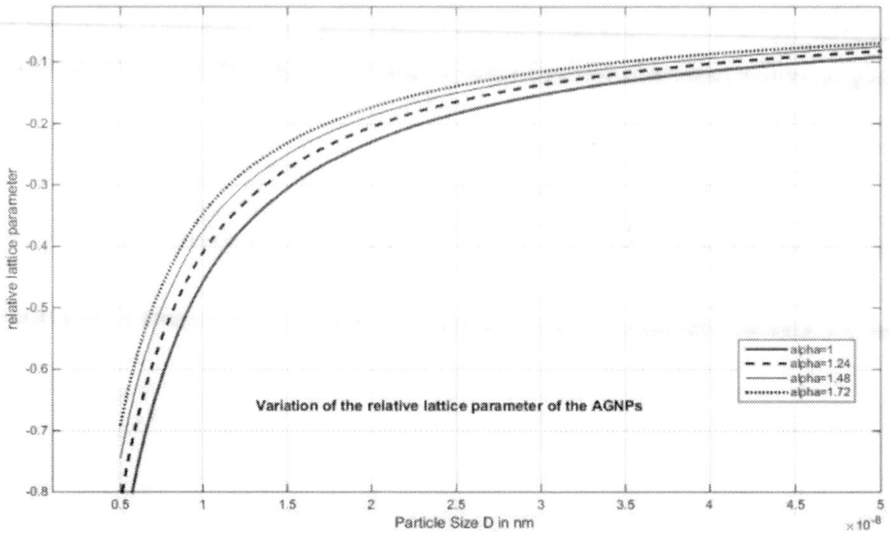

Figure 4. Variation of the relative lattice parameter as the function of the diameter of AGNPs

This model has been successful for the LP variation studies and it is being explored further to investigate theoretically complex structural characteristics of NPs and to study the variation of dielectric and elastic properties and their influence in the controlled synthesis of nanoparticles.

Acknowledgements:

This Article was presented in the International Conference Advanced Materials (ICAM2015) held at Jordan University of Science and Technology, Irbid, Jordan between April 27-29, 2015 by one of the authors Prof. Siva Prasad Peddi who expresses his sincere thanks to the Organizing Committee for their support and hospitality during the conference.
The authors are grateful for the help and support of Prof. M.V Ramana during the characterization process. The authors are thankful to their employer M/S Aljouf University, Saudi Arabia.

References:

[1] C. Richard, A. Jr. Charles, and S. Wilson, Encyclopaedia of Materials Characterization - Surfaces, Interfaces, Thin Films, Elsevier (1992).
[2] Fabrication and characterization of metal oxide nanostrutures submitted by Li Dan, University of Hong Kong, July (2007).
[3] J. B Harborne and H. Baxter, 1993. Phytochemical Dictionary: A Handbook of Bioactive Compounds from Plants. Burgress Science Press, Taylor and Francis Ltd, pp: 479-480.
[4] Sofowora A, Medicinal plants and Traditional Medicine in Africa. Spectrum Books, Ibadan. 1993.
[5] Trease GE and Evans WC, Pharmacognosy, BaillièreTindall, London, 13th edition 1989; P176-180.
[6] Bhattacharya, D.; Gupta, R. K. Nanotechnology and potential of microorganisms. Crit. Rev. Biotechnol. 2005, 25 (4), 199−204.
[7] Kim, B. S.; Song, J. Y. Biological Synthesis of Gold and Silver Nanoparticles Using Plant Leaf Extracts and Antimicrobial Applications. In Biocatalysis and Biomolecular Engineering; Hou, C. T., Shaw, J. F., Eds.; John Wiley & Sons, Inc.: Hoboken, NJ, pp 447−457, 2010.
[8] Vedpriya A. Living Systems: eco-friendly nanofactories. Digest Journal of Nanomaterials and Biostructures 2010; 5 suppl 1: 9–21
[9] Collera Zufliga O Garcia & Jimenez &Meléndez Gordillo, R. (2005). Comparative study of carotenoid composition in three mexican varieties of Capsicum annuum L. Food Chemistry, 90(1), 109-114.

[10] Jagadeesh, B. H., Prabha, T. N., & Srinivasan, K. (2004). Improved shelf life of bell capsicum fruits by manipulation of the activities of glycosidases through heat treatment. Indian journal of plant physiology, 9(2), 164-168.
[11] Wang, C., Flynn, N. T., & Langer, R. (2004). Controlled structure and properties of thermoresponsive nanoparticle–hydrogel composites. Advanced Materials, 16 (13), 1074-1079.
[12] Biswas, A., Aktas, O.C., Schumann,U., Saeed, U., Zaporjtchenko, V. and Faupel, F. (2004). Tunable multiple plasmon resonance wavelengths response from multicomponent polymer-metal nanocomposite. CORE: core.kmi.open.ac.uk/display/1015690
[13] Shipway, A. N., & Willner, I. (2001). Nanoparticles as structural and functional units in surface-confined architectures.ChemicalCommunications,(20),20352045.
[14] Nie, S and Emory, S. R. (1997) Probing single molecules and single nanoparticles by surface enhanced Raman Scattering.Science. 275, 1102 - 1106.
[15] Govindraju, K., Kiruthiga, V and Singaravelu, G. (2008) Evaluation of biosynthesized silver nanoparticles against fungal pathogens of mulberry Morus indica. Journal of Biopesticides, 1(1),101 - 104.
[16] Govindraju, K., Kiruthiga, V., Ganesh Kumar, V and Singaravelu, G. (2009), Extracellular synthesis of silver nanoparticles by a marine alga, Sargassum wightii Grevilli and their antibacterial effects. Journal of Nanoscience and Nanotechnology, 9, 5497 - 5501.
[17] Lok C, Ho C, Chen R, He Q, Tu W, Sun H, Tam P K, Chui J, Che C (2007) Silver nanoparticles: Partial oxidation and antibacterial activities. J. Biol. Inorg. Chem. 12: 527-534.
[18] Ip M, Lui SL, Poon VKM, Lung I, Burd A (2006) Antimicrobial activities of silver dressings: an in vitro comparison J. Med. Microbiol. 55:59-63.
[19] Jain D, Daima HK, Kachhwaha S, Kothari SL (2009) Synthesis of plant-mediated silver nanoparticles using papaya fruit extract and evaluation of their antimicrobial activities. Digest J. Nanomater. Nanostruct. 4(3): 557-563.
[20] Siva Prasad Peddi, Bilal Abdallah Sadeh - Study Of The Contraction Of The Lattice Parameters of Silver Nanoparticles Synthesized Using Achyranthes Aspera L Stem Extract - Proceedings of the Saudi Physical Society's, 2nd Annual Meeting Atomic, Molecular And Optical Physics (Amop) - Al-Jouf University Campus - March 31 – April 1, 2015.
[21] Thorat A, Jadhav V, Gaikwad D, Jadhav S. Phytochemical and Pharmacological potential of Eclipta alba: A review. Int Res J Phar 2010; 1(1): 77-80.
[22] Prakash K.M.M.S., Naidu P.V.S. and Muralidhar Promosing phytochemicals from medicinal plant eclipta alba International Journal Of Pharmacy&Technology. 2011; Vol. 3: 2868-2873
[23] Reddy.M, Reddy K, Reddy M. A Survey of Plant Crude Drugs of Anantapur District, Andhra Pradesh, India, Int. J. Crude Drug Res 1989; 27(3), 145-55.
[24] Elisa P, Lopes D, Da Silva A, Pimenta P, Leitao, Noel F, Costa P. Structure–activity relationship of wedelolactone analogues: Structural requirements for inhibition of Na+,K+-ATPase and binding to the central benzodiazepine receptor. Bioorg Med Chem 2006; 14: 7962-66.
[25] Miguel C, Maria G, Maria D, Thereza C, Antonio P. Assessment of luteolin (3, 4, 5, 7-tetrahydroxyflavone) neuropharmacological activity. Behav Brain Res 2008; 189: 75–82.
[26] Elumalai EK, Prasad TNVKV, Hemachandran J, Viviyan Therasa S, Thirumalai T, David E (2010) Extracellular synthesis of silver nanoparticles using leaves of Euphorbia hirta and their antibacterial activities. J. Pharma. Sci. Res. 2(9): 549-554.
[27] K. Sahayaraj and S. Rajesh, Science against Microbial Pathogens: Communicating Current Research and Technological Advances, Microbiology Book Series - Number 3, Edited by A. Méndez- Vilas Vol. 1, ISBN-13: 978-84-939843-1-1
[28] Siva Prasad Peddi, Bilal Abdallah Sadeh and M. V. Ramana - International Journal of Physics, and Research (IJPR)ISSN(P): 2250-0030; ISSN(E): 2319-4499 Vol. 5, Issue 1, Feb 2015, 7-16 © TJPRC Pvt. Ltd.
[29] Shankar S, Rai SA, Ahmad A, Sastry M (2004) Rapid synthesis of Au, Ag, and bimetallic Au core–Ag shell nanoparticles using neem (Azadirachta indica) leaf broth. Colloid Interface Sci. 275:496-502
[30] Huang J, Li Q, Sun D, Lu Y, Su Y, Yang X, Wang H, Wang Y, Shao W, He N, Hong J, Chen C (2007) Biosynthesis of silver and gold nanoparticles by novel sundried Cinnamomum camphora leaf. J. Nanotechnol. 18: 105104-105115.
[31] Balaji SD, Basavraja S, Deshpande R, Mahesh BD, Prabhakar KB, Venkatraman A (2009)

[32] LingaRao M, Savithramma N (2012) Antimicrobial activity of silver nanoparticles synthesized by using stem extract of Svensonia hydrobadensis (Walp.) a rare medicinal plant. Res. Biotechnol. 3(3): 41-47. Nalwade et al. / Research in Pharmacy 3(3): 22-28, 2013
[33] Wiley BJ, Im SH, McLellan J, Siekkinen A, Xia Y (2006) Maneuvering the surface plasmon resonance of silver nanostructures through shape-controlled synthesis. J. Phys. Chem. B110: 15666.
[34] L. Baia, S. Simon, A. Mendez-Vilas, and J. Diaz, *Modern Research and Educational Topics in Microscopy eds.* 576 (2007).
[35] G. Wang, Ch. Shi, N. Zhao, and X. Du, *Mater. Lett.* 61, 3795 (2007).
[36] D. L. V. Hyning and C. F. Zukoski, *Langmuir* 14, 7034 (1998).
A. Taleb, C. Petit, and M. P. Pileni, *J. Phys. Chem. B* 102, 2214 (1998).
[37] M. A. Noginov, G. Zhu, M. Bahoura, J. Adegoke, C. Small, B. A.Ritzo, V. P. Drachev, and V. M. Shalaev, *Opt. Lett.* 31, 3022 (2006).
[38] S. Link and M. A. El-Sayed, *Annu Rev. Phys. Chem.* 54, 331 (2003).
[39] U. Kreibig and M. Vollmer, Optical Properties of Metal Clusters, Springer, Berlin (1995).
[40] S. S. Nath, D. Chakdar, and G. Gope, *Nanotrends—A Journal of Nanotechnology and Its Application* 02 (2007).
[41] R. He, X Qian, J. Yin, and Z. Zhu, *J. Mater. Chem.* 12, 3783 (2002).
[42] S. K. Ghosh and T. Pal, *Chem. Rev.* 107, 4797 (2007).
[43] C. J. Addison and A. G. Brolo, *Langumir* 22, 8696 (2006).
[44] M. Meyer, E. C. Ru, and P. G. Etchegoin, *J. Phys. Chem. B* 110, 6040 (2006).
[45] C. S. Seney, B. M. Gutzman, and R. H. Goddard, *J. Phys. Chem. C*113, 74 (2009).
[46] K. C. Grabar, G. R. Freeman, M. B. Hommer, and M. J. Natan, *Anal. Chem.* 67, 735 (1995).
[47] R. G. Freeman, M. B. Hommer, K. C. Grabar, M. A. Jackson, and M. J. Natan, *J. Phys. Chem.* 100, 718 (1996).
[48] Z. Zhang and Y. Wu, *Langmuir* 26, 9214 (2010).
[49] P. Mulvaney, *Langmuir* 12, 788 (1996).
[50] P. Mulvaney, J. Perez-Juste, M. Giersig, L. M. Liz-Marzan, and C. Percharroman, *Plasmonics* 1, 61 (2006).
[51] G. Mie, Ann. Phys. Lpz 25, 377 (1908).
[52] Qi W H, M. P. Wang and Y. C. Su, Size effect on the lattice parameters of nanoparticles, Journal of Materials Science Letters, 21（2002）877-878.
[53] Qi W H, M. P. Wang, Size effect on the cohesive energy of nanoparticle, Journal of Materials Science Letters ，21（2002）1743-1745
[54] Qi W H, M. P. Wang, Size dependence of vacancy formation energy of nanoparticle, Physica B, 334（2003）432-435
[55] Qi W H, M. P. Wang, G. Y. Xu, The partical size dependence of cohesive energy of metallic nanoparticles, Chemical Physics Letters 372 (2003) 632-634.
[56] Qi W H, M. P. Wang, Size and shape dependent melting temperature of metallic nanoparticles, Materials Chemistry and Physics 88（2004）280-284
[57] Alymov M.I. & M.K. Shorshorov, 1999. Surface tension of ultrafine particles. Nanostruct. Mater. 12, 365–368.
[58] QI W H, M. P. Wang, Size and shape dependent lattice parameters of metallic nanoparticles，Journal of Nanoparticle Research, 7 (2005) 51–57.
[59] Miedema A.R., 1978. Surface energies of solid metals. Z. Metallkde 69, 287–292.
[60] Brands E.A., 1983. Smithells Metals Reference Book. 6th edn. Butterworths, p. 15–2.
[61] Miedema A.R., 1978. Surface energies of solid metals. Z. Metallkde 69, 287–292.

IOP Conf. Series: Materials Science and Engineering **92** (2015) 012005 doi:10.1088/1757-899X/92/1/012005

Synthesis, characterization and electrochemical properties of metal-doped nanoporous carbon

N Ben Mansour[1], M Hjiri[1,2], R Dahari[1,2], L El Mir[1,3], M Bonyani[2,4], A Mirzaei[2,4], S G Leonardi[2], G Neri[2]*

[1]Lab. of Physics of Materials and Nanomaterials Applied at Environment, Faculty of Sciences of Gabes, Tunisia.
[2]Dept. of Electronic Engineering, Chemistry and Materials Engineering, Univ. of Messina, Italy.
[3]Dept. of Physics, Al Imam Mohammad Ibn Saud Islamic University, Riyadh, Saudi Arabia.
[4]Dept. of Engineering, Materials Science and Engineering, Shiraz University, Iran.

E-mail: gneri@unime.it

Abstract. A novel nickel/nanoporous carbon (Ni-NC) composite has been synthesized via sol-gel method after the incorporation of metal oxide nanoparticles in an organic matrix based on pyrogallol-formaldehyde and successive heating under inert atmosphere. The nanocomposites have been characterized by various techniques such as XRD, SEM-EDX and TEM. The Ni-NC composite was successfully used to fabricate a modified screen printed carbon electrode (SPCE) and applied in the determination of glucose. The electrochemical behavior of the Ni-NC nanocomposite has been investigated by cyclic voltammetry performed under different experimental conditions (potential, scan rate and analyte concentration). Based on the modified Ni-NC/SPCE electrode, an amperometric sensor for glucose determination with a sensitivity of 146.4 $\mu A\ mM^{-1}cm^{-2}$and lower detection limit of 10μM, has been developed.

1. Introduction

Glucose is one of the vital substances for life activities, as it is widely distributed in blood of being livings it can provide energy to maintain the normal life activities by ingesting directly in the metabolic process [1]. In spite of this benefit, enhanced blood glucose levels could cause diabetes, one of the most serious causes of death and disability in the world [2]. So the glucose detection is one of the most popular research topics in the field of sensors both in blood and in other sources such as foods and pharmaceuticals, for clinical analysis, food industry, biotechnology, and so forth [3, 4]. Till now many techniques including UV fluorescence [5], chemiluminescence [6, 7], titrimetry [8] and electrochemical biosensors have been reported in literatures for the detection of glucose. Among them, electrochemical sensors result to be the most simples, reliable, fast, sensitives and cheaper [9].
Most previous studies on this subject involved the use of enzymatic electrochemical sensors with high sensitivity as well as excellent selectivity, but many factors such as temperature, humidity, pH, toxic chemicals and the insufficient long-term stability originating from the nature of the enzymes could

affect the activity of enzyme [10-12]. Then, many efforts have been put into producing non-enzymatic alternatives with excellent catalytic effect, reproducibility, and stability.

Recent advancements in this field have suggested that a variety of nano-structured transition metals or transition metal oxides dispersed onto a conductive matrix (carbon, graphite, carbon nanotubes) continued to receive great attention for glucose sensing due to their outstanding properties such as low cost, excellent semi-conductivity, and good biocompatibility [15-17]. As a conductive matrix, porous carbons have attracted extensively research interest all over the world because of its stable physicochemical property, high electrical conductivity, low cost, and availability which makes it one of the ideal electrode supports for electrochemical sensors [13, 14]. At present, the development of new metal-nanoporous carbon composite materials (M-NC) opened new opportunities for developing electrochemical sensors, due to their tunable electrocatalytic activity depending on the metal nature and loading.

Here, new electrode materials based on nanoporous carbon with Ni nanoparticles (Ni-NC) were proposed. In the specific, Ni-NC nanocomposites were successfully prepared by sol-gel method after the incorporation of metal nanoparticles in an organic matrix based on pyrogallol-formaldehyde (PF) and successive heating under inert atmosphere. Thus formed Ni-NC showed very good electrocatalytic ability for the oxidation of glucose in alkaline medium.

2. Experimental

2.1. Synthesis of Ni-nanoporous carbon

The synthesis of Ni-NC samples has been carried out in three steps. NiO nanoparticles were first prepared by sol-gel process using the nickel (II) chloride ($NiCl_2 \times 6H_2O$) precursor dissolved in methanol. After 15 min under magnetic stirring at room temperature, the solution was then placed in an autoclave and dried in supercritical condition of ethyl alcohol and the obtained aerogel was annealed at 500 °C in air for 2 hours. In the second one, organic wet gels were prepared by mixing formaldehyde (F) with dissolved pyrogallol (P) in water, using picric acid as a catalyst and nickel oxide (NiO) as an incorporated element with 5% of mass ratio. In the third step, the obtained wet gels were dried in humid atmosphere at 50 °C for two weeks, and in the aim to obtain a structured xerogel, the sample was transferred in an incubator and dried at 150 °C at a heating rate of 10 °C/day. The drying temperature was then maintained for two days.

In the present study, the Ni-NC sample synthesized was treated thermally in a tubular furnace under nitrogen atmosphere at 675 °C. The sample was heated up to the desired pyrolysis temperature with a heating rate of 5 °C/min. A constant selected pyrolysis temperature was then maintained for 2 h after that, the sample was cooled naturally.

2.3. Apparatus and procedures

X-ray diffraction patterns were obtained with Bruker D8 Advance X-ray diffractometer equipped with graphite monochromatized CuKα radiation (λ= 0.15418 nm) from 10° to 80° (2Θ). Transmission electron microscopy (TEM) was performed using a JEOL JEM 2100 system operating at 200 kV.

Electrochemical measurements were performed with a Dropsens µStat 400 potentiostat. A conventional three-electrode system cell was employed. The working electrodes were made by casting carbon-supported nanoparticles (M-NC nanocomposites) as a thick film onto a glass carbon. The counter electrode was carbon, and a silver pseudo-reference electrode. Cyclic voltammetry (CV) and amperometry was carried out at room temperature in 0.1 M NaOH (pH 13) in presence and absence of glucose. All measurements were performed at room temperature (25 ^{0}C).

IOP Conf. Series: Materials Science and Engineering **92** (2015) 012005 doi:10.1088/1757-899X/92/1/012005

3. Results and discussion

3.1. Characterization of the Ni-NC nanocomposite

The morphology and microstructural characterization of composite obtained after pyrolysis at 675 °C were investigated by electron microscopy. SEM image in Fig. 1a shows that at this temperature pyrogallol-formaldehyde precursors have been transformed in an amorphous carbon matrix. TEM image in Fig. 1b evidence clearly that the metal oxide precursor has been, in the same conditions, transformed in the corresponding metallic phase, as highlighted by XRD analysis reported in the inset. Ni metallic nanoparticles appears to be well dispersed on the surface of amorphous carbon matrix.

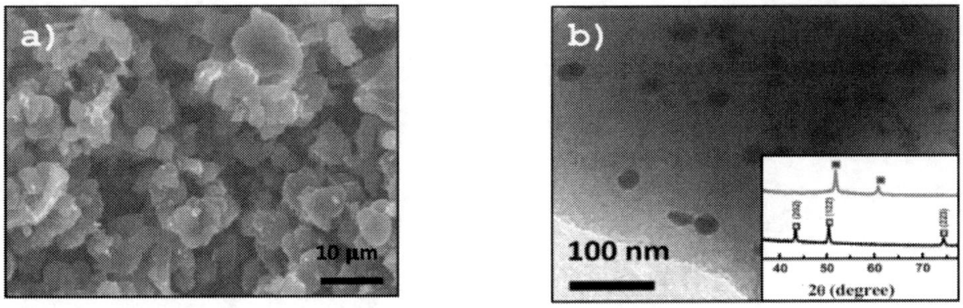

Figure 1. a) SEM and b) TEM analysis of the Ni-NC sample annealed at 675 °C. Inset shows XRD analysis of as prepared (black) and calcined (red) sample.

3.2 Electrocatalysis of glucose at the Ni-NC nanocomposite electrode

To investigate the electrocatalytic behaviour of the Ni-NC nanocomposite towards glucose reduction, cyclic voltammetry (CV) was performed in 0.1 M NaOH. The CV spectra obtained under the following conditions, potential range 0 – 0.7 V and scan rate 50 mV/s, in the absence and presence of glucose, are displayed in Fig. 1a and b. As shown in Fig. 1a, CV pattern of bare SPCE electrode exhibits no redox process in the absence of glucose. Instead, CV of 5%Ni-NC nanocomposite modified electrode, shows two peaks at the potential of 0.38 and 0.55 V, respectively, which can be assigned to the electrochemical redox reaction of Ni^{2+}/Ni^{3+}couple on the electrode surface [18].

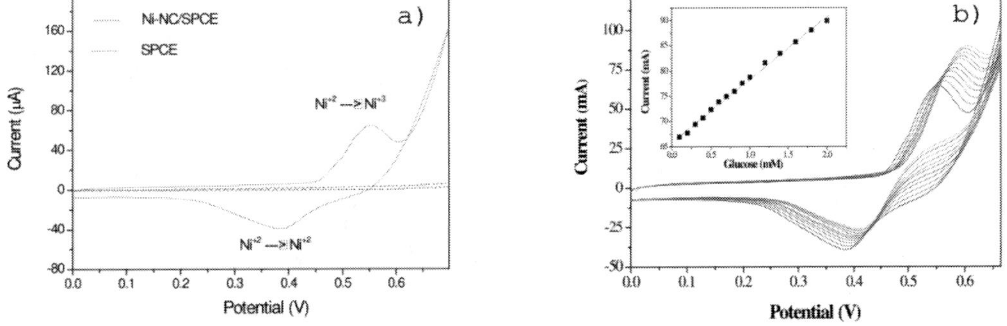

Figure 2. a) CV of SPCE and Ni-NC/SPCE electrodes in 0.1 M NaOH solution; b) CV of Ni-NC/SPCE electrode in the presence of different glucose concentrations.

IOP Conf. Series: Materials Science and Engineering **92** (2015) 012005 doi:10.1088/1757-899X/92/1/012005

CV carried in the presence of glucose (not shown) indicated that no process occurs on the bare SPCE electrode also in this condition. Vice versa, with the modified electrode, we noted a change in the CV pattern (see Fig. 1b). In the specific, the above described anodic peak at 0.55 V shows a remarkable increase in the peak current when glucose concentration increases while, correspondingly, the cathodic peak current decrease. The typical current-time curve is shown in the inset of Fig. 2b, displaying a linear trend in the range from 0 to 2mM.
Data above reported indicated that Ni species are responsible of the electrochemical behaviour observed; in particular, the Ni^{+2}/Ni^{+3} redox couple play an important role. Alkaline medium is responsible of the formation of the Ni^{+3}active sites. It can be considered indeed that, initially, Ni is electrochemically oxidized to strong oxidizing Ni (III) species such as NiOOH:

$$Ni + 3OH^- \rightarrow NiOOH + 3e^- + H_2O \quad (1)$$

Then glucose is oxidized catalytically by generated Ni^{+3} species and produce gluconic acid and/or other intermediates:

$$NiOOH + e^- + H_2O + glucose \rightarrow Ni(OH)_2 + OH^- + gluconic\ acid \quad (2)$$

As shown in Scheme 1, during glucose oxidation the electrocatalytic specie NiOOH is readily reduced to $Ni(OH)_2$. The applied oxidative potential on the electrode surface leads to the conversion to the oxidized form and the collected electrons can be directly correlated to the concentration of glucose in the solution.

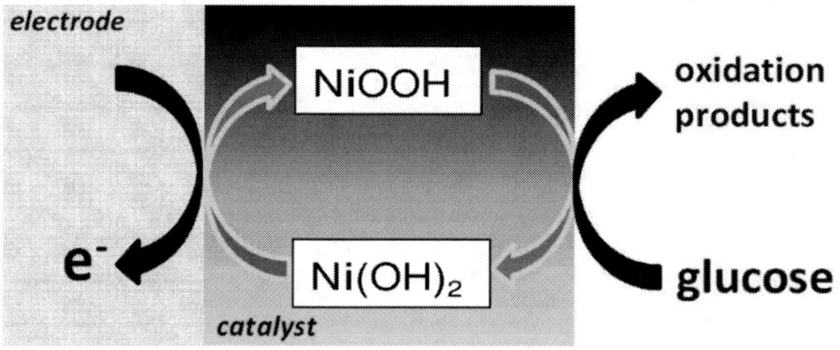

Scheme 1. Electrocatalytic behavior of modified Ni-NC/SPCE electrode.

The effect of scan rate (25-200 mVs^{-1}) on cyclic voltammetry response, at a glucose concentration of 1mM in 0.1 M NaOH, was also investigated (Fig. 3a). The result from CV shows that the redox peak current varies with the scan rate; also the oxidation and reduction potential shifts to more negative and positive values respectively with increasing scan rate. The peak current for both the oxidation and reduction processes were proportional to the square root of scan rate (Fig. 3b). Such behavior of peak current vs. square root of scan rates relationship generally appears in CV in case of diffusion of an ionic species in the vicinity of the electrode surface, which defines the rate of the reaction, i. e. under diffusional limitation.

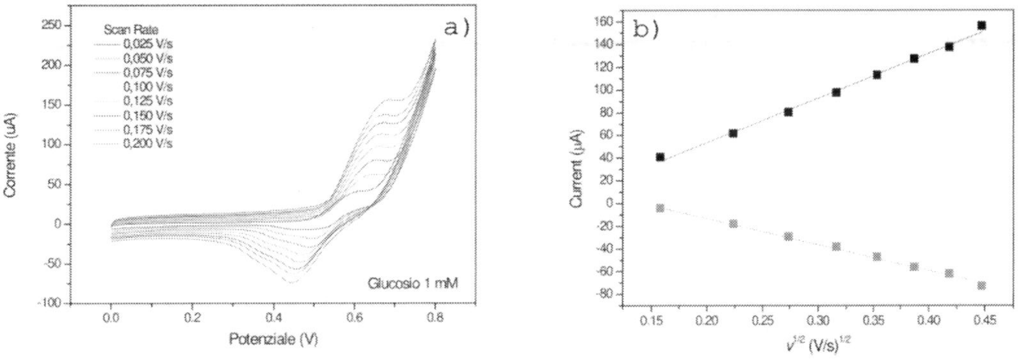

Figure 3. a) CV of Ni-NC/SPCE electrode in 1 mM glucose at different scan rate; b) trend of anodic and catodic current peak vs. scan rate.

The amperometric response of Ni-NC/SPCE electrode with successive addition of 20μM glucose in the 0.1 M NaOH solution was investigated and the results are shown in Fig. 4a-b. At the applied potential of 0.6 V it is clearly visible the remarkable enhancement in the current response upon each addition of glucose.

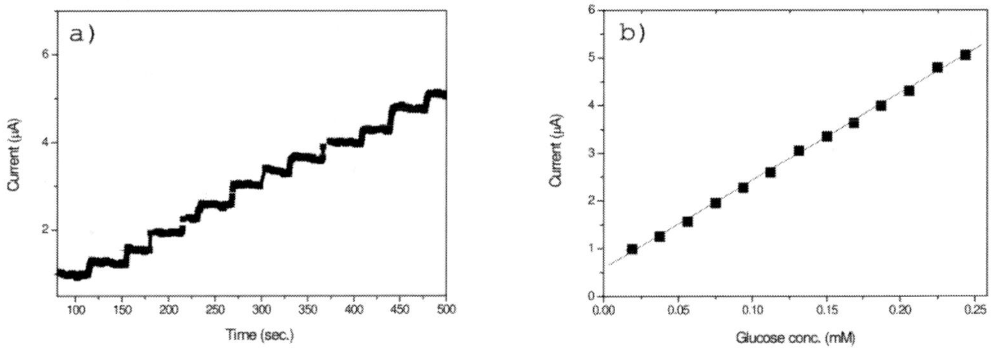

Figure 4. a) Amperometric curve of the Ni-NC/SPCE electrode at the potential of 0.6V; b) Calibration curve.

From the calibration figure, we can draw the linear regression equations as: $y=18.366x+0.5731$. The detection limit for Ni-NC/SPCE was calculated to be 10μM (S/N = 3) with a sensitivity of 146.4 μA $mM^{-1}cm^{-2}$.

4. Conclusions

In this work, a non-enzymatic glucose sensor based on Ni nanoparticles dispersed in the porous surface of carbon was developed. The electrochemical performance of the sensor in the determination of glucose in alkaline medium has been investigated. It has been demonstrated that both the nanostructured Ni particles and porous surface of carbon carrier offer particular advantages due to their large specific surface areas. Based on the modified Ni-NC/SPCE electrode, an amperometric sensor for glucose determination with a sensitivity of 146.4 μA $mM^{-1}cm^{-2}$and lower detection limit of 10μM, has been developed. These good analytical performances, coupled with low cost and straightforward preparation method make this novel electrode material promising for the development of an effective glucose sensor.

5. Reference

[1] Zhang Y C, Manuzzi L S D, de los Monteros H V E, Jia W Z, Huo D Q, Hou C J and Lei Y 2012 *Biosens. Bioelectron.* **31** 426–432.

[2] Wang J, 2008 *Chem. Rev.* **108** 814.

[3] Jin L, Shang L, Guo S, Fang Y, Wen D, Wang L, Yin J, Dong S *Biosens Bioelectron.* **26** 1965–1969.

[4] Yehezkeli O, Tel-Vered R, Raichlin S and Willner I 2011 *ACS Nano* **5** 2385–2391.

[5] Peng J, Wang Y, Wang J, Zhou X and Liu Z 2011 *Biosens. Bioelectron.* **28** 414–420.

[6] Bostick D T and Hercules D M, 1975 *Anal Chem* **47** 447–452.

[7] Lv Y, Zhang and Z Chen F 2003 *Talanta* **59** 571–576.

[8] Mazzuchin A, Thibert E J, Walton R J and Pedley E C 1971 *Microchim. Acta* **59** 285–289.

[9] Zheng B Z, Liu G Y, Yao A W, Xiao Y L, Du J, Guo Y, Xiao D, Hu Q and Choi M M F 2014 *Sens. Actuators B* **195** 431–438.

[10] Katakis I. and Dominguez E 1995, *Trends Anal. Chem.* **14** 310–319.

[11] Meng L, Jin J, Yang G, Lu T, Zhang H and Cai C *Anal. Chem.* **81** 7271–7280.

[12] Wang J, Thomas D and Chen A 2008 *Anal. Chem.* **80** 997–1004.

[13]. Zhai Y, Dou Y, Zhao D, Fulvio P F, Mayes R T and Dai S 2011 *Adv. Mater.* **23** 4828–4850.

[14] Inagakia M, Konno H and Tanaike O 2010 *Power Sources* **195** 7880–7903.

[15] Yu D, Ying W, Liang S, Heng Z and Yu L 2010 *Mater Chem* 20 9918–9926.

[16] Yang J, Jiang L -C, Zhang W-D and Gunasekaran S 2014 *Sens. Actuators B* **191** 86–93.

[17] Yang J, Jiang L-C, Zhang W-D and Gunasekaran S 2013 *Biosens. Bioelectron.* **47** 530–538.

[18] Lu P, Yu J, Lei Y, Lu S, Wang C, Liu D and Guo Q 2015 *Sens. Actuators B* **208** 90–98.

[19] Vidotti M, Cerri C D, Carvalhal R F, Dias J C, Mendes R K, Torresi S I C and Kubota L T 2009 *J. Electroanal. Chem.* **636** 18-23.

Structural and magnetic properties of Vanadium Doped M-Type Barium Hexaferrite ($BaFe_{12-x}V_xO_{19}$)

Ahmad Awadallah[1], Sami .H. Mahmood[1*], Yazan Maswadeh[1], Ibrahim Bsoul[2], Aynour Aloqaily[1]

[1]Physics Department, The University of Jordan, Amman 11942, Jordan

[2]Physics Department, Al al-Bayt University, Mafraq 13040, Jordan

*e-mail: ahmadmoh@yahoo.com

Abstract: Precursor powders of barium hexaferrite doped with vanadium, $BaFe_{12-x}V_xO_{19}$ with (x = 0.1, 0.2, 0.3, 0.4, 0.5), were prepared using the ball milling technique and then sintered at different temperatures for 2 h. The structural properties of the prepared samples were investigated by X-ray diffraction (XRD) and scanning electron microscopy (SEM), while the magnetic properties were examined by the vibrating sample magnetometry (VSM). XRD and SEM studies of the samples sintered at 1100° C indicated the presence of $Ba_3V_2O_8$ and α-Fe_2O_3 non-magnetic oxide phases in addition to BaM hexaferrite phase. The fractions of the non-magnetic oxide phases were found to increase with increasing x, and sintering the samples at temperatures higher than 1100° C was found to reduce the amounts of these non-magnetic phases only slightly. However, the addition of barium in excess of the stoichiometric ratio was found to remove the α-Fe_2O_3 oxide, and improve the saturation magnetization of the samples significantly. In addition, washing these samples with HCl was found to improve the saturation magnetization further. The effect of sintering the samples at higher temperatures was also found to reduce the coercivity due to growth of the particle size. However, the coercivity of all samples remained high enough for potential permanent magnet and magnetic recording applications.

1. Introduction

M-type hexaferrites, discovery in the early 1950s, had demonstrated to be an important class of magnetic materials. The growing interest in these ferrites is due to the suitability of their properties for a wide range of applications [1-8]. Barium-based (BaM) and strontium-based (SrM) hexaferrites are used in many technological applications such as recording media, permanent magnets, and microwave devices [9-16]. The general structural and magnetic properties of these ferrites are thoroughly discussed in the literature [1, 7, 17]. The magnetic properties of the hexaferrites are modified by different ways, such as changing the experimental conditions involved in their fabrication, substitutions of Ba or Sr by other cations, or substituting Fe by another trivalent cation or combinations of cations with different valence states [13, 18-28]. The hexaferrites were prepared using different methods such as ball milling [29-34], Sol-gel method [35-37], wet mixing method [27, 38], hydrothermal method [39-41], and coprecipitation method [42-44].

The ball milling method was used in this work to produce fine-particles powders with particle size smaller than the single domain size.
The unit cell of BaM ($BaFe_{12}O_{19}$) is composed of the stacking sequence RSR*S* of the rhombohedral (R) and spinel (S) blocks, where the star indicates rotation by 180° around the *c*-axis. The small metal ions (such as the Fe ions) occupy five different interstitial sites in the hexaferrite lattice: three octahedral sites ($12k$, $4f_2$, $2a$), one tetrahedral site ($4f_1$), and one trigonal bipyramid site ($2b$) [22-24]. Eight spin-up ferric ions per formula occupy the $2a$, $2b$, and $12k$ sites, while four spin down ferric ions per formula occupy the $4f_1$ and $4f_2$ sites [34, 45]. According to Gorter's collinear model [46, 47], the net magnetic moment per formula of BaM is equal to $4\times5\mu_B = 20\mu_B$, corresponding to a magnetization of 100 emu/g at 0 K [26] and 72 emu/g at room temperature [1]. The magnetic properties of M-type hexaferrites can be modified by appropriate substitutions of Fe^{3+} ions by other magnetic or non-magnetic ions. Such substitutions result in significant changes in the saturation magnetization (M_s) due to changes in the net magnetic moment per molecule [48-51]. Also, such substitutions could be used to tune the coercivity of the hexaferrite to fit specific applications [48, 52]. The coercivity of the hexaferrite is given by [53, 54]:

$$H_c = 0.48\left(\frac{2K_1}{M_s} - NM_s\right) \quad (1)$$

Here, K_1 is the first magnetocrystalline anisotropy constant, N (= 4π) is the demagnetizing factor along the *c*-axis for a uniaxial (BaM or SrM) crystal, and the number (0.48) arises from the random orientations of the *c*-axis. The single ion magnetocrystalline anisotropy model revealed that the contribution of Fe^{3+} ion to the first anisotropy constant is largest for ions at $2b$ site [55].
In this article, the effects of partial substitution of iron ions by vanadium ions on the structural and magnetic properties of BaM hexaferrites were discussed. Also, the effects of different heat treatments and sample stoichiometry were addressed. Sample characterization was carried out by X-ray diffraction (XRD), scanning electron microscopy (SEM), and vibrating sample magnetometer (VSM).

2. Experimental procedures

The appropriate molar ratios of analytical grade starting materials ($BaCO_3$, Fe_2O_3, and V_2O_3) were used to synthesize the desired samples. We used 1:11 as the ratio between Ba: Fe which is the optimal ratio to give a single phase of undoped $BaFe_{12}O_{19}$ [56]. The powder precursors were mixed in zirconia containers and ball-milled by zirconia balls in a Fritch Pulverissette-7 ball mill for 16 h at 250-rpm rotational speed. The wet milling was carried out in an acetone bath, with powder to ball mass ratio of 1:14, and powder to acetone ratio of 5:8. The wet powder was then left to dry in air, and then part of it was pressed into a disc of 1.2 cm in diameter and ~ 0.15 cm in thickness under a force of 5 tons. The discs were sintered in air at $T \geq 1100\,^{\circ}$C for 2 h. The structure of each sample was then investigated by the x-ray diffractometer with Cu-Kα radiation (λ = 1.5406 Å). The angular range of the x-ray diffraction (XRD) patterns was $20° \leq 2\theta \leq 70°$, and the scan was performed with angular step of 0.01°. The obtained patterns were analyzed through X'pert HighScore software, and Rietveld refinement of the patterns was carried out using FULLPROF software. Scanning electron microscopy (SEM) was used to examine the particle size and morphology of the samples using the imaging mode, and the chemical composition by the energy dispersive x-ray spectrometer (EDS) facility. Room-temperature magnetic measurements were carried out using a vibrating sample magnetometer (VSM Micro Mag 3900, Princeton Measurements Corporation) with a maximum field of 10 kOe. A needle shape sample cut from each sintered disc was used for the magnetic measurements.

3. Results and Discussion

3.1. Structural Analysis

Figure 1 shows the XRD patterns of $BaFe_{12-x}V_xO_{19}$ samples with (x = 0.0 - 0.5) sintered at 1100° C for 2 h in air. For x = 0.0, the XRD pattern shows a single phase of $BaFe_{12}O_{19}$ consistent with the standard pattern (JCPDS: 00-043-0002), indicating that the solid-state reaction at 1100° C for 2 h was suitable for the production of pure BaM hexaferrite phase [57]. The diffraction pattern for the sample with x = 0.1 indicated the presence of BaM, in addition to very small amounts of secondary phases evidenced by very weak reflections corresponding to $Ba_3(VO_4)_2$ vanadate phase (JCPDS pattern no: 00-025-1192), and α-Fe_2O_3 (JCPDS: 01-086-2368). As x increased, a large increase in $Ba_3(VO_4)_2$ and α-Fe_2O_3 peak intensities was observed. These results suggest that the substituted V reacts with barium to form $Ba_3(VO_4)_2$ rather than substituting Fe^{3+} in the hexaferrite lattice.

The lattice parameters a and c, and the cell volume V of the BaM phase in all samples were obtained from Rietveld refinement, and the results are listed in Table 1. In addition, the X-ray densities (ρ_x) for the samples were calculated using the formula:

$$\rho_x = \frac{2(M_w)}{V} \quad (2)$$

Here M_w is the molecular weight, and V is the unit cell volume. The values of X-ray densities are also listed in Table 1. The lattice parameters and cell volume of the BaM phase were found to fluctuate slightly with increasing V concentration. The slight decrease in density, and increase in cell volume for x > 0.3 may suggest that a small amount of V ions entered the hexaferrite lattice, and indicate the low solubility of V in this case.

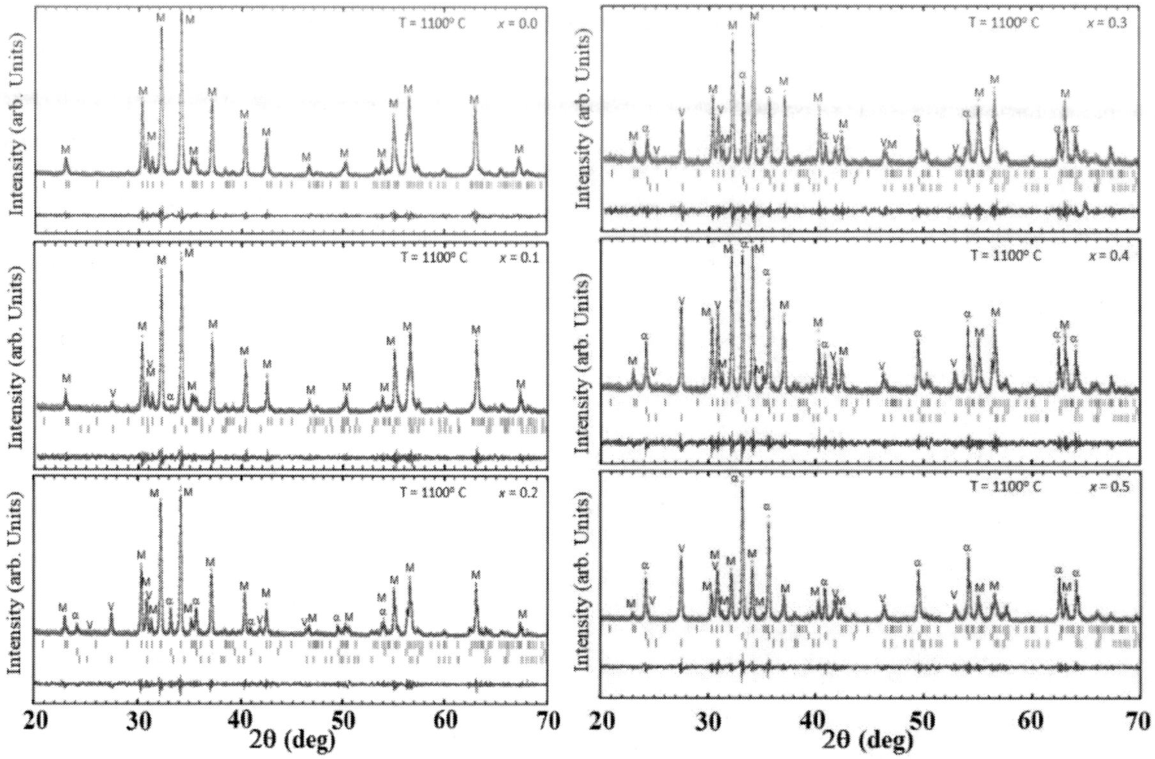

Figure 1. XRD patterns with Rietveld refinement for the samples $BaFe_{12-x}V_xO_{19}$ sintered at 1100° C.

Table 1. The lattice parameters a and c (Å), unit cell volume V (Å^3), and X-ray densities (g/cm^3) for the $BaFe_{12-x}V_xO_{19}$ samples sintered at T = 1100° C.

X	$a = b$	c	V	ρ_x	χ^2
0.0	5.89 ± 0.03	23.21 ± 0.03	697.50	5.29	1.25
0.1	5.89 ± 0.03	23.20 ± 0.03	697.05	5.28	1.21
0.2	5.89 ± 0.03	23.20 ± 0.03	697.71	5.29	1.16
0.3	5.89 ± 0.03	23.21 ± 0.03	697.34	5.29	1.17
0.4	5.89 ± 0.03	23.20 ± 0.03	697.65	5.28	1.16
0.5	5.90 ± 0.03	23.24 ± 0.03	699.34	5.27	1.16

The broadening of the diffraction peaks is mainly due to three factors [58]: instrumental effects, crystallite size (D), and lattice strain. In our system, however, the lattice strain effect was found to be very small and could be neglected. The instrumental broadening was estimated from the broadening of the peak profile of a standard silicon sample with large crystallite size (supplied by the manufacturer of the XRD instrument). The peak corresponding to a certain reflection was fitted with a Lorentzian line-shape, and integrated intensity (A) and maximum intensity (I_0) were determined from the fit. The integral breadth ($\beta = A/I_0$) was determined for the peak, and then corrected by subtracting from it the integral breadth of the standard silicon pattern. The crystallite sizes of the different samples were then calculated using the relation [38]:

$$D = \frac{\lambda}{\beta \cos \theta} \quad (3)$$

Here D is the crystallite size, λ is the wavelength of radiation (1.5406 Å), β is the corrected integral breadth, and θ is the peak position. In our analysis, the (110) peak at $2\theta = 30.3°$ was fitted to probe the crystallite size along the basal plane of the hexagonal lattice, while the approximate crystallite size along the c-direction was determined from the (107) peak at $2\theta = 32.2°$ and (20 11) peak at $2\theta = 56.6°$. In addition, the crystallite size in a direction perpendicular to the (114) crystallographic planes was determined from the most intense peak at $2\theta = 34.1°$. The results shown in Table. 2 indicate that the crystallite size along the basal plane is obviously larger than that along the c-axis, suggesting that the crystallites grow in platelet-like shapes.

Table 2. The crystallite size values for the system $BaFe_{12-x}V_xO_{19}$ calculated using different peaks.

x	D (nm)		
	(1 1 0)	**(1 0 7), (2 0 11)**	**(1 1 4)**
0.0	68	49	73
0.1	162	79	128
0.2	251	88	143
0.3	437	80	124
0.4	173	107	128
0.5	429	68	133

To shed light on the evolution and temperature dependence of the stability of the $Ba_3(VO_4)_2$ phase, the $BaFe_{12-x}V_xO_{19}$ samples were sintered at higher temperatures for 2 h, and representative XRD patterns for the sample with $x = 0.5$ are shown in figure 2. Comparing the diffraction patterns, a clear increase in the intensities of the peaks associated with BaM phase at the expense of $Ba_3(VO_4)_2$ and α-Fe_2O_3 was observed in the pattern of the sample sintered at 1300° C, while on the contrary, the vanadate $Ba_3(VO_4)_2$ phase increased upon sintering at 1200° C. This suggests that a small amount of vanadium was dissolved in the M-type phase of the sample sintered at 1100° C, and that a small fraction of this V-substituted BaM phases precipitates into $Ba_3(VO_4)_2$ and α-Fe_2O_3 upon sintering at 1200° C, while the vanadate and iron oxide phases react to form BaM phase at 1300° C. This result indicates that the vanadate phase stabilized at 1200° C, and the diffusion of higher amounts of V ions in the BaM lattice requires high temperatures (> 1200° C). This qualitative conclusion, however, needs to be confirmed by another technique, such as monitoring the magnetic properties of the samples with increasing sintering temperature; this will be done in a forthcoming section. The refined lattice parameters a and c, and the cell volume V (not shown) for the BaM phase in the samples sintered at different temperatures generally indicated that the samples had the same structure with a slight increase in the cell volume with increasing the sintering temperature. This can be associated with lattice relaxation and removal of lattice defects at higher sintering temperatures.

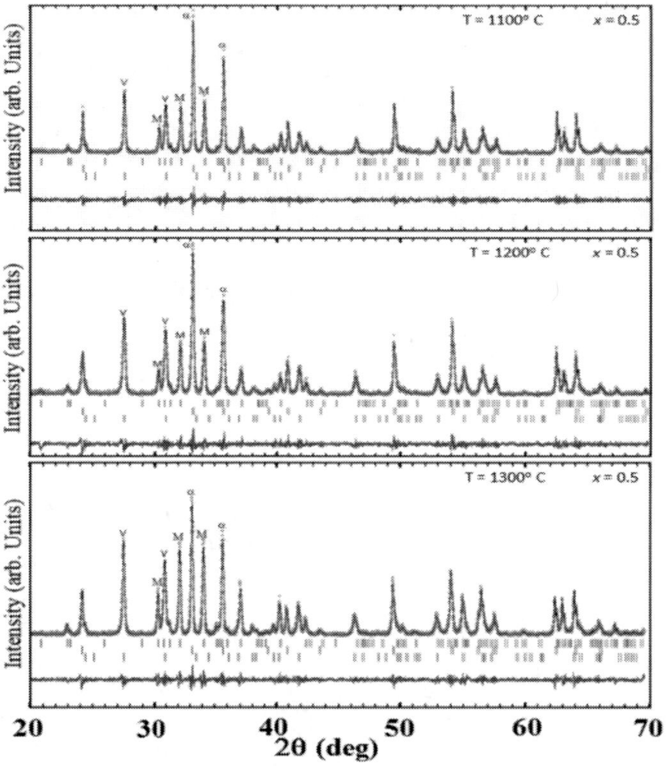

Figure 2. XRD patterns with Rietveld refinement for the sample $BaFe_{11.5}V_{0.5}O_{19}$ sintered for 2 h at different temperatures.

To examine our assumption that the vanadium has a very low solubility in BaM hexaferrite at low temperature (T = 1100° C), the percentage weight (wt. %) of each phase in each sample was calculated on the basis of the assumption that all vanadium is incorporated in the vanadate phase, with no solubility in the M-phase. Also, the precise values of the weight percent of the different phases were obtained from quantitative analysis by fitting of the X-ray patterns. The results in Table. 3 indicate that the actual weight percent of $Ba_3(VO_4)_2$ is generally lower than the calculated value. This result supports the conclusion that vanadium ions diffused into the BaM lattice, but with a very low solubility.

In light of the previous discussion, an additional amount of 75 % molar ratio of $BaCO_3$ to the precursor powders should provide the amount of Ba required for the production of the $Ba_3(VO_4)_2$ phase in the sample with $x = 0.5$, and leave an amount which is just right for the production of BaM phase with no excess amounts of α-Fe_2O_3. If no vanadium (or only a non-detectable amount of V^{3+}) inters the BaM lattice, then we expect that the peaks corresponding to the α-Fe_2O_3 should completely disappear from the XRD pattern. If, on the other hand, a small amount of vanadium ions diffuse into the BaM lattice, the amount of Ba consumed by the vanadate phase is less, resulting in excessive amounts of Ba, which leads to the production of barium spinel phase with higher Ba:Fe ratio compared to BaM; this phase should then be detected by XRD. Accordingly, we prepared a new sample of $BaFe_{12-x}V_xO_{19}$ ($x = 0.5$) using the same procedure under the same conditions, but with 75% additional molar ratio of $BaCO_3$. The XRD pattern for this sample was compared with the original one (without extra Ba addition) as shown in figure 3. The figure shows that all peaks of α-Fe_2O_3 disappeared completely, and no signature of Ba spinel was detected. This result supports our previous conclusion regarding the phases and their relative proportions in the different samples, and the low solubility level of V in the hexaferrite phase.

Table 3. Weight percent of the different phases through the different samples.

X	$BaFe_{12}O_{19}$			$Ba_3(VO_4)_2$			α-Fe_2O_3		
	# of moles	wt. % (calc.)	wt. % (fitted)	# of moles	wt. % (calc.)	wt. % (fitted)	# of moles	wt. % (calc.)	wt. % (fitted)
0.1	0.85	91.5	97.5	0.05	3.1	1.6	0.35	5.4	0.9
0.2	0.70	75.3	82.5	0.10	6.2	4.9	1.20	18.5	12.7
0.3	0.55	59.1	60.3	0.15	9.3	9.4	2.05	31.6	30.3
0.4	0.40	42.9	50.1	0.20	12.4	10.9	2.90	44.7	39.0
0.5	0.25	26.8	29.5	0.25	15.5	14.4	3.75	57.7	56.1

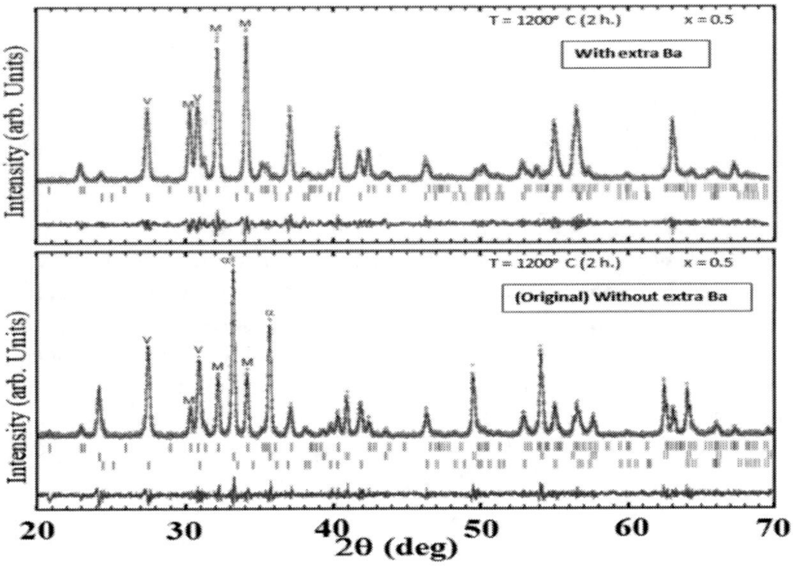

Figure 3. XRD patterns with Rietveld refinement for samples $BaFe_{12-x}V_xO_{19}$ ($x = 0.5$) sintered at 1200° C with (upper panel) and without (lower panel) extra Ba.

In order to remove the $Ba_3(VO_4)_2$ phase, the sample with $x = 0.5$ added with $BaCo_3$ and sintered at 1200° C was washed with diluted HCl solution (diluted in distilled water with HCl: H_2O equals 2:10), and figure 4 shows the XRD pattern of this sample, together with that of the un-washed sample. It is clear that the vanadate phase was completely dissolved in HCl, leading to a single BaM phase in the sample.

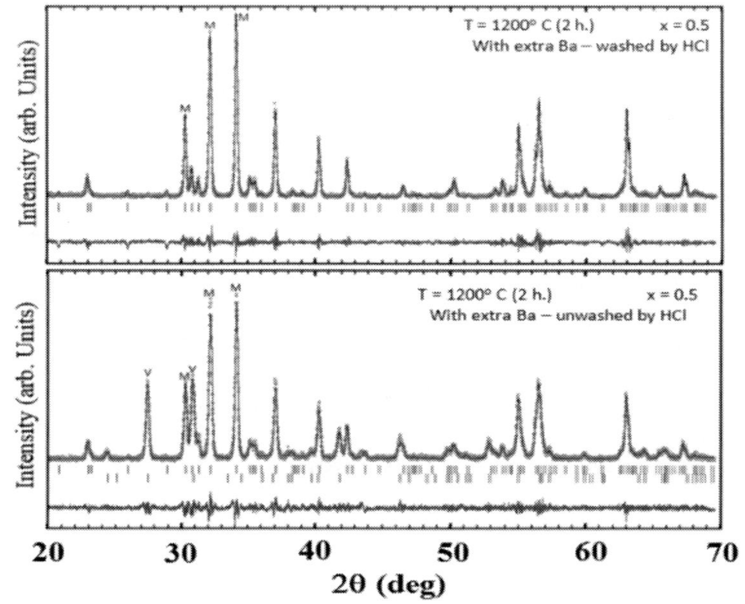

Figure 4. XRD patterns with Rietveld refinement for the samples $BaFe_{12-x}V_xO_{19}$ (x = 0.5) added with $BaCo_3$ and sintered at 1200° C for 2 h. Upper panel: washed with HCl, and lower panel: unwashed.

3.2. Electron microscopy

Figure 5 shows the SEM image for the un-doped $BaFe_{12}O_{19}$ sample, and that with $x = 0.1$, sintered at T = 1100° C for 2 h. The image of the pure sample shows evidence of hexagonal platelets with diameters ranging from 200 to 750 nm. The size distribution is, however, dominated by grains with typical diameters in the range of 400 to 500 nm. The image of the sample with $x = 0.1$ also shows evidence of hexagonal platelets with a more regular and narrower grain size distribution and grain diameters in the range from about 170 to about 500 nm. This distribution is dominated by grains with diameters in the range ~ 350 to 400 nm. The porosity in this sample is clearly lower than that in the pure sample.

Figure 5. SEM images for $BaFe_{12-x}V_xO_{19}$ with (a) $x = 0$, and (b) $x = 0.1$ samples sintered at $T = 1100^{\circ}$ C.

Typical SEM images for the sample with $x = 0.2$ are shown in figure 6. Evidence of platelet-like hexagonal grains is provided by image (a). The grain diameters range from ~ 170 to 600 nm, and the size distribution is dominated by grains with typical diameters of 350 to 400 nm. Image (b) (in the EDS mode) demonstrated the presence of lighter and darker particles in the powder, which can be associated with different phases. The chemical composition of the light particles was examined by EDS, where the observed Ba:Fe:V atomic ratios were found to be 1:6.2:0.36. The Ba:(Fe+V) ratio of 1:6.56 indicate a Ba-rich region with stoichiometry much different from the stoichiometric ratio of 1:12 for V-substituted BaM. The Ba-rich phase confirmed by XRD was the vanadate phase, and EDS measurements on the light particle, therefore, suggest that the measured region contained both $Ba_3(VO_4)_2$ and BaM phases. On the other hand, EDS measurements on a typical "dark" particle revealed that the Ba:V:Fe atomic ratios in this particle were 1:0.09:10.13, which is close to the BaM stoichiometric ratio. Although the difference of about 15% between the observed and theoretical Ba:Fe ratio of BaM is within experimental uncertainty of the technique, the observed higher concentration of Ba makes one believe that the possibility of the presence of small amounts of the Ba-rich vanadate phase in the measured region should not be excluded. The results, however, indicate clearly that while the darker particles consist mainly of barium hexaferrite phase, the lighter particles contain Ba-rich vanadate as a major phase.

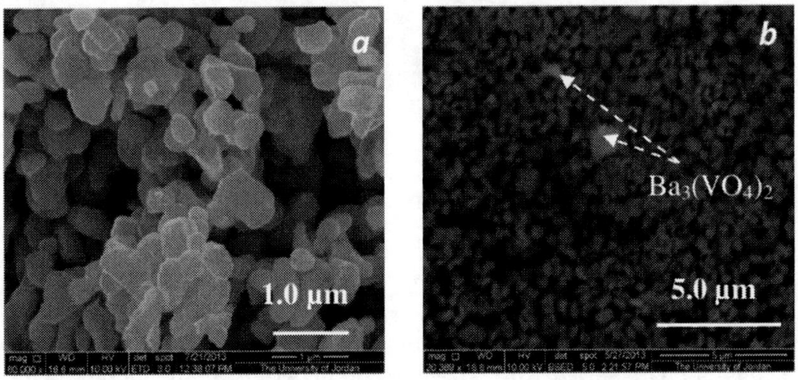

Figure 6. SEM images for $BaFe_{11.8}V_{0.2}O_{19}$ sample sintered at $T = 1100^{\circ}$ C: (a) in the imaging mode, and (b) in the EDS mode.

IOP Conf. Series: Materials Science and Engineering **92** (2015) 012006 doi:10.1088/1757-899X/92/1/012006

Figure 7 shows representative SEM images for $BaFe_{12-x}V_xO_{19}$ (x = 0.3, 0.4, 0.5) samples sintered at T = 1100° C for 2 h. The image of the sample with x = 0.3 indicates that the particle size distribution is similar to that for the samples with x = 0.1 and x = 0.2, with a particle size ranging from about 200 nm up to 650 nm. For x = 0.4, however, the image indicates the presence of larger particles with diameters ~ 1000 nm, while the size distribution is still dominated by particles with diameters in the range between 350 and 400 nm. In addition, SEM image for the sample with x = 0.5 demonstrate similar grain size distribution with no large-grain growth.

Figure 7. SEM images for the samples of $BaFe_{12-x}V_xO_{19}$ (x = 0.3, 0.4, 0.5) sintered at T = 1100° C.

3.3. VSM results

3.3.1. The system $BaFe_{12-x}V_xO_{19}$

Hysteresis loops (HL) for all samples sintered at 1100° C are shown in figure 8, and the remanence magnetization and the coercive fields were determined directly from the loops. The hysteresis loops showed a behavior characteristic of hard magnetic material for all samples. The curves did not saturate up to the upper limit of the applied field (10 kOe), so the law of approach to saturation (LAS) was used to obtain the saturation magnetization [59, 60]:

$$M = M_s\left[1 - \frac{A}{H} - \frac{B}{H^2}\right] + \chi H \qquad (4)$$

where M_s is the spontaneous saturation magnetization of the domain per unit volume, A is a constant representing the contributions of inclusions and/or micro-stress, B is a constant representing the contribution of magnetocrystalline anisotropy, and χH is the forced magnetization term. A plot of M_s vs. $1/H^2$ in the high field region (8.5kOe < H < 10kOe) for each sample gave a perfect straight line as shown in figure 9, indicating that the contributions of the forced magnetization terms and the inclusions/micro-stress are negligible. The intercept of this straight line with the M-axis gave the saturation magnetization of the sample.

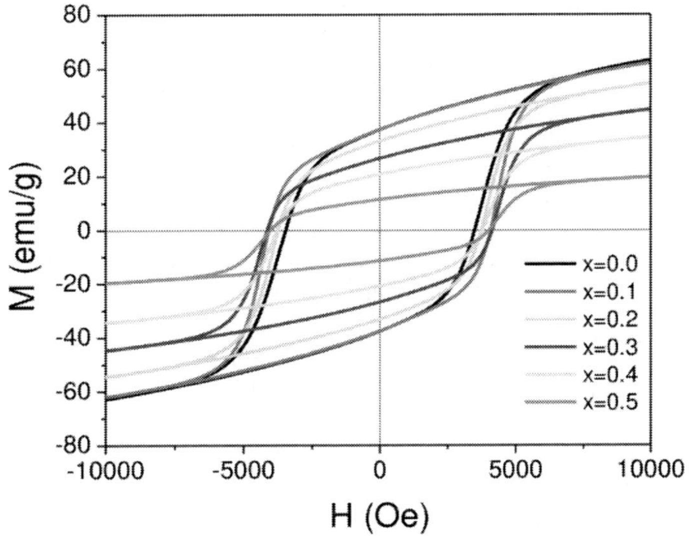

Figure 8. Room temperature hysteresis loops for the samples $BaFe_{12-x}V_xO_{19}$.

The saturation magnetization M_s of the pure sample was found to be 69.0 emu/g, which is consistent with the reported value for bulk BaM at room temperature [1, 7, 59]. The remanence (M_r) was found to be 37.4 emu/g, and the squareness ratio ($S = M_r/M_s$) was 0.54, which is close to that of random, single domain particles. This result is consistent with SEM images which showed that the particle size distribution for this sample is dominated by particles with sizes of the order or below the critical single domain size of 460 nm for hexaferrites. The saturation magnetization, remanence, and the squareness ratio for each sample are tabulated in Table 4.

IOP Conf. Series: Materials Science and Engineering **92** (2015) 012006 doi:10.1088/1757-899X/92/1/012006

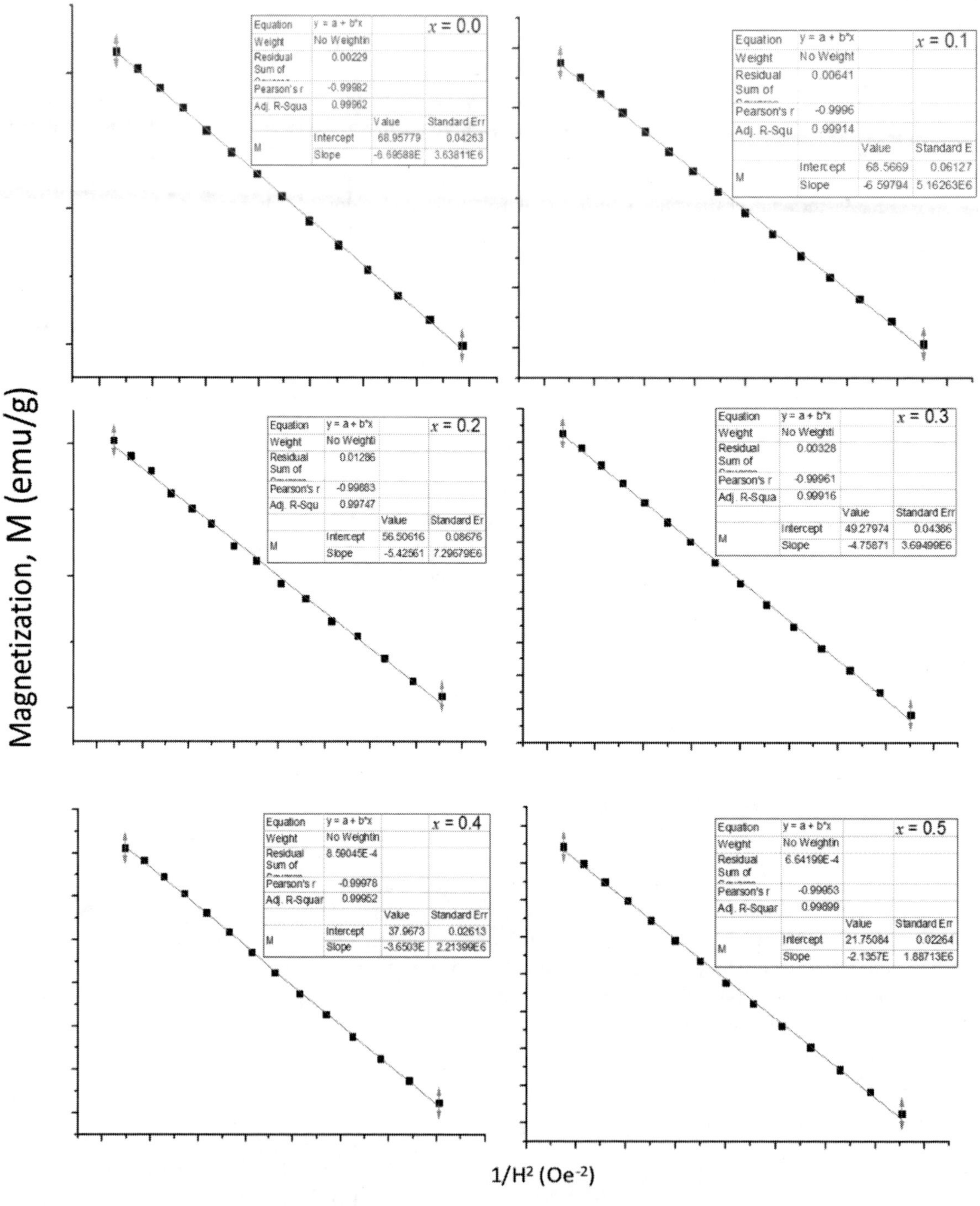

Figure 9. *M* vs. 1/H^2 relation with the straight line fit from which the saturation magnetization and the magnetocrystalline anisotropy parameters were derived.

The dependence of M_s and H_c on the concentration of vanadium is shown in figure 10. The saturation magnetization decreased monotonically with increasing vanadium concentration, recording a reduction of 68 % in the sample with x = 0.5. The monotonic decreases in saturation magnetization is associated mainly with the progressive increase of the amounts of $Ba_3(VO_4)_2$ and α-Fe_2O_3 non-magnetic phases with increasing x as confirmed by the XRD analysis.

Table 4. The saturation magnetization, remanence, the normalized remnant magnetization, coercive field, anisotropy field, and first anisotropy constant for the system $BaFe_{12-x}V_xO_{19}$ measured at room temperature.

x	M_s (emu/g)	M_r (emu/g)	S	H_c (kOe)	H_a (kOe)	K_1 (10^6 erg/cm^3)
0.0	69.0	37.3	0.54	3.5	12.1	2.2
0.1	68.6	37.4	0.54	4.1	12.1	2.2
0.2	56.8	30.8	0.54	4.4	12.3	1.8
0.3	49.3	26.7	0.54	4.1	12.1	1.6
0.4	38.0	20.8	0.55	3.8	12.0	1.2
0.5	21.8	11.4	0.52	4.0	12.3	0.7

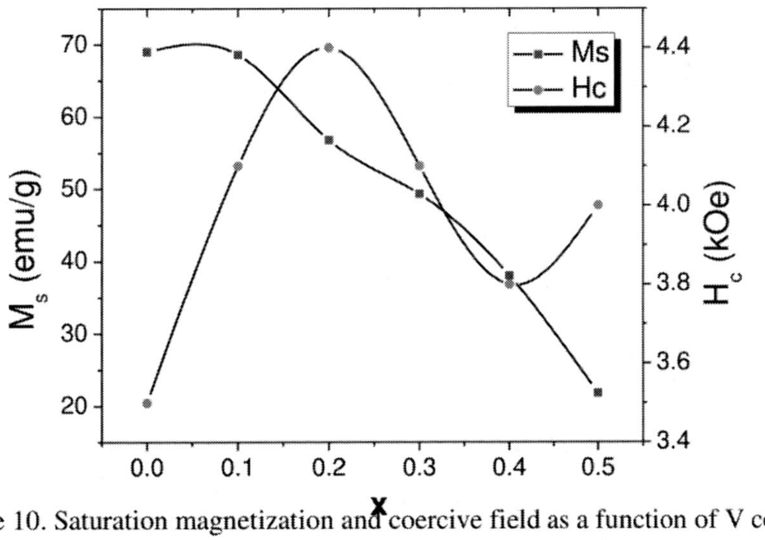

Figure 10. Saturation magnetization and coercive field as a function of V concentration.

The coercivity is known to depend on the magnetocrystalline anisotropy and saturation magnetization in accordance with equation (1). The anisotropy field (H_a) and the first anisotropy constant (K_1) were calculated from the slope of the $M - 1/H^2$ straight line, where according to equation (4), the slope = $-M_sB$, and the constant B in the LAS for the hexagonal ferrite is given by [20, 27]:

$$B = \frac{H_a^2}{15} = \frac{1}{15}\left(\frac{2K_1}{M_s}\right)^2 \quad (5)$$

The magnetic anisotropy field as well as the first anisotropy constant for each sample is listed in Table 4, and the variations of these parameters with x are shown in figure 11. It should be mentioned here that while the anisotropy field is characteristic of the BaM phase (since it is the only magnetic phase in the sample), the anisotropy constant does not have much meaning in multi-phase samples. Therefore, only the anisotropy constant in the pure, single phase sample represents magnetocrystalline anisotropy of BaM hexaferrite, whereas the constants of the other samples should be normalized by the volume fraction of the magnetic phase in the sample.

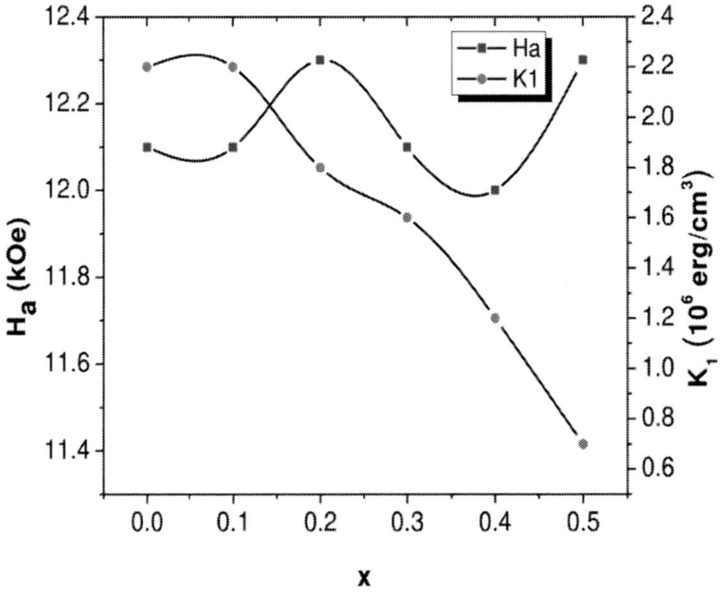

Figure 11. The magnetic anisotropy field and the first anisotropy constant as a function of V concentration.

The experimental value of H_c for the single BaM phase in the sample with $x = 0$ is 3.5 kOe (from the HL). When the first anisotropy constant (2.2×10^6 erg/cm^3) and saturation magnetization (69.0 emu/g = 365 emu/cm^3) for this sample are substituted in equation (1), the calculated coercivity is found to be 3585 Oe, in very good agreement with the experimental value. This result supports the conclusion that this sample consists of a random assembly of single domain particles, in agreement with the size distribution concluded from SEM imaging. Similar calculations cannot be carried out for the remaining samples, since these samples do not contain a single magnetic phase with well-defined saturation magnetization and first anisotropy constant. If, however, the value of the anisotropy field is used in equation (1), and the saturation magnetization is assumed to be equal to that of the pure sample (since the coercivity is characteristic of the magnetic phase only), the calculated coercivity of the samples would range between

3560 Oe and 3700 Oe. The observed variations in the experimental coercivity are significantly larger than the calculated variations, which could be attributed to differences in the microstructure of the samples. The porosity was reported to result in a decrease in the coercivity of the sample [54], which explains the lower coercivity for the pure sample, which has the highest porosity as confirmed by SEM images.

3.4.2 Effect of extra barium addition

VSM measurements were performed for the sample with x = 0.5 added with 75% extra barium and sintered at 1200° C with and without washing by HCl. As discussed previously, the additional barium reacted with the unreacted α-Fe_2O_3 phase to form BaM and remove residual hematite phase, while washing with HCl removed the vanadate phase. figure 12 shows the HL of the un-washed and washed sample, together with that of the samples with no barium addition sintered at 1100° C and 1300° C for comparison.

The saturation magnetizations of the samples were calculated from the law of approach to saturation (equation (4)) by plotting M vs. $1/H^2$. In addition, the remnant magnetization and coercivity were found directly from the HL. The values of the magnetic parameters for these samples are listed in Table 5. The saturation magnetization increased significantly from 21.8 to 56.3 emu/g with Ba addition due to the reaction of the residual α-Fe_2O_3 phase with the extra added barium, resulting in an increase in the fraction of magnetic BaM phase. The coercivity of this decreased by 48% with respect to that with no extra Ba added and sintered at 1100° C. However, this comparison should not be made since the two samples were sintered at different temperatures, and comparison between their coercivity should be addressed in the context of discussing the effect of heat treatment on the magnetic properties. The sample washed by HCl showed higher saturation magnetization due to removal of the non-magnetic vanadate phase from the sample. The coercivity of the washed sample did not change with respect to the un-washed one, which indicates that HCl did not attack the magnetic particles which provided the magnetic coercivity.

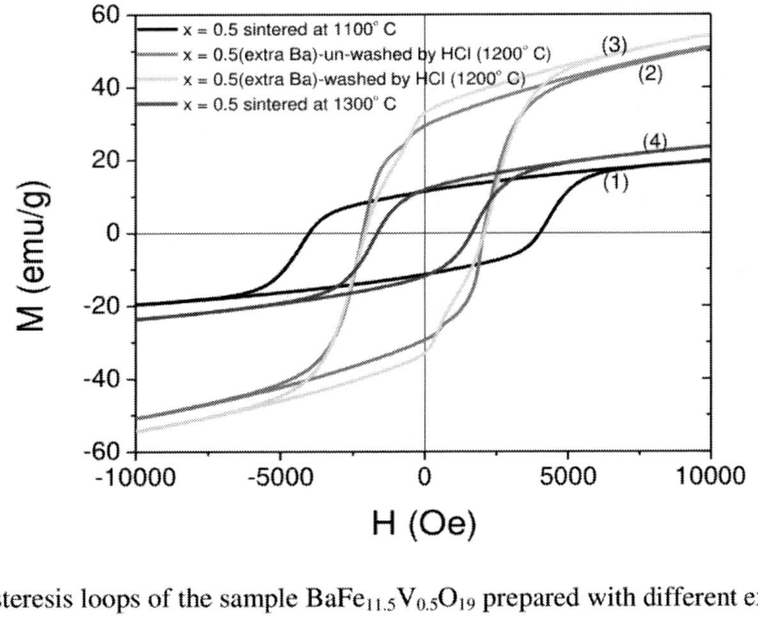

Figure 12. Hysteresis loops of the sample $BaFe_{11.5}V_{0.5}O_{19}$ prepared with different experimental conditions.

The first anisotropy constant of the sample added with extra barium increased significantly with respect to that of the sample with no barium addition, which is due to the conversion of the relatively large fraction of the α-Fe_2O_3 non-magnetic phase into BaM and the consequent significant increase in saturation magnetization. When this sample was washed with HCl, the magnetization increased slightly (~ 4%) due to removal of the extra non-magnetic vanadate phase. However, XRD analysis confirmed that the sample added with extra Ba and washed by HCl was composed of a pure BaM phase, and should have a saturation magnetization similar to that of the pure sample. Therefore, the observed decrease of M_s from ~ 69 emu/g for the sample (with x = 0.0) to M_s = 59.6 emu/g for this sample could be evidence of partial limited substitution of vanadium for iron in the BaM lattice; such a substitution should take place at spin-up sites.

Table 5. The magnetic parameters of $BaFe_{12-x}V_xO_{19}$ (x = 0.5) samples prepared under different experimental conditions: heat treatment, addition of extra Ba, and washing by HCl.

x = 0.5	M_s (emu/g)	M_r (emu/g)	S	H_c (kOe)	H_a (kOe)	K_1 (10^6 erg/cm^3)
(Sintered at 1100° C) (1)	21.8	11.4	0.52	4.0	12.3	0.7
(Sintered at 1300° C) (4)	26.6	11.9	0.45	1.6	13.0	0.9
(Sintered at 1200° C) (2) (Extra Ba)-unwashed	56.3	29.3	0.52	2.1	11.9	1.8
(Sintered at 1200° C) (3) (Extra Ba)-washed by HCl	59.6	32.8	0.55	2.1	11.7	1.8

3.4.3 Effect of sintering temperature

To examine the temperature effect on the magnetic properties, room-temperature magnetic measurements for the samples with x = 0.5 sintered at 1100° C, 1200° C, and 1300° C were compared in figure 12 (curves (1), (2), and (4), respectively). The saturation magnetization of the sample sintered at 1300° C increased by 22% with respect to that sintered at 1100° C. This could be signature of partial substitution of vanadium for iron in the BaM phase, resulting in the reduction of the fraction of Ba consumed by the vanadate phase, and providing more Ba for the production of the magnetic phase. On the other hand, the coercivity decreased significantly with increasing sintering temperature, an effect that could be associated with particle growth instigated by the higher sintering temperatures. The observed coercivity for the sample sintered at 1200° C is almost identical to the 2000 Oe coercivity reported for BaM prepared by sintering at this temperature [54]. In addition, the sample sintered at 1300° C exhibited a further decrease in coercivity (down to 1600 Oe), apparently due to further particle growth. These values of the coercivity, together with the relatively high saturation magnetization, make these materials promising for high-density magnetic recording applications.

The highest anisotropy field exhibited by the sample sintered at 1300° C could be associated with better crystallization and lower levels of crystal defects at higher sintering temperatures [54], and the lower values of the anisotropy field of the sample added with extra Ba could be indicative of the presence of more crystal defects in this sample.

4. Conclusions

Vanadium substituted BaM hexaferrites prepared by ball-milling and sintering at temperatures ≥ 1100° C were found to be multi-phase samples, consisting of the magnetic BaM phase in addition to vanadate and iron oxide non-magnetic phases. The magnetic properties of the samples sintered at 1100° C were consistent with a system of randomly-oriented single-domain magnetic particles as confirmed by SEM measurements. These samples demonstrated typical hard magnet properties with coercivity ≥ 3500 Oe. The magnetic properties of the samples improved by sintering at higher temperatures, and by addition of extra Ba and washing by diluted HCl solution.

Acknowledgements

This work was supported by a generous grant from the Deanship of Scientific Research, The University of Jordan (grant # 1404). The technical assistance of Y. Abu Salha and W. Fares, The University of Jordan, is acknowledged.

References

[1] J. Smit, H.P.J. Wijn, Ferrites: physical properties of ferrimagnetic oxides in relation to their technical applications, Philip"s Technical Library, Eindhoven, 1959.

[2] T. Fujiwara, Barium ferrite media for perpendicular recording, Magnetics, IEEE Transactions on, 21 (1985) 1480-1485.

[3] D.E. Speliotis, Performance of MP^{++} and $BaFe^{++}$ tapes in high density recording applications, Magnetics, IEEE Transactions on, 31 (1995) 2877-2882.

[4] J. Kreisel, H. Vincent, F. Tasset, P. Wolfers, The magnetic anisotropy change of $BaFe_{12-2x}Ir_xCo_xO_{19}$: a single-crystal neutron diffraction study of the accompanying atomic and magnetic structures, Journal of magnetism and magnetic materials, 213 (2000) 262-270.

[5] F. Kools, A. Morel, R. Grössinger, J. Le Breton, P. Tenaud, LaCo-substituted ferrite magnets, a new class of high-grade ceramic magnets; intrinsic and microstructural aspects, Journal of magnetism and magnetic materials, 242 (2002) 1270-1276.

[6] Ü. Özgür, Y. Alivov, H. Morkoç, Microwave ferrites, part 1: fundamental properties, Journal of Materials Science: Materials in Electronics, 20 (2009) 789-834.

[7] R.C. Pullar, Hexagonal ferrites: a review of the synthesis, properties and applications of hexaferrite ceramics, Progress in Materials Science, 57 (2012) 1191-1334.

[8] V.G. Harris, A. Geiler, Y. Chen, S.D. Yoon, M. Wu, A. Yang, Z. Chen, P. He, P.V. Parimi, X. Zuo, Recent advances in processing and applications of microwave ferrites, Journal of Magnetism and Magnetic Materials, 321 (2009) 2035-2047.

[9] H. Pfeiffer, R.W. Chantrell, P. Görnert, W. Schüppel, E. Sinn, M. Rösler, Properties of barium hexaferrite powders for magnetic recording, Journal of magnetism and magnetic materials, 125 (1993) 373-376.

[10] S. Harker, G. Stewart, W. Hutchison, A. Amiet, D. Tucker, Microwave absorption and ^{57}Fe Mössbauer properties of Ni-Ti doped barium hexaferrite, Hyperfine Interactions, 230 (2015) 205-211.

[11] N. Dishovski, A. Petkov, I. Nedkov, I. Razkazov, Hexaferrite contribution to microwave absorbers characteristics, Magnetics, IEEE Transactions on, 30 (1994) 969-971.

[12] G. Li, G.-G. Hu, H.-D. Zhou, X.-J. Fan, X.-G. Li, Attractive microwave-absorbing properties of $La_{1-x}Sr_xMnO_3$ manganite powders, Materials Chemistry and Physics, 75 (2002) 101-104.

[13] J. Qiu, Q. Zhang, M. Gu, H. Shen, Effect of aluminum substitution on microwave absorption properties of barium hexaferrite, Journal of applied physics, 98 (2005) 103905.

[14] J. Fritzsche, R. Kramer, V. Moshchalkov, Visualization of the vortex-mediated pinning of ferromagnetic domains in superconductor-ferromagnet hybrids, Physical Review B, 79 (2009) 132501.

[15] I. Ali, M. Islam, M.N. Ashiq, I. Sadiq, M.A. Khan, N. Karamat, M. Ishaque, G. Murtaza, I. Shakir, Z. Ahmad, Electrical Behavior of Tb-Mn Substituted Y-Type Hexa-ferrites for High-Frequency Applications, Journal of Electronic Materials, 44 (2015) 1054-1061.

[16] K. Praveena, M. Bououdina, M.P. Reddy, S. Srinath, R. Sandhya, S. Katlakunta, Structural, Magnetic, and Electrical Properties of Microwave-Sintered Cr^{3+}-Doped Sr Hexaferrites, Journal of Electronic Materials, 44 (2015) 524-531.

[17] S. Chikazumi, C.D. Graham, Physics of Ferromagnetism, 2 ed., Oxford, New York, 1997.

[18] E. Kreber, U. Gonser, Mössbauer spectroscopy in As-substituted $BaFe_{12}O_{19}$, Applied Physics A: Materials Science & Processing, 1 (1973) 339-341.

[19] J. Qiu, Y. Wang, M. Gu, Effect of Cr substitution on microwave absorption of $BaFe_{12}O_{19}$, Materials Letters, 60 (2006) 2728-2732.

[20] M. Awawdeh, I. Bsoul, S. Mahmood, Magnetic properties and Mössbauer spectroscopy on Ga, Al, and Cr substituted hexaferrites, Journal of Alloys and Compounds, 585 (2014) 465-473.

[21] E. Gorter, Saturation magnetization of some ferrimagnetic oxides with hexagonal crystal structures, Proceedings of the IEE-Part B: Radio and Electronic Engineering, 104 (1957) 255-260.

[22] X. Zhou, A. Morrish, Z. Li, Y. Hong, Site preference for Co^{2+} and Ti^{4+} in Co-Ti substituted barium ferrite, Magnetics, IEEE Transactions on, 27 (1991) 4654-4656.

[23] J. Williams, J. Adetunji, M. Gregori, Mössbauer spectroscopic determination of magnetic moments of Fe^{3+} and Co^{2+} in substituted barium hexaferrite, $Ba(Co, Ti)_xFe_{(12-2x)}O_{19}$, Journal of magnetism and magnetic materials, 220 (2000) 124-128.

[24] S.Y. An, I.-B. Shim, C.S. Kim, Mössbauer and magnetic properties of Co–Ti substituted barium hexaferrite nanoparticles, Journal of applied physics, 91 (2002) 8465-8467.

[25] M.J. Iqbal, M.N. Ashiq, P.H. Gomez, Effect of doping of Zr–Zn binary mixtures on structural, electrical and magnetic properties of Sr-hexaferrite nanoparticles, Journal of Alloys and Compounds, 478 (2009) 736-740.

[26] X. Gao, Y. Du, X. Liu, P. Xu, X. Han, Synthesis and characterization of Co–Sn substituted barium ferrite particles by a reverse microemulsion technique, Materials Research Bulletin, 46 (2011) 643-648.

[27] S.H. Mahmood, G.H. Dushaq, I. Bsoul, M. Awawdeh, H.K. Juwhari, B.I. Lahlouh, M.A. AlDamen, Magnetic Properties and Hyperfine Interactions in M-Type $BaFe_{12-2x}Mo_xZn_xO_{19}$ Hexaferrites, Journal of Applied Mathematics and Physics, 2 (2014) 77-87.

[28] G. Amiri, R. Mousarezaei, Structural, Magnetic, and Microwave-Absorption Properties of Nanocrystalline $Ca(MnSn)_xFe_{12-2x}O_{19}$ Ferrites, Journal of Electronic Materials, 44 (2015) 715-719.

[29] S. Wang, J. Ding, Y. Shi, Y. Chen, High coercivity in mechanically alloyed $BaFe_{10}Al_2O_{19}$, Journal of magnetism and magnetic materials, 219 (2000) 206-212.

[30] G. Mendoza-Suarez, J. Matutes-Aquino, J. Escalante-Garcıa, H. Mancha-Molinar, D. Rıos-Jara, K. Johal, Magnetic properties and microstructure of Ba-ferrite powders prepared by ball milling, Journal of magnetism and magnetic materials, 223 (2001) 55-62.

[31] V. Babu, P. Padaikathan, Structure and hard magnetic properties of barium hexaferrite with and without La_2O_3 prepared by ball milling, Journal of magnetism and magnetic materials, 241 (2002) 85-88.

[32] I. Bsoul, S. Mahmood, Structural and Magnetic Properties of $BaFe_{12-x}Al_xO_{19}$ Prepared by Milling and Calcination, Jordan Journal of Physics, 2 (2009) 171-179.

[33] I. Bsoul, S. Mahmood, Magnetic and structural properties of $BaFe_{12-x}Ga_xO_{19}$ nanoparticles, Journal of Alloys and Compounds, 489 (2010) 110-114.

[34] I. Bsoul, S. Mahmood, A.-F. Lehlooh, Structural and magnetic properties of $BaFe_{12-2x}Ti_xRu_xO_{19}$, Journal of Alloys and Compounds, 498 (2010) 157-161.

[35] J. Qiu, L. Lan, H. Zhang, M. Gu, Effect of titanium dioxide on microwave absorption properties of barium ferrite, Journal of Alloys and Compounds, 453 (2008) 261-264.

[36] M. Han, Y. Ou, W. Chen, L. Deng, Magnetic properties of Ba-M-type hexagonal ferrites prepared by the sol–gel method with and without polyethylene glycol added, Journal of alloys and compounds, 474 (2009) 185-189.

[37] Z. Lalegani, A. Nemati, Effects of Ce–Co substitution on structural, magnetic and dielectric properties of M-type barium hexaferrite nanoparticles synthetized by sol–gel auto-combustion route, Journal of Materials Science: Materials in Electronics, 26 (2015) 2134-2144.

[38] G.H. Dushaq, S.H. Mahmood, I. Bsoul, H.K. Juwhari, B. Lahlouh, M.A. AlDamen, Effects of molybdenum concentration and valence state on the structural and magnetic properties of $BaFe_{11.6}Mo_xZn_{0.4-x}O_{19}$ hexaferrites, Acta Metallurgica Sinica (English Letters), 26 (2013) 509-516.

[39] A. Ataie, I. Harris, C. Ponton, Magnetic properties of hydrothermally synthesized strontium hexaferrite as a function of synthesis conditions, Journal of materials science, 30 (1995) 1429-1433.

[40] A. Ataie, M. Piramoon, I. Harris, C. Ponton, Effect of hydrothermal synthesis environment on the particle morphology, chemistry and magnetic properties of barium hexaferrite, Journal of materials science, 30 (1995) 5600-5606.

[41] J. Wang, C. Ponton, I. Harris, A study of the magnetic properties of hydrothermally synthesised Sr hexaferrite with Sm substitution, Journal of magnetism and magnetic materials, 234 (2001) 233-240.

[42] A. Ataie, S. Heshmati-Manesh, Synthesis of ultra-fine particles of strontium hexaferrite by a modified co-precipitation method, Journal of the European Ceramic Society, 21 (2001) 1951-1955.

[43] K.S. Moghaddam, A. Ataie, Role of intermediate milling in the processing of nano-size particles of barium hexaferrite via co-precipitation method, Journal of alloys and compounds, 426 (2006) 415-419.

[44] M.J. Iqbal, M.N. Ashiq, Physical and electrical properties of Zr–Cu substituted strontium hexaferrite nanoparticles synthesized by co-precipitation method, Chemical Engineering Journal, 136 (2008) 383-389.

[45] I. Bsoul, S. Mahmood, A.-F. Lehlooh, A. Al-Jamel, Structural and magnetic properties of $SrFe_{12-2x}Ti_xRu_xO_{19}$, Journal of Alloys and Compounds, 551 (2013) 490-495.

[46] R. Arendt, Liquid-phase sintering of magnetically isotropic and anisotropic compacts of $BaFe_{12}O_{19}$ and $SrFe_{12}O_{19}$, Journal of Applied physics, 44 (1973) 3300-3305.

[47] A.A. Nourbakhsh, M. Noorbakhsh, M. Nourbakhsh, M. Shaygan, K.J. Mackenzie, The effect of nano sized $SrFe_{12}O_{19}$ additions on the magnetic properties of chromium-doped strontium-hexaferrite ceramics, Journal of Materials Science: Materials in Electronics, 22 (2011) 1297-1302.

[48] S.H. Mahmood, A.N. Aloqaily, Y. Maswadeh, A. Awadallah, I. Bsoul, M. Awawdeh, H. Juwhari, Effects of Heat Treatment on the Phase Evolution, Structural, and Magnetic Properties of Mo-Zn Doped M-type Hexaferrites, Solid State Phenomena, 232 (2015) 65-92.

[49] H. Cao, P. Lu, Z. Yu, J. Chen, S. Wang, Electronic and magnetic properties of vanadium doped AlN nanosheet under in-plane biaxial strains, Superlattices and Microstructures, 73 (2014) 113-120.

[50] P. Lu, X. Zhang, H. Cao, Z. Yu, N. Cai, T. Gao, S. Wang, Vanadium doping on magnetic properties of H-passivated ZnO nanowires, Journal of Materials Science, 49 (2014) 3177-3182.

[51] H. Pan, Electronic and magnetic properties of vanadium dichalcogenides monolayers tuned by hydrogenation, The Journal of Physical Chemistry C, 118 (2014) 13248-13253.

[52] A. Alsmadi, I. Bsoul, S. Mahmood, G. Alnawashi, F. Al-Dweri, Y. Maswadeh, U. Welp, Magnetic study of M-type Ru-Ti doped strontium hexaferrite nanocrystalline particles, Journal of Alloys and Compounds, 648 (2015) 419-427.

[53] G. Bate, Magnetic recording materials since 1975, Journal of Magnetism and Magnetic materials, 100 (1991) 413-424.

[54] G. Bate, Recording materials, Handbook of Ferromagnetic Materials, 2 (1980) 381-507.

[55] Y. Xu, G.L. Yang, D.P. Chu, H.R. Zhai, Theory of the single ion magnetocrystalline anisotropy of 3d ions, physica status solidi (b), 157 (1990) 685-693.

[56] Y. Liu, M.G. Drew, Y. Liu, J. Wang, M. Zhang, Preparation, characterization and magnetic properties of the doped barium hexaferrites $BaFe_{12-2x}Co_{x/2}Zn_{x/2}Sn_xO_{19}$, x=0.0–2.0, Journal of Magnetism and Magnetic Materials, 322 (2010) 814-818.

[57] H. Sözeri, İ. Küçük, H. Özkan, Improvement in magnetic properties of La substituted $BaFe_{12}O_{19}$ particles prepared with an unusually low Fe/Ba molar ratio, Journal of Magnetism and Magnetic Materials, 323 (2011) 1799-1804.

[58] C. Suryanarayana, M.G. Norton, X-ray diffraction: a practical approach, 1 ed., Plenum, New York, 1998.

[59] B.D. Cullity, C.D. Graham, Introduction to magnetic materials, 2 ed., Wiley, Hoboken, New Jersey, 1972.

[60] A. Alsmadi, I. Bsoul, S. Mahmood, G. Alnawashi, K. Prokeš, K. Siemensmeyer, B. Klemke, H. Nakotte, Magnetic study of M-type doped barium hexaferrite nanocrystalline particles, Journal of Applied Physics, 114 (2013) 243910.

IOP Conf. Series: Materials Science and Engineering **92** (2015) 012007 doi:10.1088/1757-899X/92/1/012007

Synthesis, Characterization and properties studies of new magnetic materials

Amel Messai[1], Dominique Luneau[2]

[1] Laboratoire des Structures, Propriétés et Interactions Inter Atomiques (LASPI2A) Institut des Sciences et Technologie Université Abbés Laghrour Khenchela ; 40000, Algeria.
[2] Groupe de Cristallographie et Ingénierie Moléculaire, Laboratoire des Multimatériaux et Interfaces UMR 5615 CNRS - Université Claude Bernard Lyon 1, Campus de la Doua, France.

E-mail: messai.amel@yahoo.ca

Abstract. We are interested in molecular polymetallic species having high spin and nuclearities in relation to the field of so call single-molecule magnets (SMMs). The goal is to find a way to synthesis metal clusters which may have application in magnetism and nano-sciences. With this purpose, we decided to investigate the coordination chemistry of the Schiff base.Along this way we were able to create cubane-like complexes and elaborate new Single Molecule-Magnets. The idea was to use Schiff base ligands and different metals to generate high nuclear complexes. Complexation of Shiff base with copper has been investigated. Tetranuclear complex with a cubane like core have been synthesised with (Sciff base), with the same base and cobalt we obtains an other single magnetic complex completely different.

1. Introduction

Among the possibilities offered by the coordination chemistry, obtaining molecules loving (Single Molecule Magnets, SMM) is probably the most significant progress. In fact, a bit of information could be attributed to a single molecular entity. Many strategies have been developed to synthesize such molecules must involve many carriers spin to have a high magnetic moment and have a strong magnetic anisotropy [1-6].

One of the strategies developed by the team based on the synthesis of ligands which are simple Schiff bases of β-diketone. Attractive, these ligands have the advantage of being shaped to perfection with a multitude of functions and chemical donating groups depending on the application required. The use of these ligands for the complexation of transition metals such as copper or cobalt, for example, yielded tetra nuclear complex [$CuII_4$] or tetra nuclear mixed-valence [$CoII_3CoIII_4$].

2. Results and discuss

2.1 Materials and methods

The unit cell determination and data reduction were performed using the CrysAlis program (Oxford Diffraction., 2006) on the full set of data.

Calculations were carried out using the WinGX software package [7].The crystal structure was solved by direct methods using SIR2004 [8] and refined by full-matrix least-squares against F^2 using all data (SHELX97) [9].

All non-H atoms were modelled with anisotropic displacement parameters. The H atoms attached to $-CH_3$ were located in difference Fourier maps refined as riding atoms (isotropically with a restrained bond distance) with distances constraints of methyl C-H = 0.96 A° and Uiso(H) =

1.5Ueq(C). Aromatic H atoms were positioned geometrically and were allowed to ride on their parent C atoms with C-H = 0.93 A° and and Uiso(H) = 1.2Ueq(C).
Crystal structure was visualized using ORTEP3 [10] and MERCURY [11]. Analyses were carried out using the program PLATON [12], as incorporated in the WinGX [13].

2.1- Crystal Structure Analysis

2.1.1 Complexe of copper (I). The complex (I). crystallizes in the monoclinic space group $P2_1/n$. The molecular structure of the complexe **(I)** and the atom-labeling scheme are shown in (Figure 1).After several cycles of refinement until convergence, we obtained factors reliabilié R = 4.5% and Rw = 7.1%. The asymmetric unit corresponds II four copper ions, two ligand type (D) and two other ligand type (L). The formula of the complex is refined [Cu4 $(L\text{-}L1)_2$ $(D\text{-}L1)_2$]. The metal cluster in turn can be described as a square Cu4 in which each copper ion (II) is coordinated to a ligand type (L) and ligand type (D) coordinate two metals at a time.
The formula of the complex is refined [Cu4 $(L\text{-}L1)_2$ $(D\text{-}L1)_2$]. The metal cluster in turn can be described as a square Cu4 in which each copper ion (II) is coordinated to a ligand type (L) and ligand type (D) coordinnent two metals at a time. Such that the two ligands of the same type, are in a position facing each other. Each copper is in a slightly distorted square-planar CuO 3N type, provided by the two oxygen atoms and the nitrogen atom of the ligand L1HK type (D) or (L) and an oxygen atom of the acid function of a second ligand of the other type L1HK [(L) or (D)].

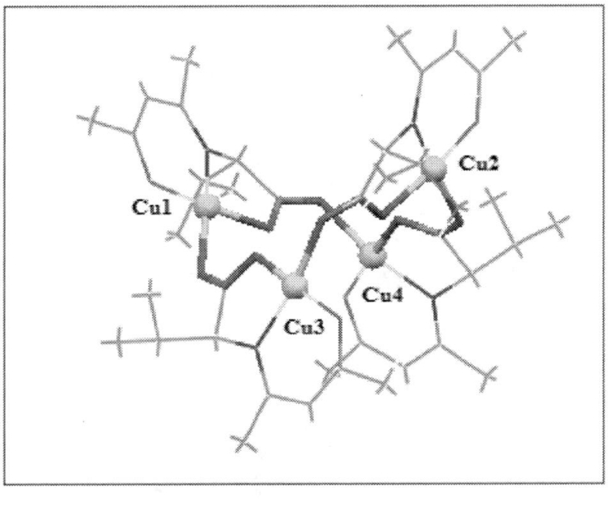

Figure 1: Asymmetric unit and atom-numbering scheme of (I)

2.1.2 The Complexe of cobalt. The complex (II); crystallizes in the triclinic, P-1 space group ,The final refinement of the structural model performed using the SHELXL97 program led to the reliability factors R = 0.13 unweighted and weighted Rw = 0.44 with a goof = 1.119. The molecular structures of **(II)** and the atom-labeling scheme are shown in (Figure 2). the four cobalt atoms form a rhombus whose center is the center of inversion of the molecule (Figure 3) . The atom Co (1) is in a hex coordinate pseudo-octahedral surrounded by oxygen of the acid function, an oxygen atom of the acac and a nitrogen atom of the same ligand (L) configuration, oxygen a methanolate type (μ3-OMe) and one oxygen methanolate another type (μ2-OMe), as well as a methanol. The base of the octahedron is formed by three atoms of oxygen and a nitrogen atom, the two peaks are the two oxygens of the

IOP Conf. Series: Materials Science and Engineering **92** (2015) 012007 doi:10.1088/1757-899X/92/1/012007

ligand. The Co atom (2) is in a second position hexacoordinate crystallographic in a distorted octahedral environment by effect of the two intramolecular hydrogen bonds. This atom (Co2) is surrounded by: two two oxygens methanolates type (μ2-OMe), a third oxygen methanolate type (μ3-OMe), an oxygen of the carboxyl function of the ligand configuration (D) and two two molecules of methanol oxygens. The base of the octahedron is formed by four oxygen atoms attributed to: methanolate type (μ2-OMe), methanolate type (μ3-OMe), methanol, and a carboxylic function and the two peaks are of a methanol oxygen and another oxygen from a methanolate(μ3-OMe).

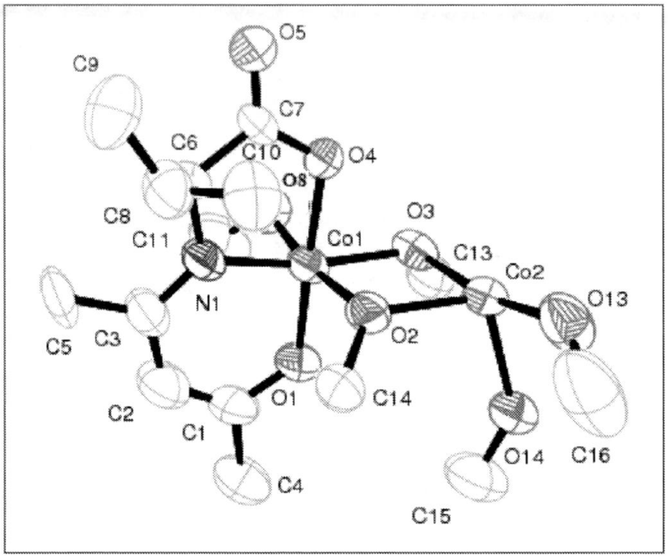

Figure 2: Asymmetric unit and atom-numbering scheme of (II) . Displacement ellipsoids are drawn at the 30% probability level. H atoms are represented as small sphere of arbitrary radii.

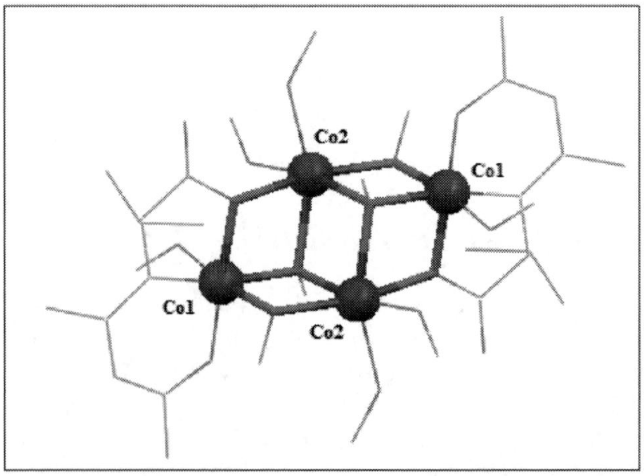

Figure 2: Complexe of Cobalt

2.2. Magnetism:

2.2.1. Complexe of copper (I). Studies of the magnetic properties were carried out for this compound; the thermal variation of the product of the magnetic susceptibility in the temperature of the complex is shown in **figure (4)**.

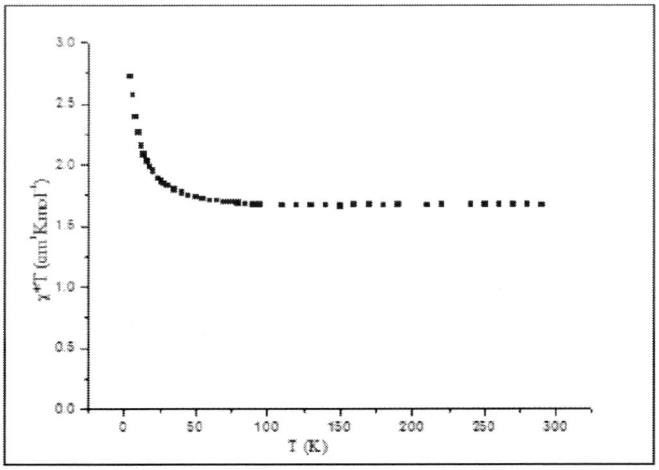

Figure (4): Evolution of temperature χT product.

At 300K, χT has a value of 1.68 cm3.K.mol-1 which is a little higher than the value expected for four copper CuII (S = 1/2) magnetically independent from each other (4x0.375cm3.K. mol-1). χT increases steadily when lowering the temperature to the value of 2.72 cm3.K.mol 4K-1 which is close to the value expected for ferromagnetically coupled four brass (S = 2, χT = 3 mol cm3.K. -1) this behavior is indicative of dominant ferromagnetic interactions within the complex which is confirmed by the value of the saturation magnetization measured at 2K **figure (5)** tends to the expected value for four copper, namely 4μB.

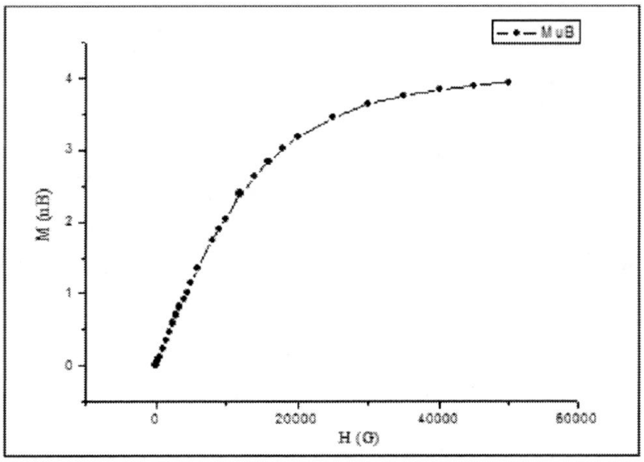

Figure (5): Dependence of the magnetization as a function of the field.

2.2.2. Complex of Cobal. In addition to its structural characteristics, the complex also has magnetic properties Figure (6) that we will discuss.

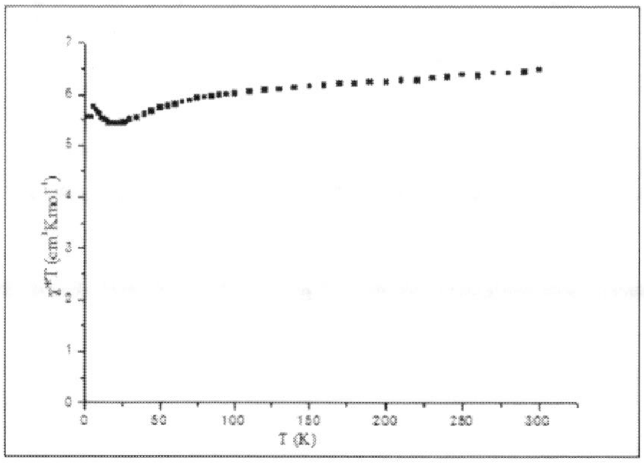

Figure (6): Evolution of temperature χT product.

At 300K, a χT value of 6.45 cm3.K.mol-1 which lower the expected value for two CoIII and CoII two magnetically independent (9.75 x3 = 2 2x1.875 cm3.K.mol-1with g = 2) . χT decreases when the temperature is lowered to a minimum and this 20K-1 5.42 cm3.K.mol then χT increases to a maximum at 6K (5.74 cm3.K.mol^{-1}), and then decreases at lower temperature. This behavior is the signature of dominant ferromagnetic interactions; the value of the saturation magnetization measured at 2K tends to 6μB **Figure (7).**

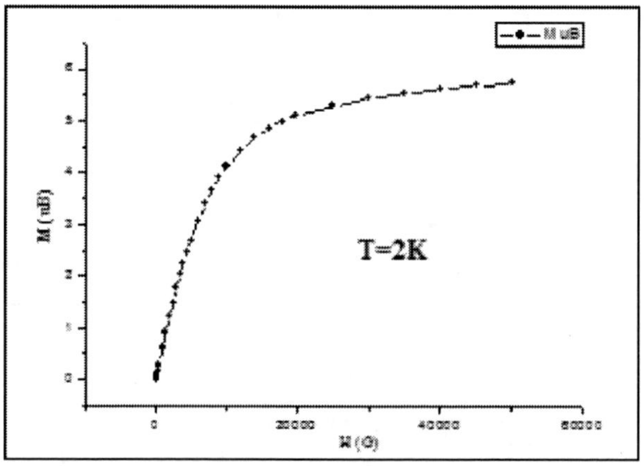

Figure (7): Dependence of the magnetization as a function of he field.

3. Experimental part

3.1 *The complexe of copper*

(1mmol, 237mg) of the ligand (L1HK) is dissolved in 20 ml of methanol, to this solution was added one equivalent of copper chloride (1 mmol, 170 mg) dissolved in 10 ml of ethanol: a green solution is obtained. After ten minutes of stirring, 3 equivalents (0.6 ml) of triéthylamine are added the solution becomes very intense green. Slow evaporation of the latter leads to the production of single crystals of good size and good quality usable in X-ray diffraction after slow evaporation of the solvent.

3.2 *The complexe of cCobalt:*
The ligand L1HK (236mg, 1mmol) was dissolved in 10 ml of methanol. To this solution is added (129mg, 1 mmol) of cobalt chloride ($CoCl_2$) previously dissolved in methanol: a brown solution is obtained. 0.6 ml of triethylamine dissolved in 10 ml of methanol was added and the solution becomes orange-brown. Slow evaporation of the latter leads to the production of single crystals after some days of acceptable size, usable in X-ray diffraction

4. Conclusion:

In general this work contributed to new approaches for the development of new molecular-based materials; We therefore conducted a study crystallographic and magnetic susceptibility measurements on a complex tetra nuclear of copper. Such complexes are currently intensively are studying because it has an interest in biological processes and in the design of inorganic materials [(4-6)]. One of our goals is the synthesis of heterometallic using tetramer as brick departure. More we can synthesize a new complex "cubane mixed valence cobalt based" rather than a wheel (similar to copper) although we used the same ligand because of the chemistry of cobalt complexes it involves complicated processes which which most crystallographically characterized compounds are not the only product present in the solution. The synthesis included the protonation / deprotonation redox chemistry and understanding of several species simultaneously. (The source of cobalt ions is a cobalt salt with the oxidation state (+ II), some of which go to the oxidation state (+ III) after partial oxidation. Process is very difficult to control because the oxidation proceeds from a part of the use of the base and the other of the water molecules.

References:

[1] C. Aronica, Y. Chumakov, E. Jeanneau,D. Luneau, P. Neugebauer, A-L Barra, B. Gillon, A. Goujon, A. Cousson, J. Tercero, E. Ruiz *Chem. 2008. Eur. J.* **14**, 9540-9548
[2] B.Le Guennic ; S.Petit ;G. Chastanet ; G. Pilet ; D. Luneau ; N. Benamor ;V. Robert , 2008, Inorg. Chem. **47**, 572-577.
[3] L. F. Chibotaru, L. Ungur, C. Aronica, H. Elmoll, G. Pilet, D. Luneau *J. Am. Chem. Soc.* 2008, **130**, 12445-12455
[4] T. Shiga, H. Oshio, *Science and techno. Adv. Mat.*, **6**, 2005, 565
[5] H. Borzel, P. Comba,H. Pritzkow, *Chem. Commun.*, **97**, 2001
[6] S.S.Y.Chui, S.M.F. Charmant, A.G. Orpen, I.G. Williams. *Science*1999, 283,1148.
[7] Farrugia LJ., 1999. J Appl Crystallogr. **32**, 837-838.
[8] Burla, M.C., Caliandro, R., Camalli, M., Carozzini, B., Cascarano, G.L., De Caro, L., iacovazzo, C., Polidori, G., Spagna, R., 2005. J Appl Crystallogr. **38**, 381-388.
[9] Sheldrick, G.M., 1997. SHELXL97. University of Göttingen, Germany.
[10] Farrugia, L.J., 1997. J Appl Crystallogr. **30**, 595.
[11] Bruno, I.J., Cole, J.C., Edgington, P.R., Kessler, M., Macrae, C.F., McCabe, P., Pearson, J., Taylor, R., 2002. Acta Cryst B. **58**, 389-397.
[12] Spek, A.L., 2003. J Appl Crystallogr. **36**,7-39.
[13] Insuasty, B., Torres, H., Cobo, J., Lowd, J.N., Gilidewell, C., 2006 .Acta Cryst C. **62**, O39-O41.

IOP Conf. Series: Materials Science and Engineering **92** (2015) 012008 doi:10.1088/1757-899X/92/1/012008

Structural and magnetic properties of Cu–V substituted M-type barium hexaferrites

Sami .H. Mahmood[1,*], Ahmad Awadallah[1], Yazan Maswadeh[1], Ibrahim Bsoul[2]

[1]Physics Department, The University of Jordan, Amman 11942, Jordan

[2]Physics Department, Al al-Bayt University, Mafraq 13040, Jordan

e-mail: s.mahmood@ju.edu.jo

Abstract: In search of magnetic materials with improved magnetic characteristics for practical applications, M-type barium hexaferrites with Fe^{3+} ions partially substituted by a mixture of Cu and V ions were prepared by ball milling and sintering at 1200° C. The structural analyses of the prepared $BaFe_{12-2x}Cu_xV_xO_{19}$ samples (x = 0.1, 0.2, 0.3, 0.4) revealed the presence of BaM phase, in addition to α-Fe_2O_3, $Ba_3V_2O_8$, and $BaFe_2O_4$ nonmagnetic phases which evolved as x increased. Scanning electron microscopy (SEM) imaging demonstrated the presence of different phases in the substituted samples, and a general trend of particle-size growth with increasing x. Energy dispersive spectroscopy was used to examine the local stoichiometry of the samples, and confirmed the different phases identified by XRD analysis. The saturation magnetization was found to be high for low substitution level (72 emu/g for the sample with x = 0.1 sintered for 2 h, and 65 emu/g for the sample sintered for 10 h), while it decreased significantly with increasing the substitution level. The coercivity (H_c) for the samples sintered for 2 h was found to decrease sharply with increasing x, even at low substitution levels ($x < 0.2$), where it decreased from about 3.5 kOe for the un-substituted sample down to about 1.6 kOe for the sample with x = 0.1, and down to below 0.3 kOe at higher substitution levels. The coercivity of the sample with x = 0.1 sintered for 10 h reduced further, down to about 677 Oe, demonstrating properties demanded for magnetic recording applications. Further, washing with HCl was found to remove some of the nonmagnetic phases, and increase the yield of the BaM phase.

1. Introduction

Magnetic materials are used as core components in a wide range of instruments, tools and devices we use in our everyday life [1-3]. Due to their low eddy current losses, low-cost processing, stability, and tunable magnetic properties, magnetic oxides had demonstrated superior properties compared with metallic magnetic materials for industrial and technological applications [4, 5]. Hexagonal ferrites discovered in the early 1950s [6, 7] are on the top list of important magnetic materials in use nowadays [8].

Due to their high magnetocrystalline anisotropy, hexagonal ferrites were recognized as important materials for microwave devices [2, 5, 9], and microwave absorbers [10-12]. Further, M-type barium-based hexaferrites (BaM or Ferroxdure) with chemical formula $BaO.6Fe_2O_3$ have found a wide range of applications in permanent magnet [3, 13, 14] and magnetic recording media [15-17]. M-type hexaferrites with Ba substituted by other divalent cations such as Sr [18-21], Pb [18, 21] and Ca [22-24] were also synthesized.

The coercivity, in addition to other important magnetic properties of the magnetic material such as saturation magnetization, remanence, and permeability, should be modified to fulfill the requirements of the specific application. This was the field of intense research work growing exponentially since the development of these materials [8]. For that purpose, different techniques were adopted to fabricate hexaferrite powders such as sol-gel [25-27], coprecipitation, [26, 28, 29] hydrothermal [30-32], wet chemical mixing [33, 34], and ball milling technique [24, 35-37]. The coercivity of M-type hexaferrites was modified by the substitution of Fe^{3+} ions by trivalent ions, or by combinations of divalent and tetravalent ions. The substitution of Fe^{3+} by Al^{3+}, Ga^{3+}, Cr^{3+} was found to enhance the coercivity of M-type hexaferrites [32, 38-40]. However, various combinations of divalent and tetravalent were found to reduce the coercivity [19, 41-44].

In a previous work conducted by our group (submitted), it was demonstrated that the substitution of Fe by V in BaM resulted in phase separation into a multicomponent powder, in which the progressive increase in V concentration resulted in a dramatic increase in the fractions of nonmagnetic phases. This led to a strong attenuation of the saturation magnetization, while the coercivity remained high. In this work, the choice of the Cu–V combination was motivated by the realization that V can assume higher valence states (V^{4+} and V^{5+}), which can be compensated for by the lower valence states of Cu ($Cu^{1+}$and Cu^{2+}) to maintain charge neutrality of the crystal when the combination replaces Fe^{3+}. The present work was therefore concerned with the fabrication of Cu–V substituted BaM hexaferrites and the investigation of their structural and magnetic properties, in search of new materials which could be candidates for useful applications.

2. Experimental

Powder precursors of $BaFe_{12-2x}V_xCu_xO_{19}$ (x = 0.1, 0.2, 0.3, 0.4) were synthesized by ball milling the stoichiometric ratios of analytical grade $BaCO_3$, Fe_2O_3, V_2O_5, Cu_2O. The appropriate amounts of these material were carefully weighed, and charged into the zirconia vessels of the Fritch Pulveressette-7 ball-mill. Wet-milling of the powder by zirconia balls in an acetone bath (Powder to acetone mass ratio = 5:8) was performed for 16 h at a rotational speed of 250 rpm, with a powder to ball mass ration of 1:14. The resulting muddy mixture was dried in air, and pressed into disc-shaped pellets 1.5 cm in diameter and ~ 1.0 mm in thickness under a force of 5 tons. The disc was then sintered at 1200° C for 2 h.

The structural characteristics of the different samples were examined by analyzing the x-ray diffraction patterns obtained using XRD 7000-Shimadzu diffractometer with Cu-Kα radiation (λ = 1.5405 Å). The samples were scanned over the angular range $20° < 2\theta < 70°$ with 0.01° scanning step and speed of 0.5°/min. The XRD patterns were analyzed using X'pert HighScore 2.0.1 software for phase identification, and Rietveld refinement of the crystal structure was performed using FULLPROF software.

The grain morphology and grain size of the samples were examined by Scanning Electron Microscope (SEM) using FEI-Inspect F50/FEG electron microscope equipped with Energy dispersive X-ray spectroscopy (EDS) facility for elemental analysis. SEM images were used to investigate the general shape and mean size of the particles, while EDX analyses were used to determine the approximate compositions of the particles of the fabricated samples. Information provided by SEM analyses is essential to complement the results of the XRD structural studies, and to examine the sample homogeneity, phase separation, and the degree of crystallinity. In addition, this information is essential to support conclusions drawn from the magnetic data, since the magnetic properties are generally sensitive to particle morphologies.

The magnetic properties determined by measurements of the magnetic parameters of the samples, such as the saturation magnetization (M_s), the coercivity (H_c), and the remnant magnetization (M_r), were investigated using a vibrating sample magnetometer (VSM MicroMag 3900, Princeton Measurements Corporation). Samples for VSM measurements were prepared by cutting small pieces from the sintered discs and gently polishing to the desired needle-shape. The magnetic measurements were carried out at room temperature with a magnetic field up to 10 kOe. The initial magnetization curves for the samples were obtained in a field range from 0 kOe to 10 kOe with field step of 10 Oe, while the hysteresis loops were performed in the field range from -10 kOe to +10 kOe.

3. Results and Discussion

3.1 Structural Analysis

Fig. 1 shows the refined XRD patterns for the samples sintered at 1200° C for 2h. The patterns demonstrated that the pure sample (with $x = 0$) is composed of a single hexagonal phase consistent with the standard pattern (JPCDS #: 00-043-0002) of BaM phase. The patterns of the substituted samples indicated, in addition, the presence of the vanadate (V: $Ba_3V_2O_8$) and α-Fe_2O_3 secondary phases, whose fractions increased with increasing the substitution level. In addition, small amounts of $BaFe_2O_4$ spinel (S) phase were detected.

To investigate the effect of the sintering time on the development of the phases, representative samples with $x = 0.1$, 0.3, 0.4 were sintered at 1200° C for 10 h, and their diffraction patterns are shown in Fig. 2. The XRD patterns for the 10 h-sintered samples showed variations which reflect changes in the fractions of the different phases, resulting from allowing enough time for reactions between the components of the powders to take place and reach the equilibrium phases.

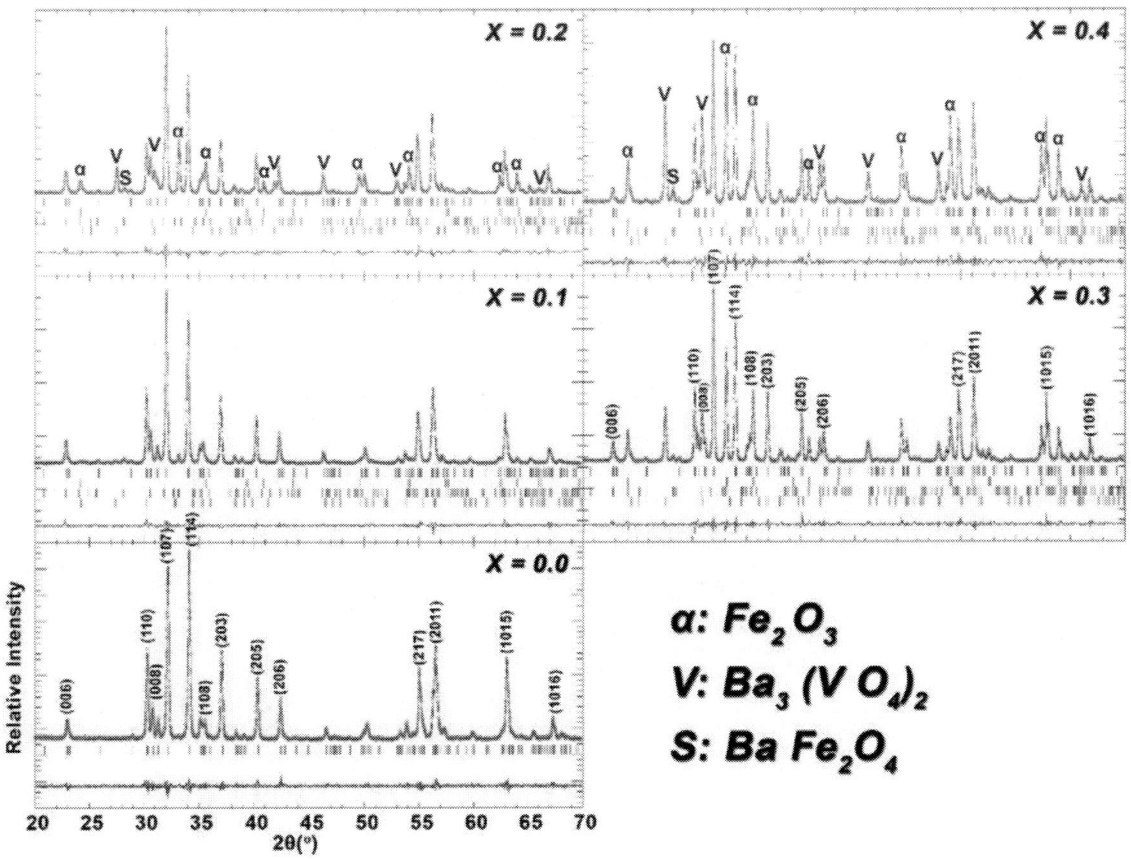

Fig. 1: XRD diffraction patterns with Rietveld refinement for the samples $BaFe_{12-2x}V_xCu_xO_{19}$ sintered at 1200° C for 2 h.

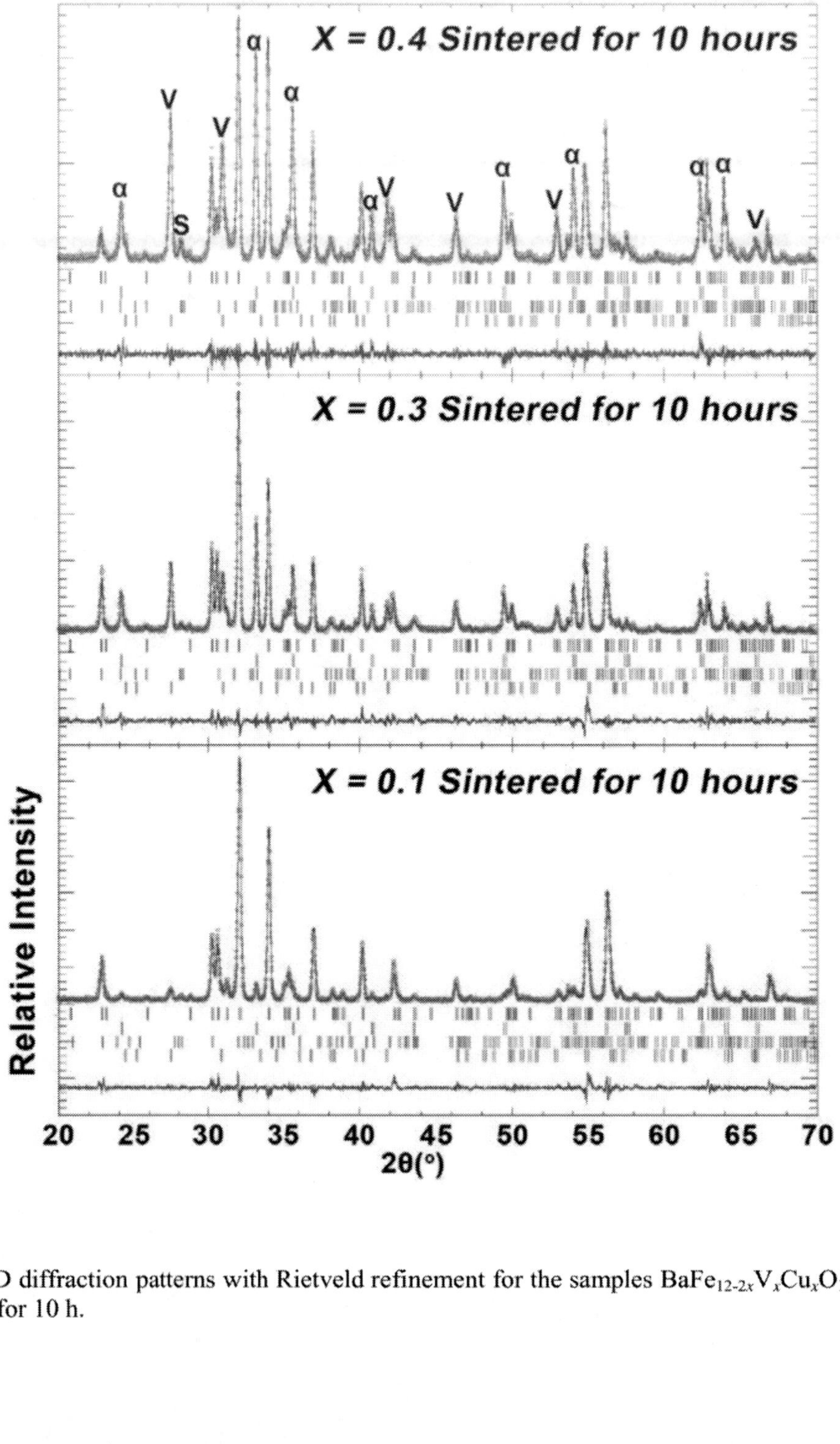

Fig. 2: XRD diffraction patterns with Rietveld refinement for the samples $BaFe_{12-2x}V_xCu_xO_{19}$ sintered at 1200° C for 10 h.

IOP Conf. Series: Materials Science and Engineering **92** (2015) 012008 doi:10.1088/1757-899X/92/1/012008

The structural analysis of each pattern was carried out using Rietveld refinement of, and full structural information related to cell parameters and volume, crystallite size, and weight fractions of the different phases were obtained. Rietveld analysis of the XRD pattern of a sample involved identification of the various phases, and obtaining their weight ratios in the sample, in addition to the refinement of the structural parameters of each phase. The output file resulting from the refinement process is huge, containing a large amount of structural data. In the present work, we considered only those structural characteristics relevant to the scope of this article.

The weight ratios of the different phases in the samples are listed in Table 1. The weight ratio of BaM phase in the samples sintered for 2 h decreased monotonically with increasing x, while the ratios of the α, and V phases increased significantly (Fig. 3). The ratio of the spinel phase also generally increased, but remained relatively low in all samples. This behavior can be well explained by the consumption of Ba in the evolving $Ba_3V_2O_8$ phase, which leaves smaller fraction of Ba available for the formation of BaM phase, and consequently higher fraction of iron which remains as unreacted α-Fe_2O_3 phase.

Sintering the samples for 10 h induced changes in the wt. % of the phases as the data of Table 1 indicated. The increase in the wt. % of the V phase indicates that the reaction involved in the formation of this phase requires prolonged sintering. The enhancement of the wt. % of the BaM phase, and the reduction in the wt. % of the α-Fe_2O_3 and S phases in the samples with $x = 0.3$ and 0.4 is attributed to the reaction of these intermediate phases to form M-type phase.

Table 1. Weight % of the different phases in $BaFe_{12-2x}V_xCu_xO_{19}$ samples sintered at 1200° C for 2 h and 10 h.

x	2 h				10 h			
	BaM	**α**	**S**	**V**	**BaM**	**α**	**S**	**V**
0.0	100	-	-	-				
0.1	95.9	2.70	0.55	0.85	93.0	5.34	0.30	1.36
0.2	76.2	18.2	0.60	5.00				
0.3	59.3	31.4	1.90	7.4	64.5	26.1	0.29	9.07
0.4	48.0	37.4	1.60	13.0	50.6	33.9	1.10	14.4
HCl Washed	62.7	36.1	1.20					

IOP Conf. Series: Materials Science and Engineering **92** (2015) 012008 doi:10.1088/1757-899X/92/1/012008

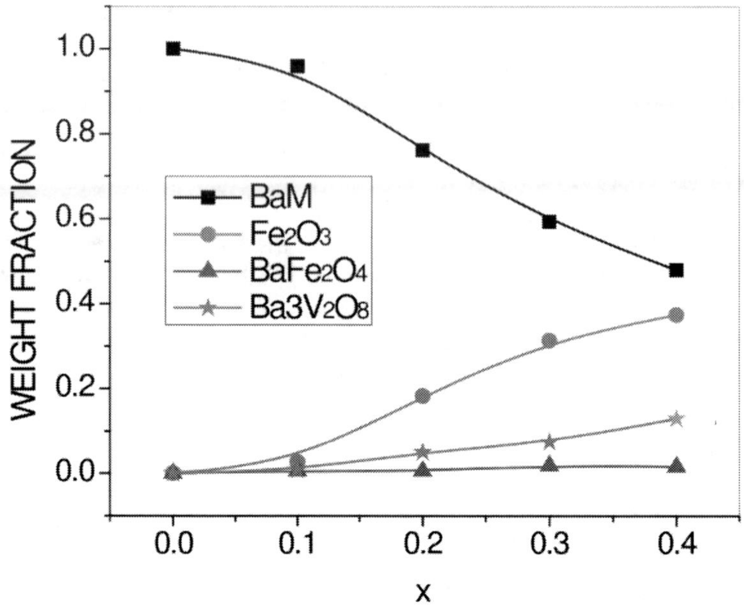

Fig. 3: Weight fractions in the samples as a function of x.

In an attempt to remove the secondary phases from the samples, the sample with $x = 0.4$ sintered for 2 h was washed with a diluted solution of HCl. XRD pattern of this sample (Fig. 4) indicated that $Ba_3V_2O_8$ phase disappeared completely, and the fractions of α-Fe_2O_3 and S phases reduced slightly with HCl washing. The weight ratio BaM:α:S = was found to be 62.7:36.1:1.2, which when compared with the ratios of 55.2:42.9:1.9 in the un-washed sample, one could deduce that the iron oxide and barium spinel phases partially dissolved in HCl, while BaM was not to be affected by the etching process. The rise in the fraction of the BaM is due to mass loss of the nonmagnetic α- and S-phases. This result indicated that HCl washing is suitable for enhancing the yield of the desired BaM phase in the sample.

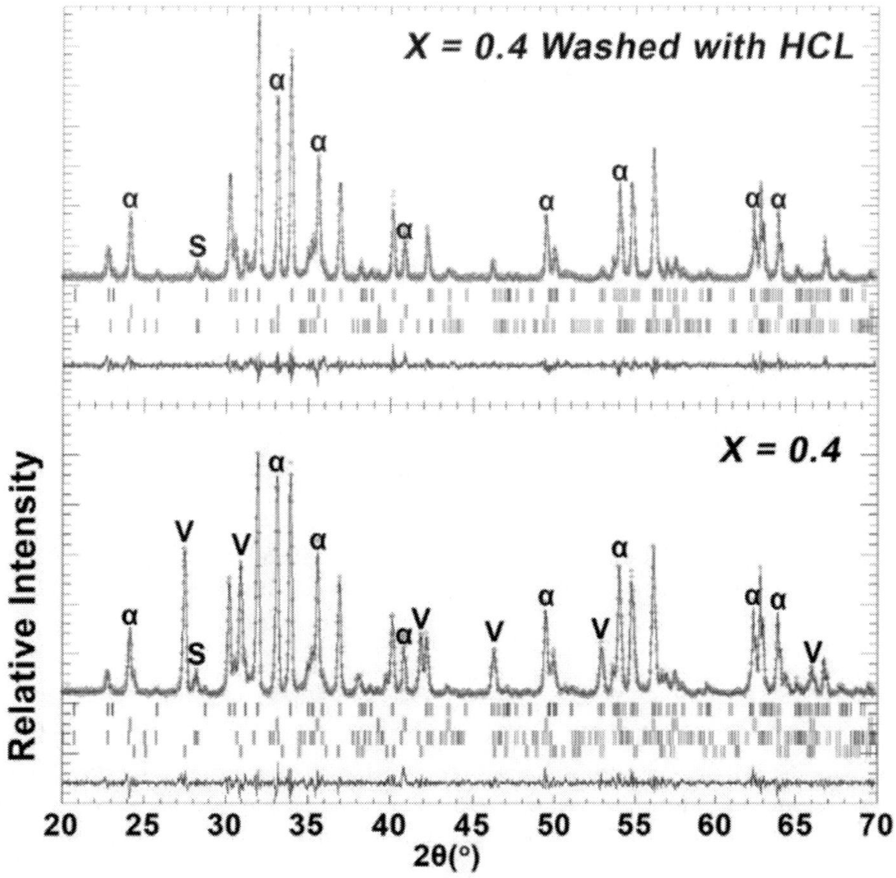

Fig. 4: XRD diffraction patterns with Rietveld refinement for the sample with $x = 0.4$ sintered for 2 h, with and without HCl washing.

The lattice parameters, a and c, of the hexagonal lattice of BaM phase in the samples sintered for 2 h increased sharply up to $x = 0.2$, and then showed tendency to level up at higher concentrations as indicated by Fig 5 A. This could be evidence of the substitution of Cu^{2+} ions for F^{3+} ions at $2b$ and $4f_1$ sites, where the radius of Cu^{2+} ion is larger than that of Fe^{3+} ions at these sites [45]. To maintain charge neutrality of the crystal, this substitution should be accompanied by the substitution of V with higher valence states, and/or the conversion of an appropriate fraction of Fe^{3+} into Fe^{4+}. Evidence of the presence of Fe^{4+} low-spin ionic state was provided by Mössbauer results recently submitted for publication by our research group.

IOP Conf. Series: Materials Science and Engineering **92** (2015) 012008 doi:10.1088/1757-899X/92/1/012008

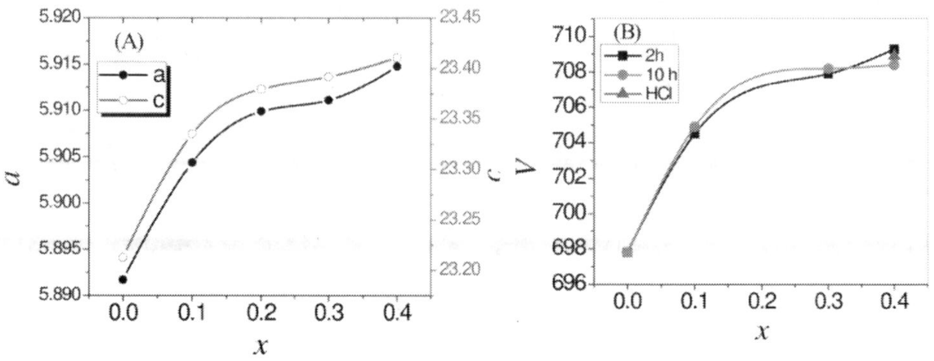

Fig. 5: (A) Lattice constants (in Å) and (B) cell volume (in $Å^3$) for the BaM phase as a function of x.

While the cell volume showed similar with increasing x, the volume did not vary significantly with increasing the sintering time up to 10 h as shown in Fig. 5 B. Also, the cell volume did not practically change with HCl washing.

The crystallite size D in a direction perpendicular to a given set of crystallographic planes (hkl) was determined for the different samples from the broadening of the corresponding peak using Stokes and Wilson formula [46]:

$$D = \frac{\lambda}{\beta \cos \theta} \quad (1)$$

where β is the integral breadth of the peak (defined as the integrated intensity of the peak divided by the maximum intensity), λ is the wavelength of the radiation (= 1.54056 Å), and θ is the peak position. The crystallite size (D_c) along the c-axis normal to the hexagonal plane was determined from the (006) and (008) reflections, whereas the size (D_p) along the a-axis in the hexagonal plane was calculated from the breadth of the (200) and (300) reflections. Using the breadth obtained from the refinement process (corrected for the instrumental broadening), the average crystallite size for each sample in the different directions is listed in Table 2. Generally speaking, the crystallization improved with the Cu-V substitution in all samples. However, the crystallites grew only in the plane of the hexagonal layers in the samples with x = 0.1 and 0.2, indicating the evolution of platelet-like crystallites in these samples sintered for 2 h. In the meantime, the crystallites in the samples with x = 0.3 and 0.4 sintered for 2 h demonstrated equal growth in both the planar and normal directions, resulting in crystallites similar to those of the pure sample in shape, but larger in size. The inhibition of crystal growth along the c-direction in the samples with x = 0.1 and 0.2 could be due to crystal defects correlated with the significant shifts of ionic positions in the hexagonal lattice for these samples (not shown for brevity). On the other hand, the samples sintered for 10 h revealed general reduction in the crystallite size along the c-axis (with respect to those sintered for 2 h), and increase in the planar size, suggesting that prolonged heat treatment resulted in crystallization in the shape of plates. Further, the noticeable decrease of the crystallite size along the c-axis for the sample washed with HCl (with respect to the unwashed sample) could be indication of partial damage of the surface of the plate.

Table 2. Crystallite size in-plane (D_p) and normal-to-plane (D_c) in $BaFe_{12-2x}V_xCu_xO_{19}$ samples sintered at 1200° C for 2 h (calculated from the full-profile fitting breadths of XRD peaks).

x	D_c (nm)		D_p (nm)	
	2 h	10 h	2 h	10 h
0.0	62		60	
0.1	66	59	70	75
0.2	64		80	
0.3	80	67	78	104
0.4	96	86	99	104
HCl Washed	78		95	

3.2 Scanning Electron Microscopy

The morphology and particle size of the samples were examined by SEM imaging, and the sample stoichiometry was determined by EDS. The representative images for the sample with $x = 0.1$ (Fig. 6) indicated the presence of large and small hexagonal-like platelet particles, in addition to semi-circular polyhedral particles. A wide distribution of particle size with particle diameters ranging from ~ 220 nm up to ~ 3.5 μm were observed, with a characteristic distribution dominated by particles with typical diameters of ~ 350 to ~ 900 nm. The significantly larger physical particle size in comparison with the crystallite size (Table 2) indicates that the particles are not single-structural domains, possibly due to internal crystal defects.

.

Fig. 6: Representative SEM images for $BaFe_{12-2x}V_xCu_xO_{19}$ ($x = 0.1$) sample sintered at $T = 1200°$ C for 2 h.

IOP Conf. Series: Materials Science and Engineering **92** (2015) 012008 doi:10.1088/1757-899X/92/1/012008

Fig. 7 shows SEM images for $BaFe_{12-2x}V_xCu_xO_{19}$ ($x = 0.2$). The images showed hexagonal platelet-like, in addition to polyhedral particles. A general growth in particle size was observed, indicating an improvement of crystallization. Some particles had a large size (> 12 μm) as indicated by Fig. 7 (a), while the majority of the particles have a smaller diameter ranging from 350 nm up to 1100 nm. The image in the EDS mode (b) showed, in addition to the regular particles, light particles with no particular shape which could be associated with different phases in the sample. The difference in particle shape is another evidence of the multi-phase nature of the sample. EDS spectrum collected on a typical "dark" particle indicated a Ba: Fe atomic ratio 1: 12.3 which is close the BaM stoichiometric ratio, indicating that the barium hexaferrite form the majority of the sample. In addition, small amounts of Cu and V were detected. The small difference between the observed and stoichiometric Ba:Fe ratio in this sample is due to experimental uncertainty of the technique.

Fig. 7: SEM images for $BaFe_{12-2x}V_xCu_xO_{19}$ ($x = 0.2$) sample sintered for 2 h: (a) imaging mode, and (b) EDS mode.

The SEM images for the sample $BaFe_{12-2x}V_xCu_xO_{19}$ ($x = 0.3$) is shown in Fig. 8. The images indicated further particle growth, with particle size ranging from ~ 1 – 10 μm or larger. Also, particles which clearly crystallized in different shapes were observed, indication that the sample may contain different phases. For brevity, in-depth analysis of the stoichiometry of the particles with different shapes will be discussed below, in the context of examining the phases in the sample with $x = 0.4$.

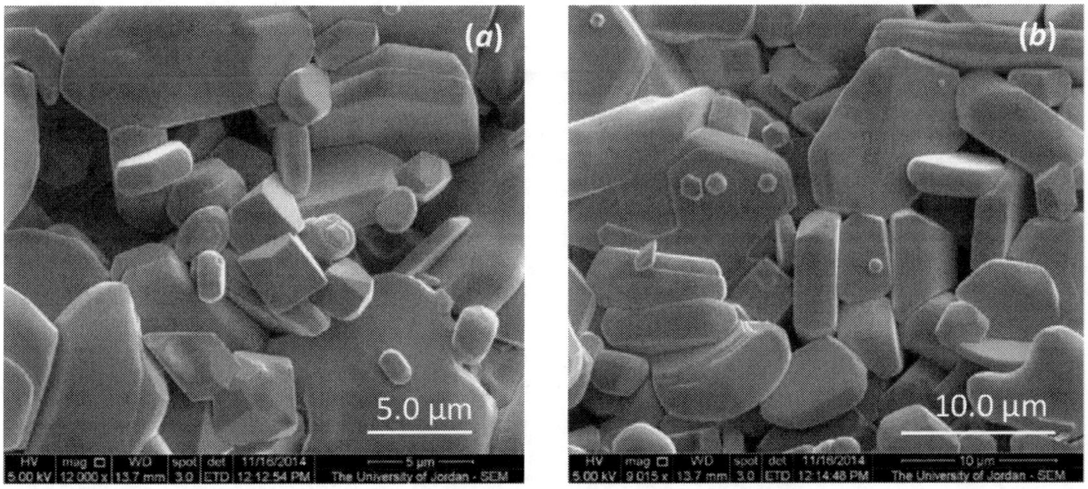

Fig. 8: SEM images for $BaFe_{12-2x}V_xCu_xO_{19}$ ($x = 0.3$) sample sintered for 2 h.

Fig. 9 shows SEM images for the sample $BaFe_{12-2x}V_xCu_xO_{19}$ ($x = 0.4$) in the imaging mode (a) and the EDS mode (b). The particle size is obviously very large in this sample, where the size of the majority of the particles is ~ 10 µm or larger. The image in the EDS mode indicated the presence of large light regions in the sample, with no particular shape.

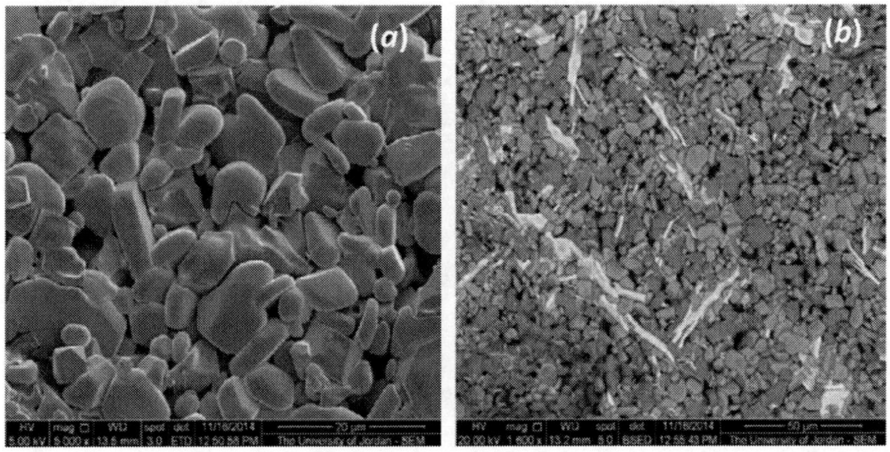

Fig. 9: SEM images for $BaFe_{12-2x}V_xCu_xO_{19}$ ($x = 0.4$) sample sintered for 2 h. (a) imaging mode, and (b) EDS mode.

To check the chemical composition of these light region, EDS spectra were collected and a typical spectrum is shown in Fig. 10. The spectrum indicated dominance of the Ba and V spectral lines with very small signature of other elements, indication that this region is mainly composed of the V-phase. On the other hand, the spectrum from an octahedral-like particle (Fig. 11) revealed the presence of Fe (in addition to oxygen) as a major element, which is an indication that this type of particles correspond to the α-Fe_2O_3 oxide. Further, EDS spectrum on hexagonal plated (Fig. 12) indicated an atomic ratio of Fe in the range 11 – 12 (with respect to Ba), and of ~ 0.5 for Cu; these stoichiometries are consistent with the expected ratios for Cu-doped BaM phase. These results confirmed the multi-phase nature of the samples as indicated by XRD results.

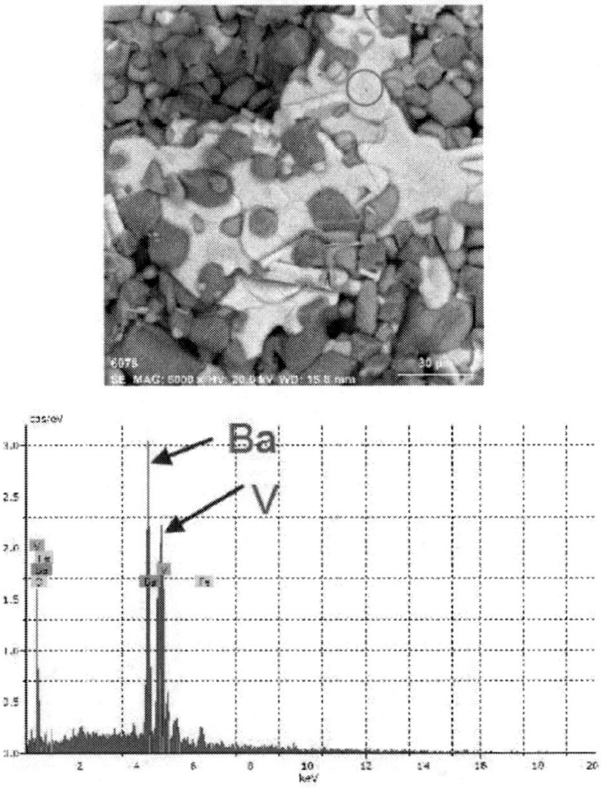

Fig. 10: EDS spectrum on the light region in the sample with $x = 0.4$.

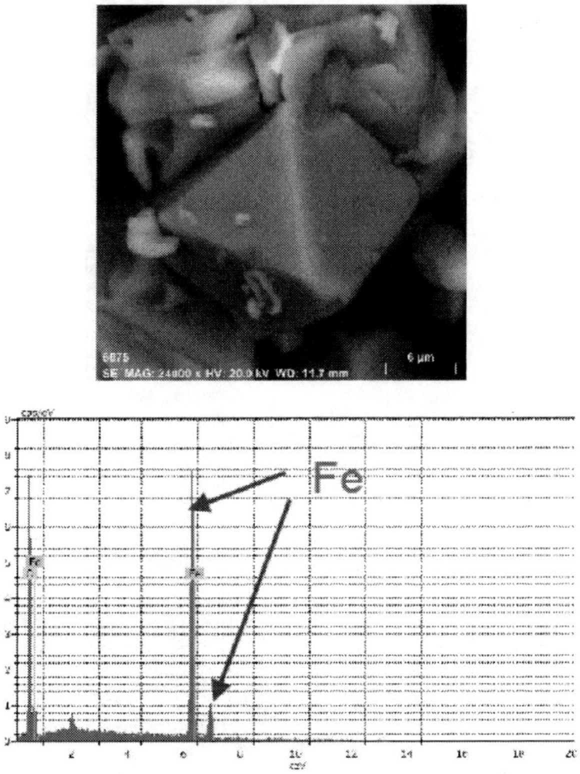

Fig. 11: EDS spectrum on the octahedral-like particle in the sample with $x = 0.4$.

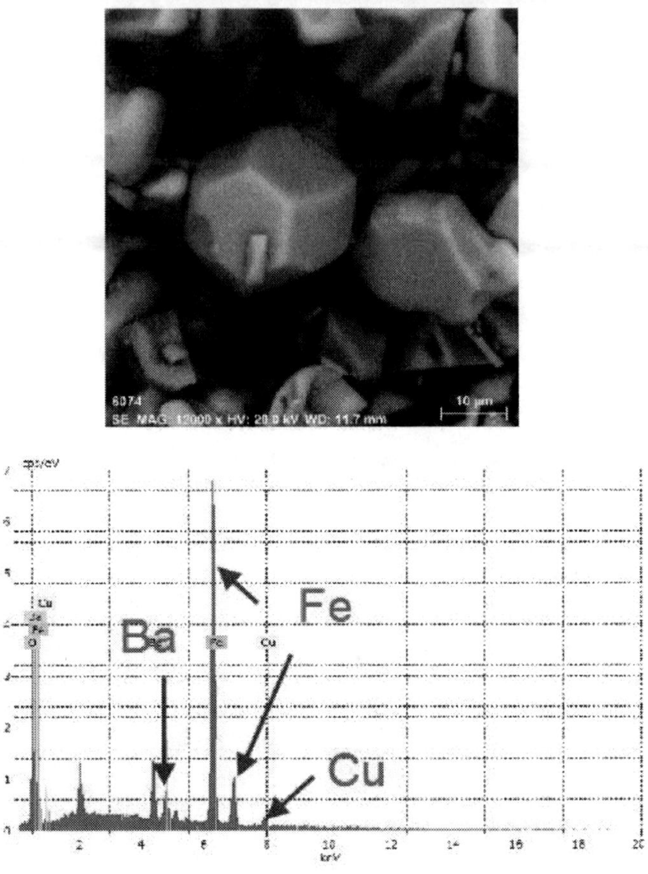

Fig. 12: EDS spectrum on the hexagonal-like particle in the sample with $x = 0.4$.

SEM images (Fig. 13) for the sample with $x = 0.4$ sintered for 10 h showed an improvement in the crystallization of the particles as indicated by the sharp-edged particles. However, well-crystallized particles with size smaller than the typical size observed in the sample sintered for 2 h seem to develop in this sample. The presence of particles with distinctly different shapes (hexagonal and irregular polyhedral particles, in addition to regions with no particular shape) is an indication of the multi-phase nature of the sample. Step-like formations in large and thick hexagonal plates similar to that reported in other multi-phase hexaferrites was also observed in this sample [8].

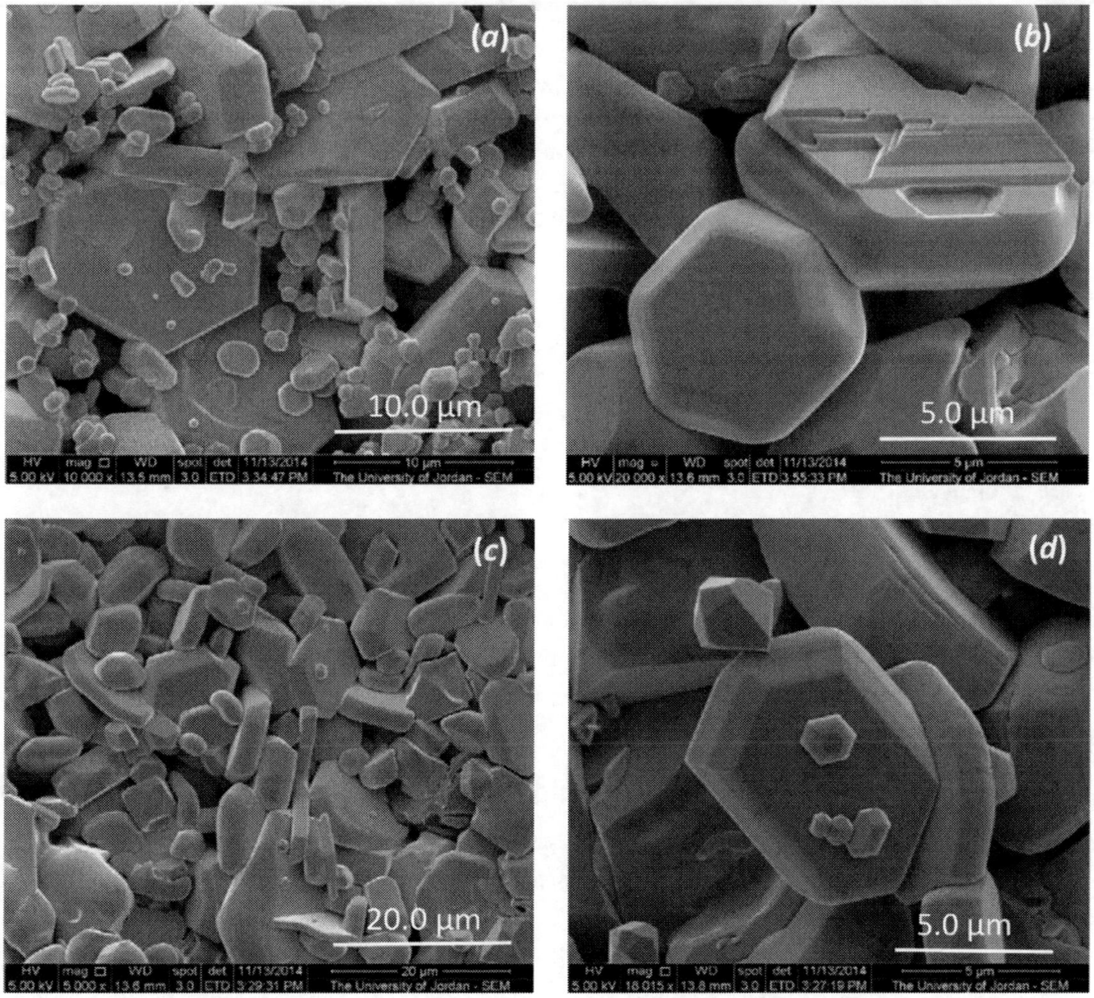

Fig. 13: SEM images for $BaFe_{12-2x}V_xCu_xO_{19}$ (x = 0.4) sample sintered for 10 h.

3.3 Magnetic Measurements

Fig. 14 shows the hysteresis loops of the samples sintered at 1200° C for 2 h, while Fig. 15 shows the hysteresis loops of the samples sintered for 10 h in air. The coercive field and remanence magnetization were determined directly from the hysteresis loops. However, since the magnetization did not saturate up to the maximum applied field of 10 kOe, the saturation magnetization was obtained from the law of approach to saturation (LAS) [34, 35, 47, 48]:

$$M = M_s\left(1 - \frac{A}{H} - \frac{B}{H^2}\right) + \chi H \qquad (2)$$

where the constant A is due to microstress/ inclusions in the sample, B arises from the magnetocrystalline anisotropy, and χH is the forced magnetization term. A plot of M vs. $1/H^2$ gave very good straight lines, indicating the dominance of the magnetocrystalline contribution in this field range (8.5 – 10.0 kOe) [34, 35, 38, 48]. The saturation magnetization is obtained from the intercept of the straight line, while the slope is used to determine the magnetocrystalline anisotropy field (H_a) and the first magnetocrystalline anisotropy constant (K_1) according to the following equations:

$$slope = -M_s B \qquad (4)$$

$$B = \frac{H_a^2}{15}\ ;\ H_a = \frac{2K_1}{M_s} \qquad (3)$$

The magnetic parameters resulting from the straight-line fit to the high-field region of the magnetization curves are tabulated in Table. 3.

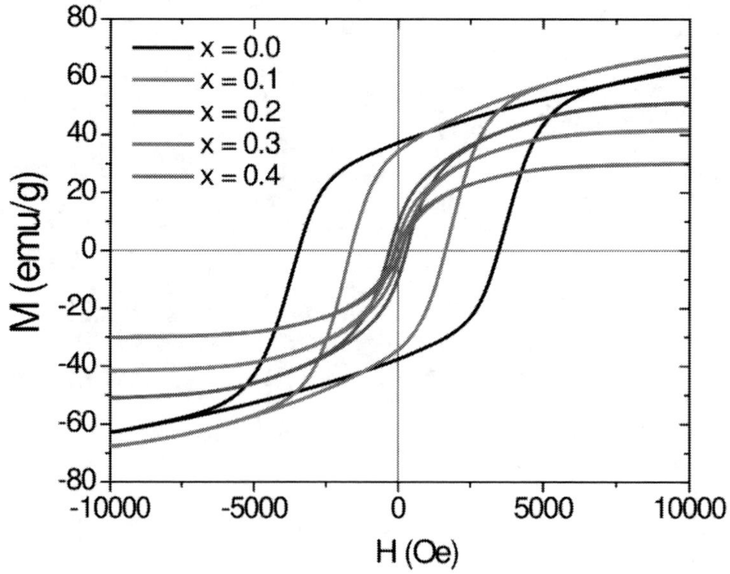

Fig. 14: Hysteresis loops for the samples $BaFe_{12-2x}V_xCu_xO_{19}$ sintered for 2 h.

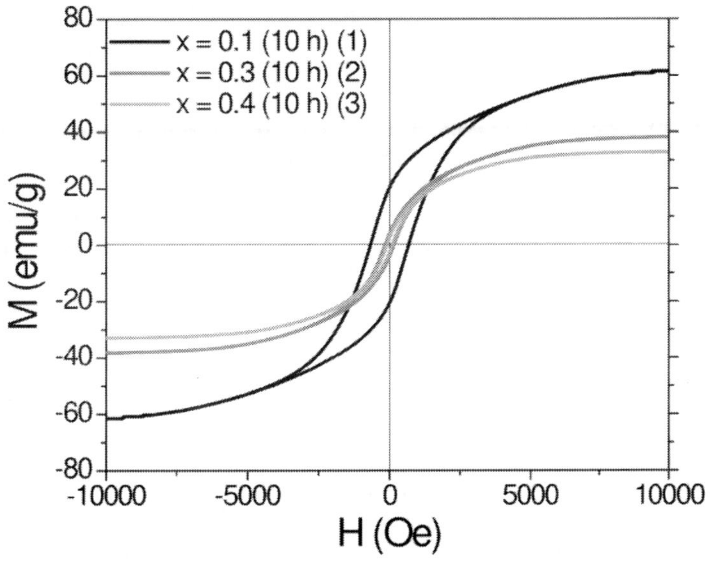

Fig. 15: Hysteresis loops for the samples $BaFe_{12-2x}V_xCu_xO_{19}$ sintered or 10 h.

Table 3 The saturation magnetization (M_s in emu/g), remanence (M_r in emu/g), the squareness ratio ($S = M_r/M_s$), coercivity (H_c in Oe), anisotropy field (H_a in Oe), and first anisotropy constant (K_1 in units of 10^6 erg/cm^3) for the systems $BaFe_{12-2x}V_xCu_xO_{19}$ samples sintered for 2 h and 10 h.

	x = 0.0		x = 0.1		x = 0.2		x = 0.3		x = 0.4	
	2 h	**10 h**	**2 h**	**10 h**	**2 h**	**10 h**	**2 h**	**10 h**	**2 h**	**10 h**
M_s	69.0	-	72.6	64.9	52.5	-	42.4	39.6	30.6	33.4
M_r	37.3	-	34.3	20.1	9.2	-	3.7	3.7	1.8	2.3
S	0.54	-	0.47	0.31	0.17	-	0.09	0.09	0.06	0.07
H_c	3455.6	-	1634.8	676.5	281.9	-	121.5	151.9	76.6	92.5
H_a	12.1	-	10.3	8.8	6.6	-	5.5	6.7	5.4	5.1
K_1	2.20	-	1.98	1.51	0.92	-	0.61	0.70	0.43	0.45

The saturation magnetization for the sample with x = 0.1 is larger than that of the pure sample by ~ 5%, which could be due to substitution of non-magnetic ions for iron at spin-down sites. However, the XRD pattern for this sample showed the presence of the vanadate phase which is evidence that the increase of M_s is due to the copper substitution. Sintering the sample for 10 h at the same temperature leads to a decrease of the saturation magnetization by 11%. This could be due to the increase in the fractions of the non-magnetic $Ba_3(VO_4)_2$ and α-Fe_2O_3 phases with increasing the sintering time as evidenced by the XRD analysis. At higher substitution levels, the saturation magnetization of the samples sintered for 2 h decreased monotonically down to 30.6 emu/g at $x = 0.4$. This decrease is also associated with the progressive increase of the fractions of the non-magnetic phases. Also, this decrease could be partially due to the onset of Cu substitution for Fe at spin-up sites at this level of substitution. In addition, the general decrease in saturation magnetization for the samples sintered for 10 h (compared to the respective values in the samples sintered for 2 h) could be associated with the enhancement of the substitution in the hexaferrite lattice promoted by the prolonged sintering process.

The coercivity dropped sharply with increasing x as shown in Fig. 16, where it was reduced from about 3.5 kOe for the sample with $x = 0$ down to about 1.6 kOe for the sample with $x = 0.1$. The coercivity was reported to proportional to the magnetocrystalline anisotropy [15], and to decrease with increasing particle size beyond the single domain critical size [15, 49]. Since the squareness ratio of 0.47 for the sample with $x = 0.1$ sintered for 2 h is very close to the theoretical ratio of 0.5 for a random assembly of single domain particles, the observed reduction in the coercivity of this particle could be mainly ascribed to the reduction in magnetocrystalline anisotropy. The further reduction of the coercivity of this sample down to about 0.68 kOe upon sintering for 10 h could, however, be associated with the increase in particle size beyond the single domain critical size as evidenced by the significant drop in squareness ratio down to 0.31. This range of coercivity, coupled with the relatively high saturation and remanent magnetization, indicate this sample could be a good candidate for high density magnetic recording.

The reduction in squareness ratio down to below 0.2 with increasing x beyond 0.1 is a clear indication of particle growth, and the multi-domain nature of particles in these samples. The coercivity in these samples dropped down to below 300 Oe, possibly as a result of the combined effect of the increased particle size and reduction in magnetocrystalline anisotropy. The values of coercivity and remanent magnetization in these samples are not suitable for magnetic recording applications. However, these samples could be promising in other applications, which call for extended studies on these samples involving the dielectric and microwave properties, which are beyond the scope of this article.

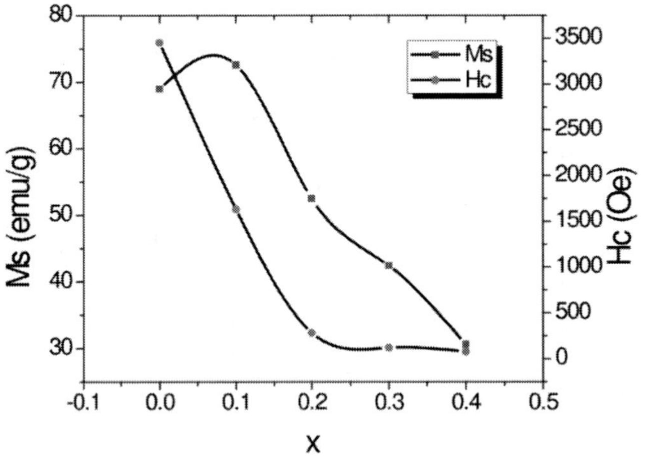

Fig. 16: Saturation magnetization and coercivity as a function of x for the samples $BaFe_{12-2x}V_xCu_xO_{19}$ sintered for 2 h.

The behavior of the coercivity with increasing x could be attributed to the behavior of the magnetocrystalline anisotropy as mentioned in the above discussion. The anisotropy field and first anisotropy constant (Fig. 17) were found to decrease with increasing the substitution level. The large drop in H_a indicates that the material is transforming into a soft magnetic material with increasing x, and this decrease in magnetic anisotropy could be responsible for the major part of the observed reduction in coercivity. The faster rate of decrease in K_1 compared to the rate of decrease of H_a is ascribed to the fact that K_1 is proportional to both H_a and M_s, both of which decrease with increasing x. At this point. It is worth mentioning that since the decrease in saturation magnetization is not completely due to changes in the intrinsic properties of the material, but partially due to the development of nonmagnetic phases, the obtained values of K_1 for the substituted samples cannot be accepted as characteristic parameter of BaM phase. However, since H_a is determined from the approach to saturation which is not influenced by nonmagnetic phases, the behavior of the magnetocrystalline anisotropy of the substituted BaM phase can only be deduced from the behavior of H_a in the present study.

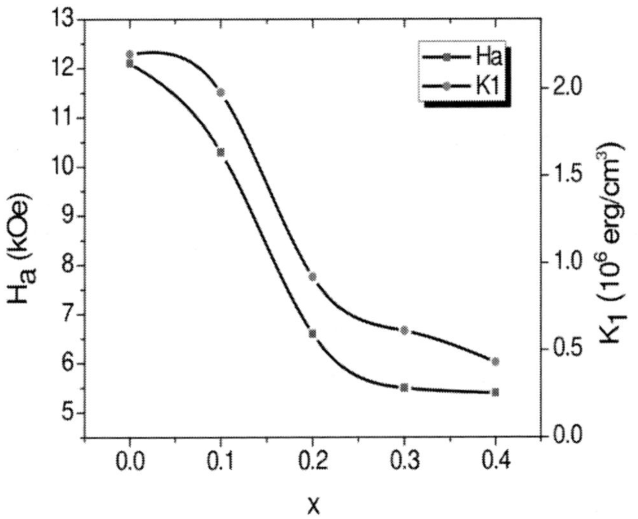

Fig. 17. The magnetic anisotropy field and the first anisotropy constant as a function of concentration for the samples $BaFe_{12-2x}V_xCu_xO_{19}$ ($x = 0.0 - 0.4$) sintered at 1200° C for 2 h.

4. Conclusions

Cu-V substituted M-type hexaferrites prepared by ball milling and sintering at 1200° C for 2 h and 10 h revealed the presence of secondary nonmagnetic phases, and the different phases existed in the form of separate particles with different sizes and shapes. The particle size was found to increase with increasing x, and sintering for 10 h was found to improve crystallization. The substitution was found to have a dramatic effect on the coercivity, even at low concentrations. The sample with $x = 0.1$ demonstrated properties promising for high density magnetic materials. The sharp decrease in the coercivity of the substituted hexaferrites resulted from a combination of the growth of the particle size, and the attenuation of the magnetocrystalline anisotropy as a result of substituting Fe^{3+} ions by nonmagnetic ions. The fast deterioration of the saturation magnetization, remanent magnetization and squareness ratio for $x > 0.1$

indicate that these samples are not good candidates for magnetic recording. Cosideration of these materials for other applications requires further investigation beyond the scope of this work

Acknowledgements

This work was supported by a generous grant from the Deanship of Scientific Research at The University of Jordan (grant # 1404). The technical assistance of Y. Abu Salha and W. Fares, the University of Jordan, is acknowledged.

References

[1] Ü. Özgür, Y. Alivov, H. Morkoç, Microwave ferrites, part 1: fundamental properties, Journal of Materials Science: Materials in Electronics 20 (9) (2009) 789-834.

[2] V.G. Harris, A. Geiler, Y. Chen, S.D. Yoon, M. Wu, A. Yang, Z. Chen, P. He, P.V. Parimi, X. Zuo, Recent advances in processing and applications of microwave ferrites, Journal of Magnetism and Magnetic Materials 321 (14) (2009) 2035-2047.

[3] U. Topal, H.I. Bakan, Permanently magnetic $BaFe_{12}O_{19}$ foams: Synthesis and characterization, Materials Chemistry and Physics 123 (1) (2010) 121-124.

[4] P. Slick, Ferrites for non-microwave applications, in: E.P. Wohlfarth (Ed.) Ferromagnetic materials, Vol. 2, North-Holland Publishing Company, New York, 1980, pp. 189-241.

[5] J. Nicolas, Microwave ferrites, in: E.P. Wohlfarth (Ed.) Ferromagnetic Materials, Vol. 2, North-Holland Publishing Company, New York, 1980, pp. 243-296.

[6] J. Smit, H.P.J. Wijn, Ferrites, Wiley, New York, 1959.

[7] J. Went, G. Rathenau, E. Gorter, G. Van Oosterhout, Ferroxdure, a class of new permanent magnet materials, Philips Tech. Rev 13 (7) (1952) 194-208.

[8] R.C. Pullar, Hexagonal ferrites: a review of the synthesis, properties and applications of hexaferrite ceramics, Progress in Materials Science 57 (7) (2012) 1191-1334.

[9] Ü. Özgür, Y. Alivov, H. Morkoç, Microwave ferrites, part 2: passive components and electrical tuning, Journal of Materials Science: Materials in Electronics 20 (10) (2009) 911-952.

[10] R. Aiyar, Microwave absorbers based on hexaferrites, Microwave and Optical Technology Letters 23 (5) (1999) 321-323.

[11] R. Meena, S. Bhattachrya, R. Chatterjee, Development of "tuned microwave absorbers" using U-type hexaferrite, Materials & Design 31 (7) (2010) 3220-3226.

[12] Y. Yang, B. Zhang, W. Xu, Y. Shi, N. Zhou, H. Lu, Microwave absorption studies of W-hexaferrite prepared by co-precipitation/mechanical milling, Journal of magnetism and magnetic materials 265 (2) (2003) 119-122.

[13] I.Y. Gershov, Barium ferrite permanent magnets, Soviet Powder Metallurgy and Metal Ceramics 1 (5) (1964) 386-393.

[14] G. Asti, P. Cavallotti, R. Roberti, Metal-barium hexaferrite composites as permanent magnets, Ceramurgia International 3 (2) (1977) 70-77.

[15] G. Bate, Recording materials, in: P. E, Wohlfarth (Ed.) Ferromagnetic materials, Vol. 2, North-Holland Publishing Company, New York, 1980, pp. 381-508.

[16] G. Bate, Magnetic recording materials since 1975, Journal of Magnetism and Magnetic materials 100 (1) (1991) 413-424.

[17] M.N. Ashiq, M.J. Iqbal, M. Najam-ul-Haq, P.H. Gomez, A.M. Qureshi, Synthesis, magnetic and dielectric properties of Er–Ni doped Sr-hexaferrite nanomaterials for applications in High density recording media and microwave devices, Journal of magnetism and magnetic materials 324 (1) (2012) 15-19.

[18] S. Sanghi, A. Agarwal, N. Ahlawat, Structure refinement and dielectric relaxation of M-type Ba, Sr, Ba-Sr, and Ba-Pb hexaferrites, Journal of Applied Physics 112 (1) (2012) 014110.

[19] I. Bsoul, S.H. Mahmood, A.F. Lehlooh, A. Al-Jamel, Structural and magnetic properties of $SrFe_{12-2x}Ti_xRu_xO_{19}$, Journal of Alloys and Compounds 551 (2013) 490-495.

[20] F. Pereira, M. Santos, R. Sohn, J. Almeida, A. Medeiros, M. Costa, A. Sombra, Magnetic and dielectric properties of the M-type barium strontium hexaferrite ($Ba_xSr_{1-x}Fe_{12}O_{19}$) in the RF and microwave (MW) frequency range, Journal of Materials Science: Materials in Electronics 20 (5) (2009) 408-417.

[21] S. Singhal, T. Namgyal, J. Singh, K. Chandra, S. Bansal, A comparative study on the magnetic properties of $MFe_{12}O_{19}$ and $MAlFe_{11}O_{19}$ (M= Sr, Ba and Pb) hexaferrites with different morphologies, Ceramics International 37 (6) (2011) 1833-1837.

[22] M.N. Ashiq, R.B. Qureshi, M.A. Malana, M.F. Ehsan, Synthesis, structural, magnetic and dielectric properties of zirconium copper doped M-type calcium strontium hexaferrites, Journal of Alloys and Compounds 617 (2014) 437-443.

[23] G. Amiri, R. Mousarezaei, Structural, Magnetic, and Microwave-Absorption Properties of Nanocrystalline $Ca(MnSn)_xFe_{12-2x}O_{19}$ Ferrites, Journal of Electronic Materials 44 (2) (2015) 715-719.

[24] Y. Yang, X. Liu, Synthesis and Magnetic Properties of $Sr_{0.65-x}Ca_xLa_{0.35}Fe_{11.32}Co_{0.28}O_{18.435}$ Hexaferrites Prepared by Solid State Reaction Method, IEEE Transactions on Magnetics 50 (9) (2014) 2201606.

[25] S. Chawla, R. Mudsainiyan, S. Meena, S. Yusuf, Sol–gel synthesis, structural and magnetic properties of nanoscale M-type barium hexaferrites $BaCo_xZr_xFe_{(12-2x)}O_{19}$, Journal of Magnetism and Magnetic Materials 350 (2014) 23-29.

[26] J.-P. Wang, L. Ying, M.-L. Zhang, Y.-j. QIAO, X. Tian, Comparison of the Sol-gel Method with the Coprecipitation Technique for Preparation of Hexagonal Barium Ferrite, Chemical Research in Chinese Universities 24 (5) (2008) 525-528.

[27] W. Abbas, I. Ahmad, M. Kanwal, G. Murtaza, I. Ali, M.A. Khan, M.N. Akhtar, M. Ahmad, Structural and magnetic behavior of Pr-substituted M-type hexagonal ferrites synthesized by sol–gel autocombustion for a variety of applications, Journal of Magnetism and Magnetic Materials 374 (2015) 187-191.

[28] J. Matutes-Aquino, S. Dıaz-Castanón, M. Mirabal-Garcıa, S. Palomares-Sánchez, Synthesis by coprecipitation and study of barium hexaferrite powders, Scripta materialia 42 (3) (2000) 295-299.

[29] G. Litsardakis, I. Manolakis, C. Serletis, K. Efthimiadis, High coercivity Gd-substituted Ba hexaferrites, prepared by chemical coprecipitation, Journal of Applied Physics 103 (7) (2008) 07E501.

[30] D. Primc, D. Makovec, D. Lisjak, M. Drofenik, Hydrothermal synthesis of ultrafine barium hexaferrite nanoparticles and the preparation of their stable suspensions, Nanotechnology 20 (31) (2009) 315605.

[31] A. Ataie, I. Harris, C. Ponton, Magnetic properties of hydrothermally synthesized strontium hexaferrite as a function of synthesis conditions, Journal of materials science 30 (6) (1995) 1429-1433.

[32] K. Praveena, M. Bououdina, M.P. Reddy, S. Srinath, R. Sandhya, S. Katlakunta, Structural, Magnetic, and Electrical Properties of Microwave-Sintered Cr^{3+}-Doped Sr Hexaferrites, Journal of Electronic Materials 44 (1) (2015) 524-531.

[33] G.H. Dushaq, S.H. Mahmood, I. Bsoul, H.K. Juwhari, B. Lahlouh, M.A. AlDamen, Effects of molybdenum concentration and valence state on the structural and magnetic properties of $BaFe_{11.6}Mo_xZn_{0.4-x}O_{19}$ hexaferrites, Acta Metallurgica Sinica (English Letters) 26 (5) (2013) 509-516.

[34] S.H. Mahmood, G.H. Dushaq, I. Bsoul, M. Awawdeh, H.K. Juwhari, B.I. Lahlouh, M.A. AlDamen, Magnetic Properties and Hyperfine Interactions in M-Type $BaFe_{12-2x}Mo_xZn_xO_{19}$ Hexaferrites, Journal of Applied Mathematics and Physics 2 (05) (2014) 77-87.

[35] A.M. Alsmadi, I. Bsoul, S.H. Mahmood, G. Alnawashi, K. Prokeš, K. Siemensmeyer, B. Klemke, H. Nakotte, Magnetic study of M-type doped barium hexaferrite nanocrystalline particles, Journal of Applied Physics 114 (24) (2013) 243910.

[36] I. Bsoul, S.H. Mahmood, A.F. Lehlooh, Structural and magnetic properties of $BaFe_{12-2x}Ti_xRu_xO_{19}$, Journal of Alloys and Compounds 498 (2) (2010) 157-161.

[37] S. Ketov, Y.D. Yagodkin, A. Lebed, Y.V. Chernopyatova, K. Khlopkov, Structure and magnetic properties of nanocrystalline $SrFe_{12}O_{19}$ alloy produced by high-energy ball milling and annealing, Journal of Magnetism and Magnetic Materials 300 (1) (2006) e479-e481.

[38] M. Awawdeh, I. Bsoul, S.H. Mahmood, Magnetic properties and Mössbauer spectroscopy on Ga, Al, and Cr substituted hexaferrites, Journal of Alloys and Compounds 585 (2014) 465-473.

[39] S. Katlakunta, S.S. Meena, S. Srinath, M. Bououdina, R. Sandhya, K. Praveena, Improved magnetic properties of Cr^{3+} doped $SrFe_{12}O_{19}$ synthesized via microwave hydrothermal route, Materials Research Bulletin 63 (2015) 58-66.

[40] I. Bsoul, S.H. Mahmood, Magnetic and structural properties of $BaFe_{12-x}Ga_xO_{19}$ nanoparticles, Journal of Alloys and Compounds 489 (1) (2010) 110-114.

[41] D. Han, Z. Yang, H. Zeng, X. Zhou, A. Morrish, Cation site preference and magnetic properties of Co-Sn-substituted Ba ferrite particles, Journal of magnetism and magnetic materials 137 (1) (1994) 191-196.

[42] A. González-Angeles, G. Mendoza-Suárez, A. Grusková, J. Sláma, J. Lipka, M. Papánová, Magnetic structure of $Sn^{2+}Ru^{4+}$-substituted barium hexaferrites prepared by mechanical alloying, Materials letters 59 (14) (2005) 1815-1819.

[43] A. González-Angeles, G. Mendoza-Suarez, A. Grusková, M. Papanova, J. Slama, Magnetic studies of Zn–Ti-substituted barium hexaferrites prepared by mechanical milling, Materials letters 59 (1) (2005) 26-31.

[44] I. Bsoul, S. Mahmood, A.-F. Lehlooh, Structural and magnetic properties of $BaFe_{12-2x}Ti_xRu_xO_{19}$, Journal of Alloys and Compounds 498 (2) (2010) 157-161.

[45] R.t. Shannon, Revised effective ionic radii and systematic studies of interatomic distances in halides and chalcogenides, Acta Crystallographica A 32 (5) (1976) 751-767.

[46] B.E. Warren, X-ray Diffraction, Addison-Wesley, Reading, Massachsetts, 1969.

[47] B.D. Cullity, C.D. Graham, Introduction to magnetic materials, John Wiley & Sons, 2011.

[48] S. Mahmood, A. Aloqaily, Y. Maswadeh, A. Awadallah, I. Bsoul, H. Juwhari, Structural and Magnetic Properties of Mo-Zn Substituted ($BaFe_{12-4x}Mo_xZn_{3x}O_{19}$) M-Type Hexaferrites, Material Science Research India 11 (1) (2014) 09-20.

[49] S. Charles, J. Popplewell, Ferromagnetic liquids, Handbook of Ferromagnetic Materials 2 (1980) 509-559.

A prototype Ultraviolet Light Sensor based on ZnO Nanoparticles/Graphene Oxide Nanocomposite Using Low Temperature Hydrothermal Method

M. Al-Fandi[1*], R. Oweis[2], B. A. Albiss[3] T. AlZoubi[4], M-Ali Al-Akhras[3], H. Qutaish[3], H. Khwailah[2], S. Al-Hattami[2], and E. Al-Shawwa[2]

[1]Department of Mechanical Engineering, Jordan University of Science & Technology (JUST), P.O. Box 3030, Irbid 22110, Jordan
[2]Department of Biomedical Engineering, Jordan University of Science & Technology (JUST), P.O. Box 3030, Irbid 22110, Jordan
[3]Department of Physics, Jordan University of Science & Technology (JUST), P.O. Box 3030, Irbid 22110, Jordan
[4] College of Engineering, American, University of the Middle East (AUM), P.O. BOX: 220 Dasman, 15453 Kuwait.

Corresponding Author E-mail: mohamed_alfandi@just.edu.jo

Abstract. A new prototype UV nanosensor using ZnO nanoparticles (NPs)/graphene oxide (GO) nanocomposite (ZnO-NP/GO) on silicon substrate is reported in this paper. The hybrid nanocomposite structure has been developed by an optimized hydrothermal process at low growth temperature (~50 °C). In this hybrid nanosensor, the ZnO nanoparticles act as UV-absorbing and charge carrier generating material, while graphene with its superior electrical conductivity has been used as a charge transporting material. Various nanostructure characterization techniques were intensively utilized including SEM, EDX, XRD, FTIR and UV-VIS. Also, the I-V measurement was employed to evaluate the prototype sensor. The morphological SEM analysis showed that the ZnO-NPs (average diameter of 20 nm) were dispersed evenly on the GO sheets. As well, the EDX spectra confirmed the exact chemical composition of the intended structure. The room temperature UV-VIS measurement revealed an enhanced optical absorption of UV-light at an absorption band centered on 375 nm. The improved optical and electrical properties were observed at an optimum relative concentration of 1:10. Under UV light illumination, the measured I-V characteristic of the prototype detector exhibited a considerable photocurrent increase of the ZnO-NP/GO nanocomposite compared to pristine ZnO nanostructure. These results can be promising for future enhanced UV- sensing applications.

1. Introduction

Zinc Oxide (ZnO) is a II-VI semiconductor with a wide band gap (3.37 eV) and a large excitation binding energy of 60 meV at room temperature [1]. It is a non-toxic material, environmentally friendly and has low production cost [2]. These properties make ZnO a desirable material for many applications such as conductive oxides [1], solar cells [3], nanogenerators [4-7], displays and sensors [8-11]. Various UV sensors based on ZnO nanostructures have been reported in literature because this extraordinary nanostructure can enhance the performance of the devices due to the large surface to volume ratio. This enhancement attracted many application fields including flame detection, environmental studies, medical and communication equipment and space exploration [12,13]. Nevertheless, pure ZnO suffers from low operating speed and weak photoresponsivity drawbacks [14,15]. Meanwhile research in graphene is booming due to many remarkable properties comprising good electrical and thermal conductivity, large specific surface area, high mobility of charge carriers, excellent chemical stability and mechanical strength [16]. However, many applications of materials depend greatly on their intrinsic properties. Hybridization of different materials offers a powerful way to enhance the application of graphene by enabling versatile and tailor-made properties with high performance far beyond those of the individual materials. It is expected that ZnO-graphene or graphene oxide nanocomposite will provide a potential for the development of UV detectors with superior response, flexibility and improved stability.

Development of modern high-performance flexible optoelectronics devices such as UV-photodetectors is subjected to the enhancement of several sensor internal key parameters such as: the sensitivity defined ($I_{photo} - I_{dark}/ I_{dark}$) as percentage, spectral responsivity (R_o) defined as, the photocurrent through the device per unit power of irradiation per active area, external quantum efficiency (EQE) defined as, the number of electron detected per light photon and the response speed, which include two values: rise time (the time taken by the sensor to reach 90 % of the max. photocurrent) and the decay time (the time taken to reach 10 % of the max. photocurrent) [17].

Up to now, there have been a few reports on the development of UV detectors based on ZnO nanocomposites [18]. However, the performance improvement of the UV detector made of ZnO-graphene nanocomposite is still restricted, which is likely due to the absence of effective morphological and interfacial control between ZnO nanostructures and graphene.

In the nanocomposite reported here, ZnO nanoparticles act as UV absorbing and charge carrier generating materials, while graphene is used as a charge transporting and highly conductive material. Many methods have been developed for the preparation of ZnO nanoparticles including chemical vapor deposition (CVD) [19], vapor phase transport process [20], metal-organic vapor phase epitaxy [21], thermal evaporation [22]. These growth techniques are complicated and the operating growth temperatures are comparatively high (> 350 °C).

The hydrothermal process [23-24] has attracted considerable attention because of its unique advantages, such as simplicity, low cost, low temperature (60-100 °C), high yield and more controllable process [25-27] than previously mentioned methods. In this paper, the fabrication of high quality UV nanosensor made of ZnO nanoparticles onto graphene oxide sheets surface, by a simple hydrothermal process, on silicon substrates is reported. The optical, electrical, morphological and structural properties were intensively investigated to assess the improved physical properties of the fabricated UV sensor. The approach presented herein utilizes the superior optical properties of ZnO nanoparticles with the superior electrical and mechanical properties of graphene nano-sheets targeting large values of sensitivity, spectral responsivity and external quantum efficiency.

2. Experimental

2.1 Hydrothermal Synthesis of ZnO nanoparticles

All chemicals used to prepare ZnO nanoparticles were of analytical reagent grade and used as received without any further purification. The ZnO nanoparticles was synthesized using the traditional hydrothermal process. In a typical initial step of the hydrothermal synthesis, a stock solution of zinc acetate dehydrate $Zn(CH_3CO_2)_2.2H_2O$ (5 mM) was prepared in 20 ml ethanol under magnetic stirring. Subsequently, 20 ml of NaOH with concentration varying from 0.2 mM to 0.5 mM in ethanol was added to the solution under stirring for 30 minutes. The solution was then transferred into a water path and maintained at a stable temperature in the range of 50-60 °C for 2-3 hours. The obtained ZnO nanoparticles (ZnO-NP) were afterward washed three times with ethanol and distilled water in order to remove any impurities and then dried in air at 50 °C for 24 hours.

2.2 Preparation of ZnO-NP/GO nanocomposite sensor

ZnO-NP/GO nanocomposite was synthesized using a customary hydrothermal method. The synthesis was started by dispersing of (1 mg/mL) graphene in 10 mL of sulfuric acid and sonicated at room temperature for 10 minutes. Similarly, 0.1 g of ZnO was dispersed and sonicated in 10 mL of distilled water. After that, 10 mL of graphene solution was added to 10mL of ZnO solution and the mixture was sonicated for 5 minutes. A layer of the prepared mixture was then deposited on a cleaned n-Si substrate at room temperature to fabricate the sensor. Finally, a thin layer of Ag was printed over the mixture in a specific pattern as shown in the Figure 1. The Cu electrodes were fixed over Ag using heat and a duct tape.

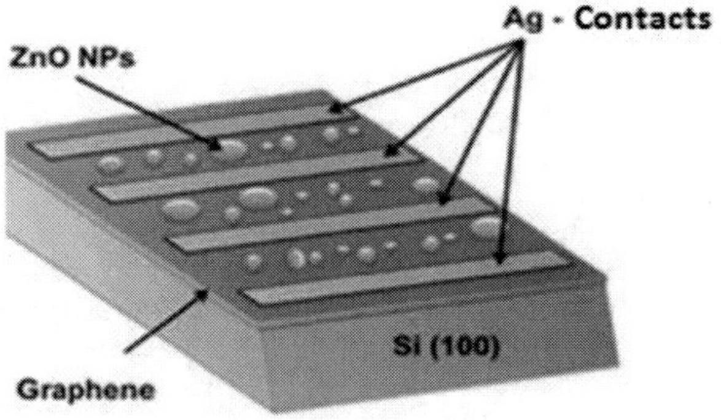

Figure 1. Schematic diagram of the ZnO-NP/GO UV sensor

3. Results and discussion

The successful preparation of the ZnO-NPs/GO hybrid structure was evaluated using various standard characterization techniques including scanning electron microscope (SEM), electron dispersive X-ray (EDX), ultra-violet visible (UV–VIS) and Fourier transform infra-red (FTIR). These techniques can assess the morphological, optical as well as the structural properties of the hybrid structure. Likewise, the I-V characteristics were utilized to evaluate the behavior of the newly developed nanosensor.

3.1 UV-VIS Spectroscopy

Room temperature UV–vis absorption spectra were carried out for diverse range of ZnO(NPs)/GO concentrations to consider the finest optical properties of the hybrid structure, see Figure 2a. The first UV-vis measurement was performed for the unmodified ZnO-NPs and GO. Typical ZnO UV-vis spectrum revealed two absorption bands centered near the wavelengths of 265 nm and 375 nm. Conversely, the transparent graphene oxide exhibits a weaker absorption band centered nearby 290 nm. On the other hand, the hybrid ZnO-NP/GO with relative concentrations of 1:1000, 1:100 and 1:50 showed no absorption peaks at the considered wavelength range, as shown in Figure 2a. Most probably this is due to the very low concentration of the ZnO-NPs, which acts as an inefficient UV absorber. Oppositely, a clear enhanced absorption band centered at a wavelength close to 375 nm was observed for the samples with ZnO-NPs/GO relative concentrations of 1:10 and 1:1. Among the diverse studied samples, it was observed that theses samples with concentrations of 1:10 and 1:1 have the intended optical behavior with a single absorption band centered at a wavelength of approximately 375 nm. Consequently, further investigation on the sample with a concentration of 1:10 was executed and analyzed. Figure 2b displays the spectrum of the hybrid ZnO-NP/GO sample with the concentration of 1:10 accompanied by the unmodified samples of ZnO NPs and GO. The figure revealed a considerable quench in the 375 nm and the 265 nm peak intensities as compared with the unmodified ZnO-NPs. This quench may be attributed to the effective electron transfer from the conduction band of ZnO to GO during the nanocomposite combination, which indicates that the effective charge transfer can reduce the charge recombination and increase the photocatalytic activity of ZnO nanoparticles.

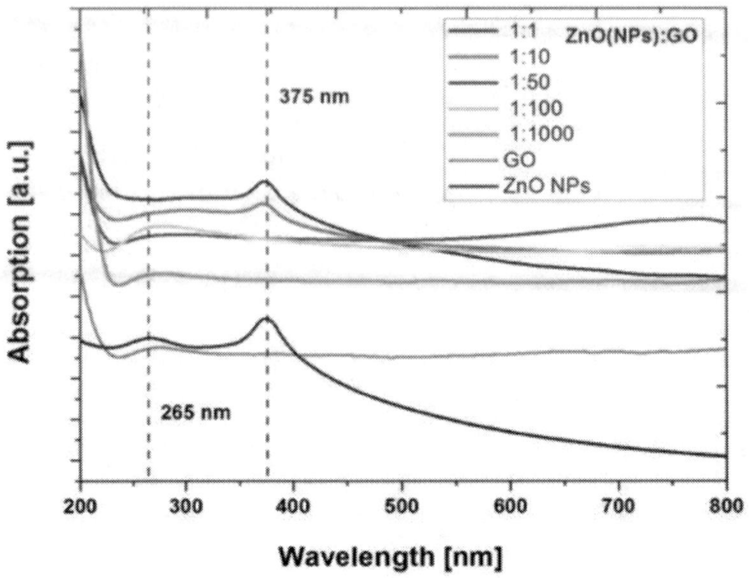

Figure 2a: UV-Vis absorption spectra of all samples with different ZnO-NP/GO ratios

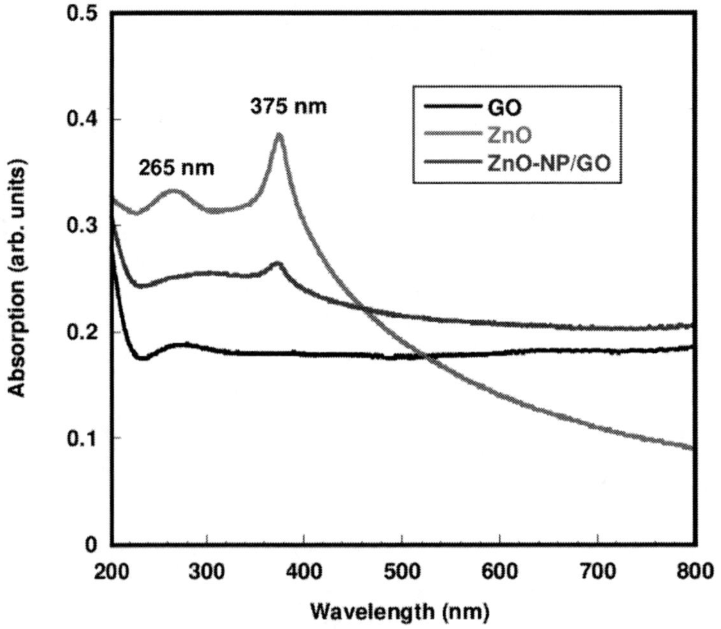

Figure 2b: UV-Vis absorption spectra for GO, ZnO and ZnO-NP/GO (ratio1:10)

IOP Conf. Series: Materials Science and Engineering **92** (2015) 012009 doi:10.1088/1757-899X/92/1/012009

3.2 XRD Analysis

The XRD spectra of ZnO-NP/GO nanocomposite films are shown in Figure 3. All major ZnO reflections are clearly observed at 2θ = 31.6°, 34.4°, 36.4°, 47.3°, 56.8°, 62.7°, 68.3, 72.5 and 76.8°, which correspond, respectively, to the (100), (002), (101), (102), (110), (103), (201), (004) and (202) planes of ZnO (Joint Committee on Powder Diffraction Standards 36-1451). All peaks in the ZnO-NPs can be indexed to hexagonal wurtzite structure with space group P63mc and lattice parameters of a = 0.3251 nm and c= 0.5208 nm, which indicates that ZnO-NPs are oriented to some extent. Moreover, a quite strong peak appeared at a 2θ value of about 26.5°, which is corresponded to the (002) plane of the GO. Because only one GO peak was observed, this confirms the partial reduction of the GO to a Gr sheet. Additionally, the absence of any extra diffraction peaks of other phases on both XRD patterns demonstrating the phase purity of the nanocomposite. The maximum diffraction peak of the GO was marked with asterisk in Figure 3.

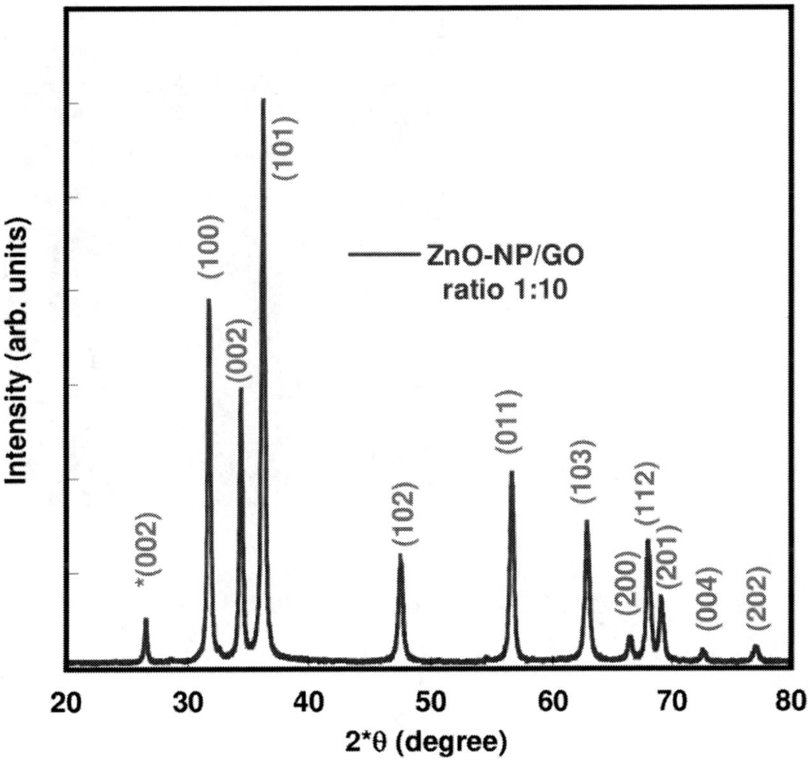

Figure 3. XRD pattern of ZnO-NP/GO nanocomposite

3.3 Morphology and size Analysis

The morphologies and compositions of the ZnO-NP/GO nanocomposite for the sample with a concentration of 1:10 were characterized by a scanning electron microscope SEM equipped with an energy dispersive X-ray (EDX). The EDX analysis of the chemical composition, as-prepared ZnO-NPs/GO nanocomposite on Si substrate, shows that only Zn, O, C and Si signals were detected, see Figure 4a. The existence of only these signals indicates that the hybrid structure indeed made up of Zn, O, C and Si elements. Also, since there is no any signals of the secondary phase or impurity were detected, this may suggest the high-purity of the grown structure. In addition, the morphological SEM analysis showed that the ZnO-NPs with an average diameter of almost 20 nm were dispersed evenly on the GO sheets, see Figure 4b.

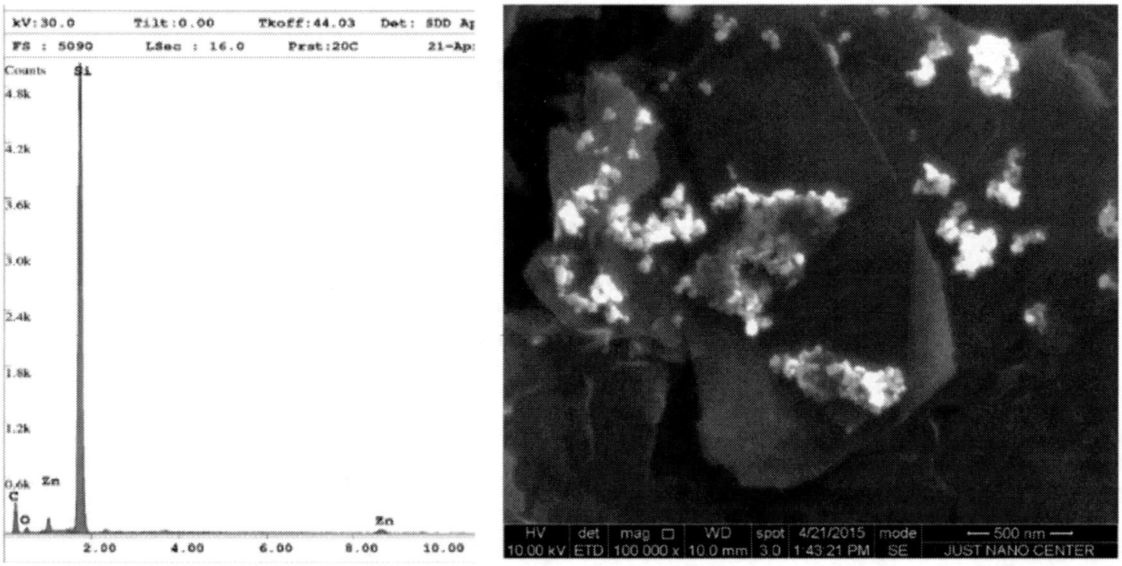

(a) (b)

Figure 4. EDX pattern (a) and SEM image (b) for ZnO-NP/GO sample (ratio1:10)

3.4 FTIR Spectroscopy

The surface modifications of the functional groups of ZnO-NPs/GO after the hydrothermal process were investigated by FTIR spectroscopy. Figure 5 shows the FTIR spectrum of the absorption bands of ZnO-NPs/GO nanocomposite with the relative concentration of 1:10. All oxidized samples exhibit an absorption band centered at 440 cm^{-1}; this is related to the stretching mode of Zn-O bond vibration [1]. The peaks at 3450, 1740, 1680 and 720 cm^{-1} are assigned to O-H stretching vibration, C=O stretching, C=C bond stretch in alkenes and C-H rock, respectively [2-4]. The acquired FTIR spectrum for the sample confirms the existence of ZnO-NPs/GO, which is in agreement with the identical contents obtained by the XRD and EDX analysis.

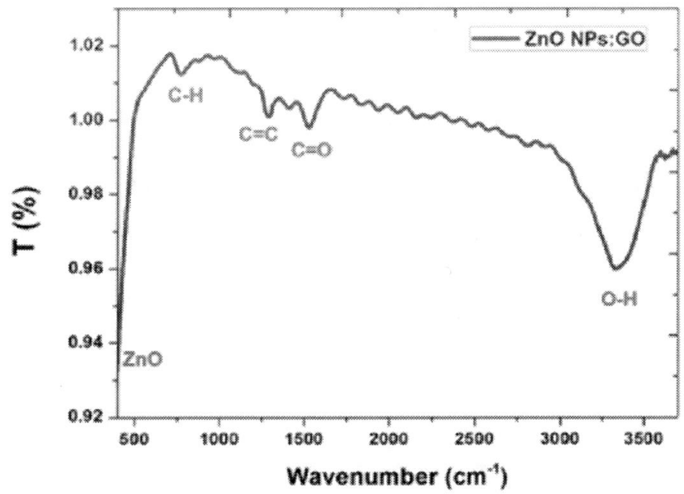

Figure 5: FTIR spectrum of ZnO-NP/GO nanocomposite (sample with ratio 1:10)

3.5 I-V characteristics

Figure 6 shows the typical I-V characteristics of ZnO-NPs/GO sensor for the sample with the concentration of 1:10 under dark and UV light illumination. Usually, for the UV detectors, the dependences of photocurrent is defined as the difference between the photoexcited and dark current ($I_{ph} = I_{light} - I_{dark}$). The UV light employed for characterization has a wavelength of 365 nm and power density of 10 mW/cm^2 at the bias of 6 V. As observed in Figure 6, the sensor shows an increase in the photocurrent when exposed to the UV light as compared with the dark condition. The sensor response exhibited power-law behavior as a function of the biased voltage with quite good hysteresis properties. The increased in the *n*-type ZnO conductivity may be attributed to the role of intrinsic defects (i.e. native defects), which are a result of the growth conditions and impurity atoms of the ZnO nanostructures under study. Moreover, it is well known that the weak photon absorption and fast carrier kinetics of the GO limit its application for photodetection. This limitation can be overcome by introducing photosensitive nanostructures such as ZnO-NPs to GO for a hybrid nanostructure. However, in spite of extensive research for many years, the role of these intrinsic defects is still not perfectly clear at present. In the dark, the oxygen molecules adsorbed on the surface of ZnO-NPs capture free electrons of the *n*-type semiconductor. As a result, a depletion layer near the nanoparticle surface is produced, resulting in an upward band bending near the surface. Due to the large surface-to-volume ratio, the adsorption of oxygen significantly reduces the conductivity of the nanowires. UV light illumination with photon energy higher than the energy gap of ZnO generates electron–hole pairs in the ZnO. Subsequently, the holes tend to migrate to the nanoparticle surface along the band bending line and recombine with O_2-trapped electrons, thus releasing oxygen from the surface. Therefore, adsorbed oxygen and the photogenerated hole contribute to local charge neutrality and free up the photogenerated electrons, which enhances the conductivity of the sensor.

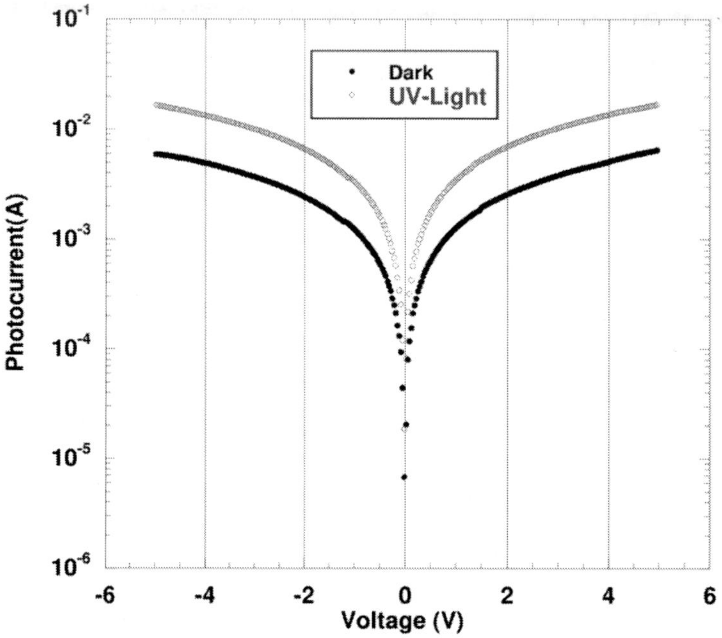

Figure 6: I–V curves of the ZnO-NR/GO (ratio 1 detector in darkness and in UV illumination of 10 mW/cm^2.

4. Conclusions

The synthesis of high quality hybrid ZnO-NPs/GO nanocomposites via simple and cost-effective method by mixing of GO solution and ZnO-NPs solution was achieved and characterized successfully. The operation of an efficient prototype sensor based on ZnO-NPs/GO nanocomposite at room temperature was also demonstrated confidently. The UV sensor of the hybrid nanocomposite was considerably enhanced under the UV light illumination. This enhancement is mainly attributed to the adsorbed oxygen on the ZnO-NPs surface and the photogenerated holes, which ultimately can contribute to the local charge neutrality and free up the photogenerated electrons thereby enhancing the conductivity of the sensor. The sensitivity of the fabricated detector is comparable to the conventional metal oxides-based detectors in the literature. The present study provides a distinctive platform for designing metal/wide band gap semiconductor/graphene oxides nanocomposites to enhance the UV detection.

References

[1] D.G. Thomas, J. Phys. Chem. Solids 15 (1960) 86.
[2] Tam. Kai-hang, thesis, HKU Scholar Hub 92007), hub.hku.hk/handle/10722/52370
[3] M. Law, L. E. Greene, J. C. Johnson, R. Saykally and P. D. Yang, Nat. Mater., 2005, 4, 455.
[4] R. Yang, Y. Qin, C. Li, G. Zhu and Z. L. Wang, Nano Lett., 2009, 9,1201.
[5] X. D. Wang, J. H. Song, J. Liu and Z. L. Wang, Science, 2007, 316,102.
[6] X. D. Wang, J. Zhou, J. H. Song, J. Liu, N. S. Xu and Z. L. Wang, Nano Lett., 2006, 6, 2768.
[7] Z. L. Wang and J. H. Song, Science, 2006, 312, 242.
[8] Z. L. Wang, Annu. Rev. Phys. Chem., 2004, 55, 159.
[9] Q. Wan, Q. H. Li, Y. J. Chen, T. H. Wang, X. L. He, J. P. Li and C. L. Lin, Appl. Phys.Lett., 2004, 84, 3654.
[10] S. Bai, W. W. Wu, Y. Qin, N. Y. Cui, D. J. Bayerl and X. D. Wang, Adv. Funct. Mater., 2011, 21, 4464.
[11] L. Luo, Y. F. Zhang, S. S. Mao and L. W. Lin, Sens. Actuators, 2006, 127, 201.
[12] E. Munoz, E. Monroy, J. L. Pau, F. Calle, F. Omnes and P. Gibart, Phys.: Condens. Matter, 2001, 13, 7115.
[13] S. Liang, H. Sheng, Y. Liu, Z. Huo, Y. Lu and H. Shen, J. Cryst. Growth, 2001, 225, 110.
[14] Xu T, Zhang L, Cheng H, Zhu Y (2011) , Appl Catal B 101(3):382–387
[15] Shao D, Qin L, Sawyer S (2012), Photonics J IEEE 4(3):715–720
[16] Geim AK, Novoselov KS., Nat Mater 2007;6(3):183–91.
[17] Wei Tian, Hao Lu, Liang Li Nanoreseach. 8 .(2015) 382.
[18] B. A. Albiss, M-Ali AL-Akhras and Ihab Obaidat, International Journal of Environmental Analytical Chemistry, 2015, 95(4) 339-348.
[19] T. Hirate, T. Kimpara, S. Nakamura, T. Satoh, Superlattices Microstruct. 42(2007) 409.
[20] J.Grabowska,K.K.Nanda,K.McGlynn,J.P.Mosnier,M.O.Henry,A. Beaucamp,A. Meaney, J. Mater. Sci.: Mater. Electron. 16 (2005) 397.
[21] K. Ogata, K. Maejima, S. Fujita, S. Fujita, J. Cryst. Growth 248 (2003) 25.
[22] Q. Wan, K. Yu, T.H. Wang,Appl. Phys. Lett. 83 (2003) 2253.
[23] C.X. Xu, A. Wei, X.W. Sun, Z.L. Dong, J. Phys. D: Appl. Phys. 39 (2006) 1690.
[24] J. Song, S. Baek, S. Lim, Phys. B: Condens. Matter 403 (2008) 1960.
[25] L. Vayssieres, K. Keis, S. Lindquist, A. Hagfeldt, J. Phys. Chem. B 105 (2001) 3350.
[26] Liu, C.H. Zeng, J. Am. Chem. Soc. 125 (2003) 4430.
[27] Z. Qiu, K.S. Wong, M. Wu, W. Lin, H. Xu, Appl. Phys. Lett. 84 (2004) 2739.

Interfacial interactions between polyethylene matrix and clay layers in polyethylene/clay nanocomposites

R Abu-Zurayk

Hamdi Mango Centre for Scientific Research (HMCSR) - The University of Jordan, 11942 Amman, Jordan.

e-mail: r.abuzurayk@ju.edu.jo

Abstract: Polyethylene/clay nanocomposites were prepared as blown films using different formulae (clay contents (4 and 6 wt%) and compatibilizer/clay ratio (1/2, 1.0, 2.0)). Structure and mechanical behaviour were tested. It was found that blown film extrusion process decreased the tactoids size and consequently enhanced the exfoliation degree of the clay layers inside the polymer matrix, which is due to the elongational stress during extrusion. Addition of clay had some effects on mechanical behaviour. There was an increase of yield strength (max 32%). Yield strength is related to the interfacial interaction between the polymer and the clay layers in the nanocomposites, which would be enhanced by enhancing the compatibility between polymer and clay layers. Correlation analysis showed good correlation between compatibility and interfacial interaction parameters, and between parameters of interfacial interaction, structure and yield strength.

1.Introduction

Polymer nanocomposites have been studied extensively since the successful production of nylon6/ clay nanocomposite by Toyota [1]. However polymer nanocomposites are still not in widespread use in industrial applications and many polymer systems do not achieve the property enhancements reported by Toyota with Nylon 6 nanocomposites. In order to be able to achieve that, relations between processing, structure and performance of nanocomposites were studied.

Regarding the effect of processing on structure, one study on Polypropylene (PP)/organoclay nanocomposites investigated the effect of screw profile in the extruder on the state of clay intercalation/exfoliation [2]. It was found that at low feed rates (long residence time), the screw profiles that are more restricted produced a smaller interlayer distance, however at high feed rates the interlayer distance was not affected by screw profile. The same authors found that neither the screw speed nor the feed rate in the extruder had any significant effect on the degree of intercalation [3]. Jiang et. al. [4] studied the effect of mixing on structuring. They found that chaotic mixing facilitates the dispersion of clay in the PP matrix compared to high shearing/mixing. Kalendova et. al [5] investigated the effect of different types of plasticizer and compounding conditions on the structure of polyvinyl chloride (PVC)/clay nanocomposites. They found that clay modified by plasticizer exhibits fine dispersion of partial to nearly full exfoliated clay. They also concluded that the processing conditions played an important role in nanocomposite production. Abu-Zurayk et. al.[6] investigated the effect of post processing on performance of Polyethylene (PE)/clay nanocomposites. They found that there is an interaction between polymer/clay compatibility and the processing route of the polymer/clay nanocomposites.

Regarding the effect of processing and structure on nanocomposite performance, Du et. al. [7] and Shi et. al. [8], have found that the addition of clay up to a certain point enhances the mechanical properties but further increases lead to a reduction in mechanical performance. This was attributed to the formation of clay aggregates, which act as stress concentrators, at high loading [8]. There is disagreement in the literature regarding the state of exfoliation necessary to maximize composite modulus. Boo et. al. [9] found that in a poorly exfoliated system the modulus improved by only 8%, in a well exfoliated system the enhancement was 50%, while in a moderately exfoliated system had the tensile properties lay between the poorly exfoliated and the well exfoliated systems. Bousmina [10] on the other hand published quite different results. He showed that the best enhancement in modulus was achieved in a system with an intermediate level of exfoliation. He attributed this to the fact that the clay particles are flexible that can bend under mechanical stress and that it is better to have the clay layers as 10-20 layers stacks in order to act as rigid fillers. Such stacks would resist the bending much more than the individual layers of a fully exfoliated system. Costantino et. al. [11] investigated the deformation and fracture properties of PP/clay nanocomposites. They concluded that the overall crack propagation energy values varied as a function of the processing conditions.

Many questions need to be answered regarding the processing of nanocomposites in order to produce useful products. For example, what is the best structure to have in a plastic sheet that is subsequently to be extruded into a blown film? What are the best processing conditions/formulae to get this structure? What are the interfacial interactions in such system? How would these interactions affect the performance? Not all questions have been answered so far in the literature. This study was performed in order to examine the link between processing and interfacial interactions in HDPE/Clay nanocomposites in a sequence of processing steps that constitute a typical processing route (i.e. compounding of nanocomposite followed by blown film extrusion) and their relations with structure parameters and mechanical performance.

2. Materials

HDPE (HYA 600)-ExxonMobil was used. It has density of 0.954g/cm^3 and MFI of 0.35g/10 min (at 190°C and 2.16 kg). Cloisite 20A was supplied by Southern Clay Products. This clay is treated with dimethyl, dehydrogenated tallow, quaternary ammonium ions. Maleated polyethylene DuPontTM Bynell 4033 was used as the compatibilizer. This has an MFI of 2 g/10 min (at 190°C and 2.16 kg).

3. Methodology

3.1 Nanocomposites preparation

A twin screw Brabender compounder was used for compounding seven batches with different formulae as shown in table 1.

Table1. Formulae for compounded batches

Batch number	HDPE (wt%)	PE-g-MA (compatibilizer) (wt%)	Clay (wt%)	Compatibilizer/clay ratio
1	100	0	0	NA
2	94	2	4	1/2
3	92	4	4	1/1
4	88	8	4	2/1
5	91	3	6	1/2
6	88	6	6	1/1
7	82	12	6	2/1

The compounded batches were extruded using blown film extrusion at temperature profile from the feeder of the single screw extruder to the die were 185, 195, 200, 205, 210, 210, 215, and 215 °C, and blow up ratio of 5.0 and final thickness of 0.03mm.

3.2 XRD

A 7000 Shimadzu 2 kW model X-ray diffractometer with a nickel-filtered copper X-ray radiation (Cu Ka 1.5418A) was used to carry out XRD on compounded batches and blown films. Data were

IOP Conf. Series: Materials Science and Engineering **92** (2015) 012010 doi:10.1088/1757-899X/92/1/012010

collected from 1 to 60°. Samples of 2 cm diameter were tested. Data were used to calculate the d-spacing of the crystal lattice of clay.

3.3 TEM

All samples were ultra-microtomed using a 45° diamond knife. Images of the microtomed sections were taken using TEM microscope. A magnification of 22K was used. Images produced were analyzed for tactoid size (thickness, length) using JMicroVision v1.25 software, which was also used for Aspect ratio (AR), exfoliation number (N) and interparticle distance (IPD). Five images of each sample were taken for analysis.

3.4 Tensile Tests

Tensile test was performed on an Instron 5564 Universal tester according to (ISO 527-3) at a crosshead speed of 200 mm/min in machine direction. Each test was repeated on seven samples for each batch.

4. Results and Discussions

4.1 XRD and TEM

XRD was performed on two systems of nanocomposites, the compounded batches which were introduced into shear stress during compounding, and the blown films batches which were introduced into elongational stress during extrusion. Figures 1 and 2 show XRD diffraction patterns. d-spacing values are shown in table 2.

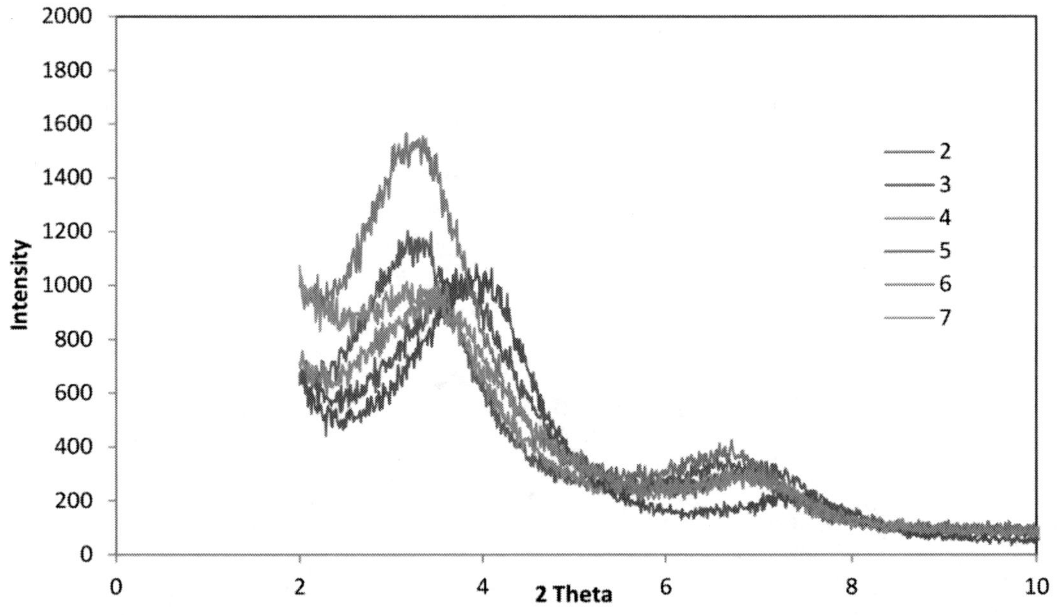

Figure 1. XRD diffraction patterns of compounded batches

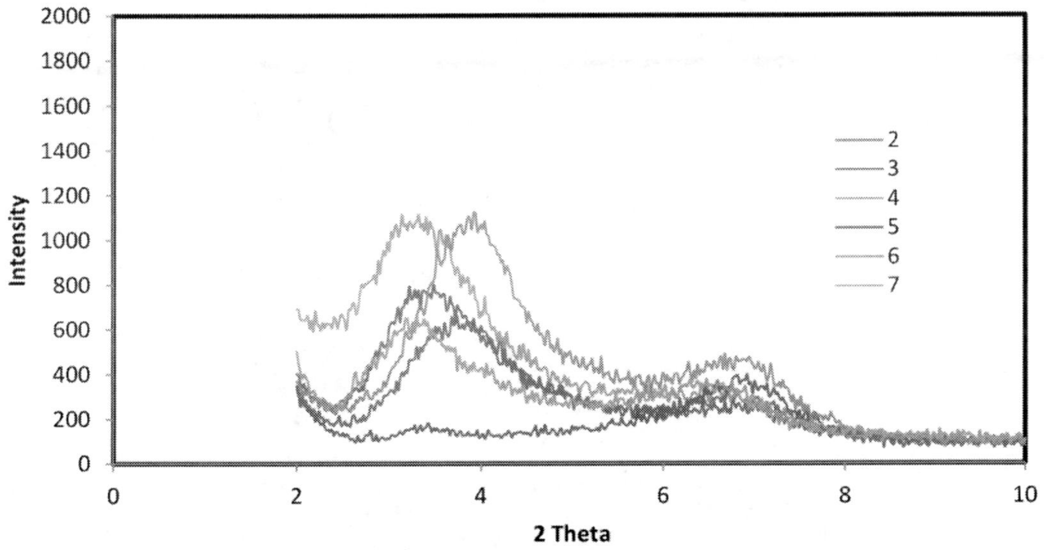

Figure 2. XRD diffraction patterns of blown film batches

Table 2. XRD parameters

d-spacing (A)	2	3	4	5	6	7
Compounded nanocomposites	23.9	23.8	26.8	22.5	26.4	27.9
Blown films nanocomposites	24.02	25.4	26.6	25.8	22.55	26.8

It is shown that at 4wt% clay loading, d-spacing was increased with the increase in compatibilizer concentration but this is not clear at all 6wt% clay loading. Based on the peak shifting between different batches as shown in figures 1 and 2, the highest d-spacing was found in batches 4 and 7, which have highest compatibilizer/clay ratio (2.0) for both clay contents (4 wt% and 6wt%). This is expected, because the higher the compatibilizer content the better the bond between clay and polymer which makes it easier for clay layers that are in contact with the polymer to peel off the clay tactoids during shear stress and elongational stress), which would increase the average d-spacing.

TEM analysis was performed to check the detailed information regarding clay tactoids dispersion and degree of exfoliation. Examples of TEM images are shown in figures 3 and 4.

IOP Conf. Series: Materials Science and Engineering **92** (2015) 012010 doi:10.1088/1757-899X/92/1/012010

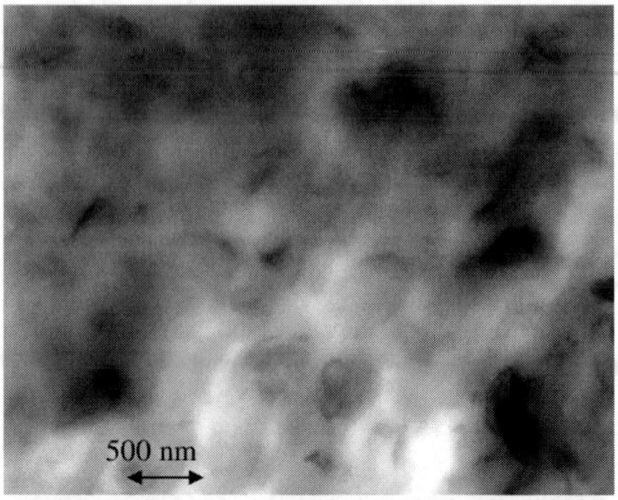

Figure3. TEM for batch 5, compounded nanocomposite, magnification of 22K

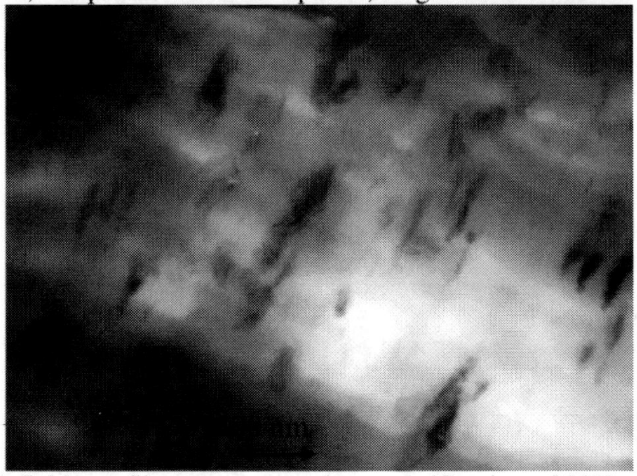

Figure 4. TEM for batch 5, blown film nanocomposite, magnification of 22K

Analyses were performed on TEM images for average tactoids thickness, length, aspect ratio (AR) for all batches, in addition to exfoliation number (N) and interparticle distance (IPD) for blown films. The exfoliation number (N) is defined as the percentage of the total clay interfacial area that is exposed to the polymer matrix. Interparticle distance (IPD) is defined as the mean of all possible straight-line distances in 3D, at arbitrary angles, between arbitrary points on the surfaces of clay particles [12]. Results of analyses are shown in table 3

Table 3 – TEM analysis parameters

Batch	Clay (wt%)	Compatibilizer /clay	Compounded Nanocomposites			Blown films Nanocomposites				
			Length (nm)	Thickness (nm)	AR	Length (nm)	Thickness (nm)	AR	N	IPD
2	4	0.5	339.0	34.1	9.9	267.4	30.3	8.8	6.8	1712.9
3	4	1.0	237.8	12.4	19.2	202.6	19.0	10.6	12.4	1164.8
4	4	2.0	241.7	14.4	16.8	189.7	12.4	15.3	18.4	818.5
5	6	0.5	241.4	22.2	10.9	281.2	27.6	10.2	8.1	1076.7
6	6	1.0	413.0	25.8	16.0	247.0	22.5	11.0	7.0	911.5
7	6	2.0	229.0	12.2	18.8	172.6	12.6	13.7	13.2	520.5

TEM results show that tactoids size in blown films have lower size (length, thickness) compared to compounded batches, which is an indication that blown film extrusion enhances the exfoliation degree of the clay layers inside the polymer matrix regardless of its formula (clay content, compatibilizer/clay ratio). There is a general trend for AR. The higher the compatibilizer/clay ratio, the higher the AR. Same trend is found for N and IPD with compatibilizer/clay ratio. Also, for blown films, at low clay content and high compatibilizer/clay ratio, both AR and N are high, which means that these conditions are best for exfoliation degree based on TEM. Similar results were found on PE/clay nanocomposites systems with different compatibilizer/clay ratios [6], where increasing the compatibility between the polymer and the clay would increase the aspect ratio of the clay layers in the polymer matrix as a result of higher degree of exfoliation. This is expected to affect the mechanical performance

4.2 Tensile test and interfacial interactions

As seen in figure 5, results of tensile tests in machine direction showed enhancement in most nanocomposites yield strength. The strength of nanocomposites relies on the effectiveness of stress transfer between matrix and filler. For well-bonded particles, the applied stress can be efficiently transferred to the particles in the matrix, which clearly improves the strength [13], and this would explain the enhancement of yield strength as result of enhancement of compatibility and bonding between the polymer and the clay.

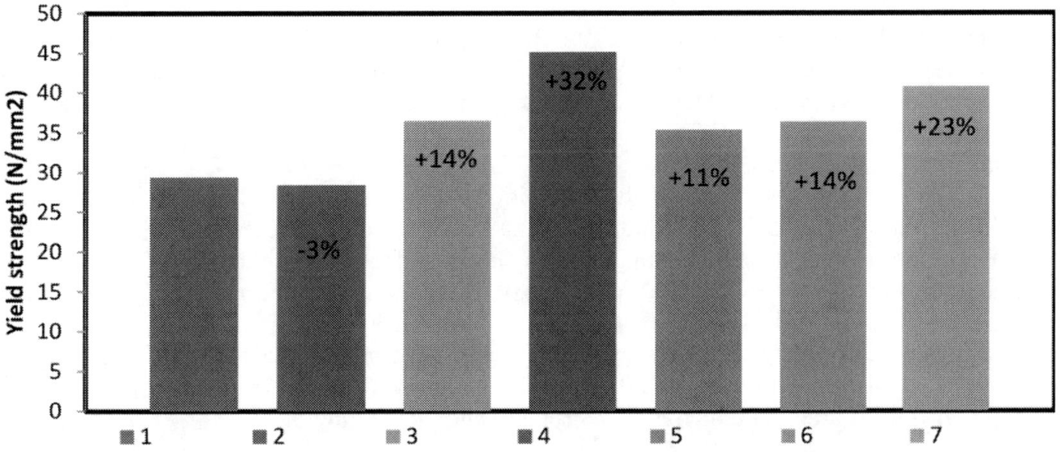

Figure 5. Yield strength of Blown films nanocomposites

Dependence of yield strength of polymer/clay nanocomposite with homogenous matrices on the interfacial interactions, clay tactoid size, and clay fraction can be modeled using the following equation [14]:

$$\sigma R = 1 + \varphi f\left(\frac{AR\,\tau}{\sigma p} - 1\right) \quad (1)$$

where τ is the interfacial stress transfer parameter; σR and σP are the relative yield strength and the yield strength of the polymer, respectively; AR is the aspect ratio of clay tactoid, and φf is the volume fraction of clay in the nanocomposites.

Applying equation (1) for each batch, using relative AR calculated using TEM analysis and yield strength data, τ was calculated. In order to check correlations between processing parameters and interfacial interactions and between interfacial interactions and structure and performance, a bootstrapping analysis via MATLAB was performed at confidence interval of 90%. Significant correlations were found between different parameters, with good correlation coefficients as shown in table 4.

Table 4- Correlation coefficients

Parameter		Processing parameter	Interfacial interaction parameter	Strucutre parameters		
		Compatibilizer/clay ratio	**τ**	**AR**	**N**	**IPD**
Interfacial interaction parameter	**τ**	+0.72	1.00	---	+0.85	---
Strucutre parameters	**AR**	+0.75	---	1.00	+0.89	-0.80
	N	+0.85	+0.85	+0.88	1.00	-0.55
	IPD	-0.79	---	-0.80	-0.55	1.00
Mechanical parameters	**Yield Strength**	+0.88	+0.94	+0.96	+0.87	-0.85

As seen in table 4, there is a significant correlation between interfacial interaction parameter and compatibility; the better the compatibility, the stronger the interaction. And this would affect the structure (better exfoliation) and consequently the yield strength of the system. Highest correlations of yield strength were found with interfacial interaction parameter (τ), and aspect ratio of clay tactoids.

5. Conclusions

Blown film extrusion process enhanced the exfoliation degree of the clay layers inside the polymer matrix, due to the elongational stress during extrusion. Regarding effects of clay on mechanical performance of nanocomposites, yield strength was enhanced for different batches of nanocomposites with a maximum enhancement of 32%. Yield strength is related to the interfacial interaction between the polymer and the clay layers in the nanocomposites, which would be enhanced by enhancing the compatibility between polymer and clay layers. This was confirmed by results of correlation analysis which showed good correlation between compatibility and interfacial interaction parameters, and between parameters of interfacial interaction, structure and yield strength.

6. Acknowledgement

The author acknowledges the Scientific Research Support Fund (SRF) at Ministry of Higher Education, Jordan, for funding this research.

7. References

[1] Hussain F, Hojjati M, Okamoto Y and Gorga R 2006 Journal of Composite Material, 40(17) 1511.
[2] Lertwimolnun W and Vergnes B 2007 Polymer Engineering and Science 47 2100.
[3] Lertwimolnun W and Vergnes B. 2006 Polymer Engineering and Science 46 314.
[4] Jiang G and Huang H 2011Polymer Engineering and Science 51 2345.
[5] Kalendova A, Zykova J, Kovarova L, Slouf M. and Gerard J 2010 AIP Conference Proceeding 1255 181
[6] Abu-Zurayk R and Harkin-Jones E 2012 Polymer Engineering and Science 52 2360.
[7] Du K, He A, Liu X and Han C 2007 Macromolecular Rapid Communication 28 2294.
[8] Shi X and Gan Z 2007 European Polymer Journal 43 4852.

[9] Boo W, Sun L, Liu J, Moghbelli E, Clearfield A, Sue H, Pham H, and Verghese N 2007 Journal of Polymer Science: Part B: Polymer Physics 45 1459.
[10] Bousmina M 2006 Macromolecules 39 4259.
[11] Costantino A, Pettarin V, Viana J, Pontes A, Pouzada A and Frontini P 2013 Journal of Mechanical Engineering 59697.
[12] Abu-Zurayk R, Harkin-Jones E, McNally T, Menary G, Martin P and Armstrong C 2009 Composites Science and Technology 69 1644.
[13] Tanasa F, Zanoaga M, Darie R 2014 International Conference of Scientific Paper Brasov, 22-24 May 2014.
[14] Durmus A, Kasgoz A and Macosko C W 2008 Journal of Macromolecular Science, Part B: Physics 47 608

IOP Conf. Series: Materials Science and Engineering **92** (2015) 012011 doi:10.1088/1757-899X/92/1/012011

Investigating Negative Magnetization and Blocking Temperature in Aggregates of Ferrite Nanoparticles

I. M. Obaidat[1,*], B. Issa[1], B. A. Albiss[2], and Y. Haik[3,4]

[1]Department of Physics, United Arab Emirates University, Al-Ain 17551, United Arab Emirates
[2]Superconductivity & Magnetic Measurements Laboratory, Physics Department, Jordan University of Science and Technology, Irbid 22110, Jordan
[3]Center for Research Excellence in Nanobiosciences, University of North Carolina-Greensboro, NC 27412, USA
[4]Department of Mechanical Engineering, Qatar University, Qatar

*e-mail:iobaidat@uaeu.ac.ae

Abstract. Blocking temperatures of aggregates of $Mn_{0.5}Zn_{0.5}Gd_xFe_{(2-x)}O_4$ ferrite nanoparticles, with x = 0.02, 0.05, 0.11, 0.15, and 0.2, were obtained from the zero-field-cooled (ZFC) magnetization measurements. We found a nonmontonic behavior of the blocking temperature with increasing size of the particles. The effective magnetic anisotropy was calculated and found to have two distinct rates of increase with decreasing the size of the particles. These results were attributed to the strong inter-particle interactions in the aggregated nanoparticles and to the enhanced role of surface anisotropy with the decrease of the size of the particles. In three samples, the ZFC magnetization was found to exhibit a significant negative magnetization in a considerable part of the low temperature region. To our knowledge, this is the first time that negative magnetization is reported in such nanoparticles. These peculiar results are discussed and are currently under investigation.

1. Introduction

It is well-established that, due to large surface-to-volume ratio, surface effects of magnetic nanoparticles (MNPs) strongly influence their magnetic properties [1-3]. The magnetic surface effects could result of several factors such as the expansion or contraction of the lattice structure near the surface [4, 5], the missing of crystallographic symmetry at the surface [6], roughness, reduction in coordination number, charge transfer, and other factors [7-9]. The impact of each of these surface effects cannot be separated from that of the other effects, and hence all of these effects contribute to the surface anisotropy of the NPs. Technological applications of MNPs demand their thermal stability [10]. If thermal energy (k_BT, where k_B is Boltzmann's constant) exceeds the magnetic anisotropy energy (magnetic energy barrier) between stable magnetic directions, the nanoparticles (NPs) relax from the blocked state to the superparamagnetic state, where the magnetic moment of the NP starts flipping randomly and will not be fixed in a specific direction. The blocking temperature, T_b is the temperature at which the magnetization direction starts flipping. The blocking temperature of MNPs depends on the magnitude of the energy barrier which is determined by several factors such as the size, shape and magnetic anisotropy of MNPs, and inter-particle interactions. Exchange-bias effect can also impact the energy barrier of MNPs [7, 11, 12]. For single-domain particles with small inter-particle interactions, the relaxation time of the magnetic moment of the NPs is given by [13]

$$\tau = \tau_0 e^{\Delta E/k_B T} \quad (1)$$

where τ_0 is an attempt time (10^{-9} – 10^{-13}s), ΔE is the magnetic energy barrier of spin reorientation the MNPs, and T is the absolute temperature. At the blocking temperature, T_b, the NPs become superparamagnetic where the relaxation time, τ becomes equal to the measurement time, τ_m. In a typical magnetization experiment, $\tau_m = 100s$. Hence, $\ln \tau_m/\tau_0 = \Delta E/k_B T = 25$.
Thus the magnetic energy barrier, $\Delta E = 25k_B T_b$. For single-domain MNPs with uniaxial anisotropy, ΔE is given by

$$\Delta E = K_{eff} V \quad (2)$$

where V is the average volume of the NPs, and K_{eff} is the effective anisotropy energy density constant of the NPs. Thus for NPs with small inter-particle interactions, the effective anisotropy constant, K_{eff} is given by [14, 15]

$$K_{eff} = 25k_B T_b/V \quad (3)$$

The effective anisotropy constant of MNPs, K_{eff} is usually given by the phenomenological formula [16],

$$K_{eff} = K_V + \frac{K_S}{d} \quad (4)$$

where K_V, is the bulk anisotropy energy density constant, K_S, is the surface anisotropy constant, and d, is the diameter of the MNP.

Reversal magnetization or negative magnetization effect appears when the direction of magnetization (in the FC or ZFC branches) in a material becomes opposite to the direction of the applied magnetic field to that material. The temperature at which magnetization becomes zero and changes sign (from negative to positive in the ZFC branch and from positive to negative in the FC branch) is called the compensation temperature, T_{comp}. Reversal magnetization is a challenging field of research and has a potential applications in magnetic devises [17- 21]. This effect was first observed by Néel in ferrimagnets that consist of two magnetic ions coupled antiferromagnetically each other [22]. The cause of negative magnetization was suggested to be due to the difference in the temperature dependent magnetization of two magnetic ions. Negative magnetization has been observed in several magnetically ordered materials, such as intermetallic alloys and multilayers, perovskites, spinel ferrites, molecular magnets, and garnets [21]. However, the origin of the observed negative magnetization in some materials is not well-known yet [21, 22].

In this paper we show interesting results about the blocking temperature behavior as function of size of the NPs. We also show the occurrence of negative magnetization in some of these NPs.

1. Experimental Methods

The ferrite nanoparticles with several Gd^{3+} concentrations were obtained via chemical co-precipitation and feritization. Details on the synthesize method can be found in [24]. The X-ray diffraction (XRD) was performed using Agilent/Oxford Gemini Diffractometer using molybdenum (Mo) $K\alpha$ radiation. The magnetic measurements were conducted using Quantum Design Superconducting Quantum Interference Device (SQUID).

2. Results and Discussion

As reported in [25], the XRD diffraction patterns for $Mn_{0.5}Zn_{0.5}Gd_xFe_{(2-x)}O_4$ nanoparticles with x = 0.02, 0.05, 0.11, 0.15, and 0.2 indicate a single-phase FCC spinel structure for the samples with no detectable impurity phases. From the XRD patterns and using Scherrer's formula [26, 27], the sizes of the particles were estimated to be in the range 4 - 10 nm indicating a single domain structure. For x = 0.02, 0.05, 0.11, 0.15, and 0.2, the diameters are 9.2, 7.4, 7.1, 5.0, and 4.2 nm respectively. Using a SQUID, the measurements of magnetization versus temperature (*M-T*) were conducted on all samples from 5 K to 400 K in an applied magnetic field of magnitude 100 Oe.

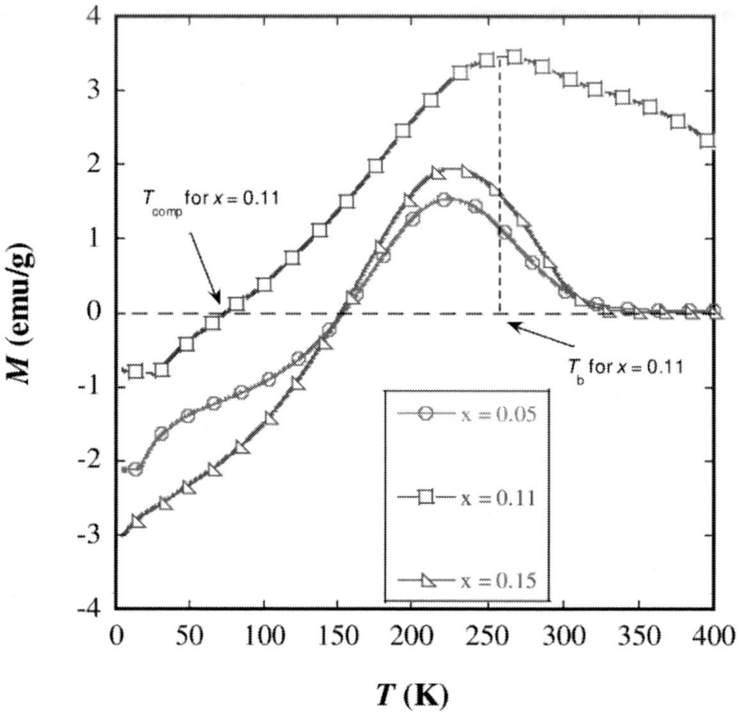

Figure 1: ZFC magnetization curve for $Mn_{0.5}Zn_{0.5}Gd_xFe_{(2-x)}O_4$ nanoparticles with x = 0.05, 0.11, and 0.15.

Figure 1 displays the zero-field-cooled (ZFC) magnetization measurements as function of temperature for x = 0.05, 0.11, and 0.15. In the ZFC branch, the samples were cooled down to the lowest temperature (5 K) in zero applied field. A magnetic field of magnitude 100 Oe was then applied, and the magnetic measurements were performed during the heating process up to 400 K. It can be seen that the peaks of the ZFC curves occur at large temperatures, indicating large particle-particle interactions [28]. On the other hand, the broad peaks indicate large size distribution of the NPs. The blocking temperature T_b is obtained from the peaks of the ZFC curves. An arrow in figure 1 shows T_b for x = 0.11, the other values were obtained similarly. Figure 2 displays the blocking temperature as function of the NPs sizes. For MNPs, T_b is expected to increase with increasing the size of particles [29, 30]. However, figure 2 displays a nonmonotonic behavior of T_b with the size of NPs. This behavior could be attributed to large inter-particle interactions. In [25], the SEM images show large aggregates of these NPs which supports the idea of large inter-particle interactions. It was reported that decreasing the inter-particle spacing (increasing inter-particle interactions) leads to an increase in the blocking

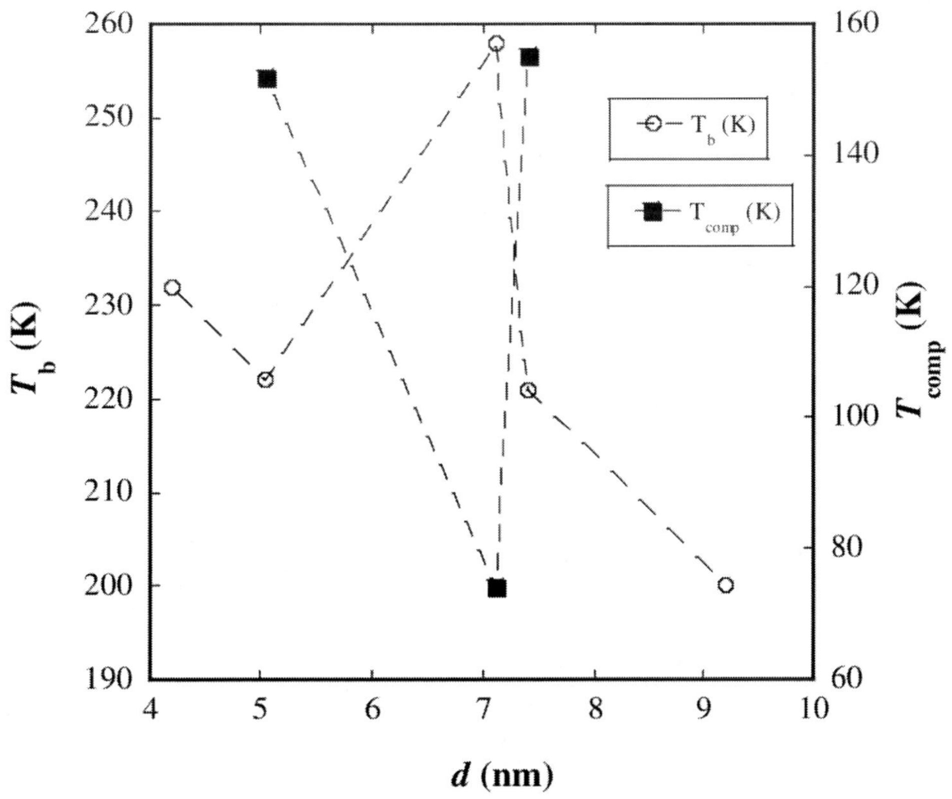

Figure 2: The blocking temperature, T_b and the compensation temperature, T_{comp} as function of the diameter of the NPs. The dashed lines are guide for the eye.

temperature [31-33]. This result was also reported using Monte Carlo simulations [34]. It is well-known, but not well-understood, that the magnetic properties of MNPs are sensitive to particle-particle interactions. Not only intrinsic properties, but also the extrinsic properties play significant roles in determining the blocking temperature. Thus, T_b not only depends on the size of NPs but also on the magnetic inter-particle interactions. Hence, the nonmonotoinc behavior of T_b as function of the size in our ferrite NPs could be explained by assuming different strength of the inter-particle interactions among the aggregated samples.

There are several experimental methods that could be used to obtain the effective anisotropy constant, K_{eff} [35]. One of these methods is to use the Zero Field Cooled (ZFC) magnetization measurements. K_{eff} for a particular MNP with volume V, and the blocking temperature, T_b was calculated using equation (3) [36]. Any experimental sample contains large numbers of NPs with different sizes, shapes, surface effects, and particle-particle interactions. Hence, $K_{eff}$is a result of many effects. Figure 3 displays the effective anisotropy constant as function of the size of the NPs. In obtaining this figure we used equation (3) as an approximation (knowing that it might not yield accurate results in the case of our aggregated samples). As can be seen in this figure, K_{eff}increases with two distinct rates as the size of NPs decreases.

For NPs with diameters above 7 nm, K_{eff} increases slowly with decreasing the diameter. However, for diameters smaller than 7 nm, K_{eff} increases steeply with decreasing diameter. This behavior could be attributed to the enhanced role of surface anisotropy with decreasing the size of NPs.

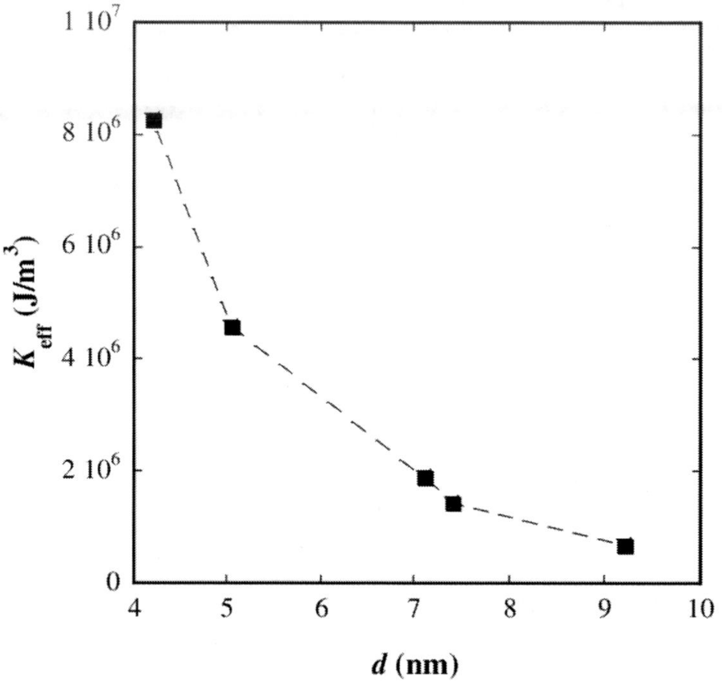

Figure 3: The effective anisotropy constant as function of the diameter of the NPs. The dashed line is a guide for the eye.

As can be seen in figure 1, for the samples with $x = 0.05$, 0.11, and 0.15, the initial magnetization (at 5 K) starts with significant negative values and increases monotonically upon heating the samples up the compensation temperatures. Reversal magnetization was not observed in the ZFC branches of the samples with $x = 0.02$, and 0.2 (not shown in figure 1). For the samples with $x = 0.05$, and 0.15 the negative magnetization persists up to temperatures above 150 K while for the sample with $x = 0.11$, the magnetization direction flips positive at nearly 74 K. An arrow in figure 1 shows T_{comp} for $x = 0.11$, the other values were obtained similarly. This indicates a dependence of the compensation temperature, $T_{\rm comp}$ on the composition of the samples (among other possible factors). In figure 2, $T_{\rm comp}$ (for $x = 0.05$, 0.11, and 0.15) is plotted as function of the size of NPs, where is evident that $T_{\rm comp}$ has a nonmonotonic behavior that is opposite to the behavior of $T_{\rm b}$. Several parameters are suggested to play roles in the appearance of negative magnetization, these include, temperature dependence of sublattice magnetization, magnetic anisotropy, magnetic exchange interactions, and crystal structure [21]. In [37], the negative magnetization was observed in polycrystalline Co_2CrAl compound (in both bulk and thin films) in the ZFC branch at fields below 100 Oe. The compensation temperature decreases with the increase of magnetic field. This behavior was suggested to be initiated by Landau diamagnetism, and then preserved by the peculiar electronic structures of the compound. In [38], negative magnetization was observed in the ZFC branch in polycrystalline $SrRuO_3$ at low applied fields.

Several factors such high magnetocrystalline anisotropy constant, intrinsic negative spin polarization, and random orientation of easy axis were suggested to be key points in this behavior. Negative magnetization in the ZFC branch, in low applied field, was observed at low temperatures in $CoCr_2O_4$ nanoparticles of sizes 30 - 40 nm [39]. The origin of the negative magnetization in those $CoCr_2O_4$ nanoparticles is not well-known.

It is important to state that to our knowledge, this is the first time that negative magnetization is observed in such ferrite NPs. We believe that these results are not due to experimental errors. We are aware of the fact that residual field of the SQUID magnetometer during the cooling process could cause negative magnetization. However, in our study, the applied field was 100 Oe and the negative magnetization persists up to more than 150 K in some samples, making the possibility of residual field very small. The origin of the negative magnetization observed in our samples could be attributed to the difference in temperature dependence of the magnetic moments at the tetrahedral and octahedral sites coupled with surface spin structures which change with the x value. However, detailed investigation is needed to understand this effect in our NPs.

3. Conclusions

Blocking temperatures of aggregates of $Mn_{0.5}Zn_{0.5}Gd_xFe_{(2-x)}O_4$ ferrite nanoparticles (with several Gd^{3+} cations concentrations) were obtained from the ZFC magnetic measurements. The blocking temperature was found to have nonmonotonic behavior as function of the size of NPs. This behavior was attributed to significant inter-particle interactions between the NPs. Reversal magnetization was observed in the low temperature ZFC branches of some samples while it did not appear in other samples. This behavior was attributed to the difference in temperature dependence of the magnetic moments at the tetrahedral and octahedral sites coupled with surface spin structures which change with the Gd^{3+} cations concentrations. The steep increase of the effective anisotropy with decreasing the size of NPs is suggested to be due to the enhanced role of surface effects as the NPs become smaller.

ACKNOWLEDGMENT
This work was financially supported by the UAEU Program for Advanced Research (UPAR) under the Grant no. 31S162.

References

1. D. Fiorani (ed) Surface Effects in Magnetic nanoparticles, Springer. New York, 2005.
2. I. M. Obaidat, B. Issa, and Y. Haik, Nanomaterials **5**, 63 (2015).
3. B. Issa, I. M. Obaidat, B. A. Albiss, and Y. Haik, Int. J. Mol. Sci. **14**, 21266 (2013).
4. F. Dorfbauer, R. Evans, M. Kirschner, O. Chubykalo-Fesenko, R. Chantrell, T. Schrefl, J. Magn. Magn. Mater. **316**, E791 (2007).
5. R. Evans, F. Dorfbauer, O. Chubykalo-Fesenko, T. Schrefl, R.W. Chantrell, IEEE Trans. Magn. **43**, 3106 (2007).
6. M. Jamet, W. Wernsdorfer, C. Thirion, V. Dupuis, P. Mélinon, A. Pérez, D. Mailly, Phys. Rev. B **69**, 24401 (2004).
7. R.F.L. Evans, R. Yanes, O. Mryasov, R.W. Chantrell, O. Chubykalo-Fesenko, Europhys. Lett. **88**, 57004 (2009).
8. F. Garcia-Sanchez, O. Chubykalo-Fesenko, O. Mryasov, P. Asselin and R.W. Chantrell, J. Appl. Phys. **103**, 07F505 (2008).
9. G.R Aranda, O. Chubykalo-Fesenko, R. Yanes, J. Gonzalez, J.J. del Val, R.W. Chantrell, Y.K. Takahashi, K. Hono, J. Appl. Phys. **105**, 07B514 (2009).
10. S. Sun, C.B. Murray, D. Weller, L. Folks, A. Moser, Science **287**, 1989 (2000).

11. R. Yanes, O. Chubykalo-Fesenko, R.F.L. Evans, and R.W. Chantrell, J. Phys. D: Appl. Phys. **43**, 474009 (2010).
12. V. Skumryev et. al., Nature **423**, 850 (2003).
13. L. Néel, Ann. Geophys. **5**, 99 (1949).
14. P. Guardia, B. Batlle-Brugal, A. G. Roca, O. Iglesias, M. P. Morales, C. J. Serna, A. Labarta and X. Batlle, J. Magn. Magn. Mater. **316**, e756 (2007).
15. R. Venugopal, B. Sundaravel, W. Y. Cheung, I. H. Wilson, F. W. Wang and X. X. Zhang, Phys. Rev. B **65**, 014418 (2001).
16. F. Bodker, S. Mörup and S. Linderoth, Phys. Rev. Lett. **72**, 282 (1994).
17. Y. Ren, T. T. M. Palstra, D. I. Khomskii, E. Pellegrin, A. A. Nugroho, A. A. Menovsky, and G. A. Sawatzky, Nature (London) **396**, 441 (1998).
18. K. Yoshii and A. Nakamura, J. Solid State Chem. **155**, 447 (2000).
19. S. M. Yusuf, A. Kumar, and J. V. Yakhmi, Appl. Phys. Lett. **95**, 182506 (2009).
20. B. Sarkar, B. Dalal, and S.K. De, Appl. Phys. Lett. **103**, 252403 (2013).
21. A. Kumar, and S.M. Yusuf, Physics Reports **556**, 1 (2015).
22. S. Gu, W. He, M. Zhang, T. Zhuang, Y. Jin, H. ElBidweihy, Y. Mao, J. H. Dickerson, M. J. Wagner, E. D. Torre, and L. H. Bennett, Scientific Reports **4**, 6267 (2014).
23. L. Néel, Ann. Phys. (Leipzig) **3**, 137 (1948).
24. I. M. Obaidat, B. Issa, and Y. Haik, J. Nanosci. Nanotech. **11**, 3882 (2011).
25. I. M. Obaidat, B. Issa, B. A. Albiss, G. Qabaja, N. E. Al Khaili, Z. A. Karam, A. M. Hefaity, S. Qadri, A. Al-Okour, and Y. Haik, J. Nanoeng. Nanomanuf. **2**, 325 (2012).
26. P. Scherrer, Nachr. Ges. Wiss. Göttingen **26**, 98 (1918).
27. A. Patterson, Phys. Rev. **56**, 978 (1939).
28. D. Predoi, V. Kuncser, G. Filoti, G. Schinteie, J. Optoelectron. Adv. Mater. **5**, 211 (2003).
29. M Hanson, C Johansson, M *S* Pedersen and *S* Mørup, J. Phys.: Condens. Matter **7**, 9269 (1995).
30. T.C. Monson, E.L. Venturini, V. Petkov, Y. Ren, J.M. Lavin, D.L. Huber, J. Magn. Magn. Mater. **331**, 156 (2013).
31. H. T. Yang, D. Hasegawa, M. Takahashi, and T. Ogawa, J. Appl. Phys. **94**, 013103 (2009).
32. J. Dai, J-Q. Wang, C. Sangregorio, J. Fang, E. Carpenter, and J. Tang, J. Appl. Phys. **87**, 7397 (2000).
33. K. Rumpf, P. Granitzer, P. M. Morales, P. Poelt, and M. Reissner, Nanoscale Res. Lett. **7**, 445 (2012).
34. J. García-Otero, M. Porto, J. Rivas, and A. Bunde, Phys. Rev. Lett. **84**, 167 (2000).
35. S. Yoon, J. Korean Phys. Soc. **59**, 3069 (2011).
36. T. Hyeon, Y. Chung, J. Park, S. S. Lee, Y-W. Kim and B. H. Park, J. Phys. Chem. B **106**, 6831 (2002).
37. K.W. Kim, J.Y. Rhee, Y.V. Kudryavtsev, V.G. Prokhorov, J.M. Kim, and Y.P. Lee, J. Korean Phys. Soc. **59**, 3064 (2011).
38. B. Sarkar, B. Dalal, and S.K. De, Appl. Phys. Lett. **103**, 252403 (2013).
39. D.P. Dutta, J. Manjanna, and A.K. Tyagi, J. Appl. Phys. **106**, 043915 (2009).

Temperature Dependence of Saturation Magnetization and Coercivity in $Mn_{0.5}Zn_{0.5}Gd_{0.02}Fe_{1.98}O_4$ Ferrite Nanoparticles

I. M. Obaidat[1,*], B. Issa[1], B. A. Albiss[2], and Y. Haik[3,4]

[1]Department of Physics, United Arab Emirates University, Al-Ain 17551, United Arab Emirates
[2]Superconductivity & Magnetic Measurements Laboratory, Physics Department, Jordan University of Science and Technology, Irbid 22110, Jordan
[3]Center for Research Excellence in Nanobiosciences, University of North Carolina-Greensboro, NC 27412, USA
[4]Department of Mechanical Engineering, Qatar University, Qatar

*e-mail:iobaidat@uaeu.ac.ae

Abstract. The influence of temperature on coercivity, H_c and saturation magnetization, M_s were investigated experimentally in $Mn_{0.5}Zn_{0.5}Gd_{0.02}Fe_{1.98}O_4$ ferrite nanoparticles (average size 35 nm). Isothermal magnetization curves M (H) were obtained in the field range from -5 kOe to +5 kOe at different temperatures after the zero field cooling (ZFC) process. The temperature dependence of the coercivity, $H_c(T)$ deviated slightly from the classical Kneller's law. The temperature dependence of saturation magnetization, $M_s(T)$ was found to have an excellent agreement with the Bloch's law. These results are discussed in terms of several factors such as the size and size distribution of the particles, inter-particle interactions and the surface spin.

1. Introduction

When the radius of the magnetic nanoparticles (MNPs) decreases, larger fraction of the atoms resides on the surface compared to those in the core of the particle. This is because of the increase in the ratio of the surface area to the volume of the particle. This gives the nanoparticles (NPs) interesting properties that could be different from those of the bulk material. At very small sizes, nanoparticles display superparamagnetic behavior above the so-called blocking temperature, T_B [1]. Superparamagnetic particles (at temperatures above the blocking temperature) show a quick response to the variations in the applied magnetic field. In the typical case, these responses occur with zero remnant magnetization, M_R and coercivity, H_c.

Ferrimagnetic nanoparticles such as $Mn_{0.5}Zn_{0.5}Gd_xFe_{(2-x)}O_4$ are attractive because of their biomedical applications as contrast agents in magnetic resonance imaging (MRI) and their promising application in self-controlled magnetic hyperthermia [2, 3]. Although many interesting studies were conducted on ferrite nanoparticles, many investigations are being conducted to understand their magnetic behavior [4-17].

Due to finite size effects, surface spin effects (such as spin canting, spin freezing, spin disorder), the saturation magnetization of nanoparticles was found to be different from that of the corresponding bulk material. In ferromagnetic particles, these effects become more important because of the superexchange interaction [18, 19]. Surface spin disorder can occur at the surface of the nanoparticle because of the existence of impurities such as oxygen ion vacancy at the surface of the nanoparticle which could lead to the breaking of superexchange bonds between the magnetic cations [20]. Hence, the saturation magnetization in ferrite nanoparticles is noticeably small compared with that of bulk material [21, 22]. The decrease of saturation magnetization in nanoparticles becomes more pronounced with the decrease of the size of nanoparticles where the surface to volume ratio becomes larger [23]. On the other hand, the temperature dependence of coercivity and saturation magnetization in nanoparticles [24] showed the importance of surface spin and finite size effects in nanoparticles. The temperature dependence of coercivity is described by Kneller's law [25, 26]

$$H_c(T) = H_o\left[1 - \left(\frac{T}{T_B}\right)^{\alpha}\right] \quad (1)$$

Here T_B is the superparamagnetic blocking temperature of the nanoparticles and H_o is the coercivity at $T = 0$ K. Kneller law was derived in the temperature range (0 - T_B), where the nanoparticles were considered to be single-domain, non-interacting and with uniaxial anisotropy. For such system, α was found be equal to ½.

For bulk ferromagnetic or ferrimagnetic materials at low temperatures, the saturation magnetization, M_s as function of temperature is governed by the Bloch's law [27]

$$M_S(T) = M(0)\left[1 - \left(\frac{T}{T_o}\right)^{\alpha}\right] \quad (2)$$

Here $M(0)$ is the saturation magnetization at 0 K and T_o is the temperature at which the M_s becomes zero. The exponent α is known as the Bloch's exponent where α = 3/2 for bulk materials. Bloch derived the $T^{3/2}$ law by considering the magnon excitation of long wave-length spin-waves at low temperatures. However, in nanoparticles and due to finite size effects, magnons could have wavelengths larger than the size of the particle. The confinement of the wavelength of the spin wave by the size of the magnetic nanoparticle results in a gap in the spin-wave energy spectrum [24]. On the other hand, at temperatures larger than the spin-wave energy gap, Bloch's law seems to be applicable in nanoparticles. Several studies [28-42] reported the deviation of the saturation magnetization from Bloch's law in nanoparticles at low temperatures. The modified Bloch's law for nanoparticles was shown to have similar form to the Bloch's law for bulk materials but with Bloch's exponent α which could have values larger and smaller than 3/2 [34, 35] at intermediate temperatures and decays exponentially at low temperatures [31]. Various interpretations were proposed for the deviation from the Bloch's law in nanoparticles [24, 31-40]. For example, the additional magnetization due to shell spins was suggested to induce the deviation from the Bloch law [38]. In [40], the deviation of the saturation magnetization from the Bloch law at temperatures below 50 K was suggested to be due to the canted surface spins which freeze into a disordered state.

In this paper we studied the role of temperature on the coercivity and saturation magnetization in $Mn_{0.5}Zn_{0.5}Gd_{0.02}Fe_{1.98}O_4$ ferrite nanoparticles in temperatures between 5 and 300 K. We compare our data with Block's and Kneller's laws.

2. Experimental Methods

The aim of this synthesis process is to produce Gd-substituted Mn-Zn ferrite nanoparticles of the desired stoichiometry, exhibiting the form $Mn_{0.5}Zn_{0.5}Gd_xFe_{(2-x)}O_4$ with controlled values for the Gd concentration. Chemical co-precipitation and ferritization (formation of ferrite) method was used here. In order to obtain Fe^{3+}, Gd^{3+}, Mn^{2+} and Zn^{2+} ions in aqueous form different metal solutions were prepared in the synthesis process. Solutions such as $FeCl_3$ $6H_2O$, $MnCl_2$ $4H_2O$, $ZnSO_4$ $7H_2O$, $GdCl_3$ $6H_2O$, and $FeSO_4$ $7H_2O$ were mechanically stirred in deionized water. Controlled Argon pressure was implemented. A thick crystalline precipitate which is formed after addition of the co-precipitating base (sodium hydroxide base), was stirred at room temperature for half hour, and then heated for one hour at 90 °C. Ferritization of this precipitate is ensured by heating in the precipitation alkaline solution. The synthesized nanoparticles then were filtered and washed several times with doubly distilled water and acetone. The nanoparticles then were filtered and dried at room temperature. Finally, the nanoparticles were prepared as powders or colloids. The magnetic measurements were conducted using Superconductor Quantum Interference Device (MPMS SQUID VSMdc Magnetometer). The average size of the nanoparticles and their size distribution were extracted using scanning electron microscopy (SEM) analysis (EDX, Jeol Model JSM-5600).

3. Results and Discussion

Tunnelling electron microscopy (TEM) image of the nanoparticles is displayed in figure 1. It is seen that there are some agglomerations of the particles. In figure 2, the size histogram of the nanoparticles obtained from SEM image is displayed. The histogram revealed a non-uniform dispersion of nanoparticles with average diameter of 35 nm. The magnetization hysteresis measurements were conducted at several temperatures from 5 K up to room temperature 300 K. For each hysteresis measurement, the sample was cooled to the specific temperature and the magnetization measurement was done under the applied magnetic field from -5 kOe to +5 kOe.

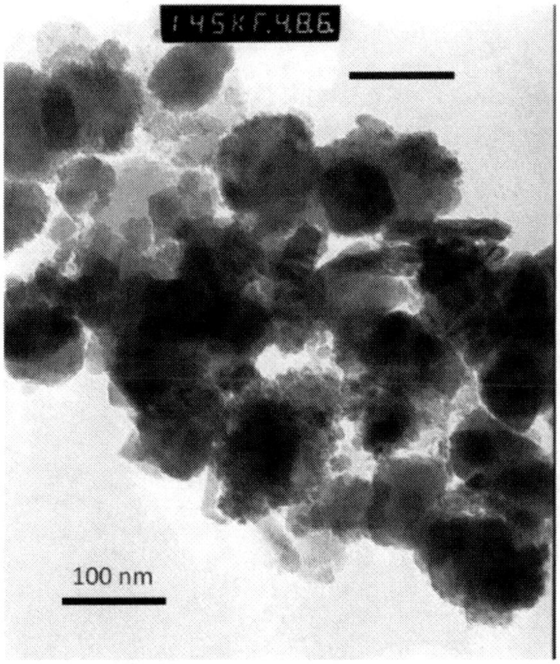

Figure 1. TEM image of the nanoparticles.

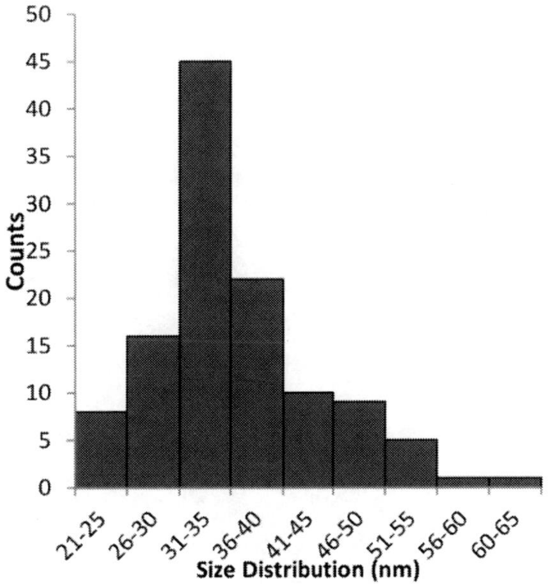

Figure 2. Size histogram of the nanoparticles obtained from a SEM image. The average diameter obtained at room temperature is 35 nm

Figure 3 displays the magnetization hysteresis loops at several temperatures (5, 50, 100, 200, and 300 K). We can see that the saturation magnetization at 200 and 300 K was achieved while it was not fully achieved at temperatures below 200 K. This could be a result of surface effects where disordered surface spins of nanoparticles are not easily aligned along the direction of the applied magnetic field leading to the incomplete saturation magnetization. The value of the magnetization at the largest applied field of 5 kOe is referred to as the saturation magnetization. In figure 4(a), we plotted the coercive fields, $H_c(T)$ as a function of temperature. $H_c(T)$ which were obtained from the magnetization hysteresis loops. It is clear that the coercivity decreases monotonically with increasing temperatures.

IOP Conf. Series: Materials Science and Engineering **92** (2015) 012012 doi:10.1088/1757-899X/92/1/012012

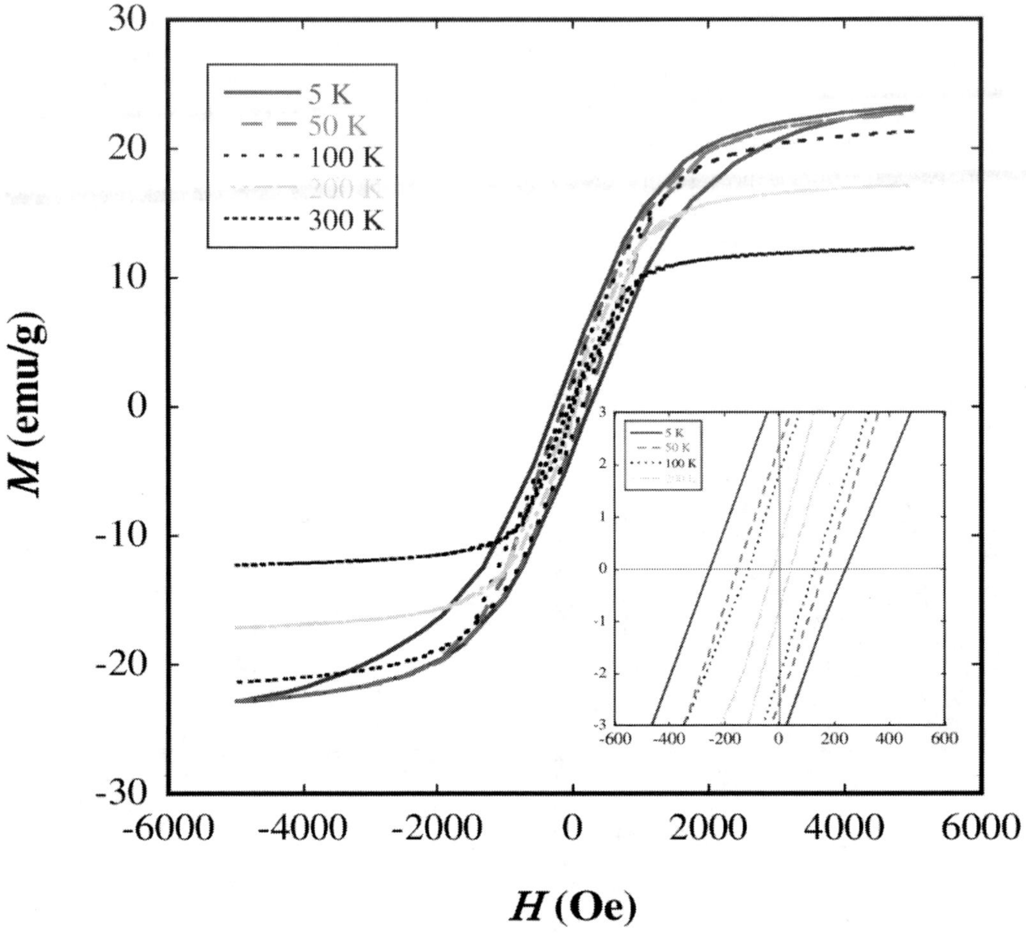

Figure 3. $M(H)$ hysteresis curves of the ferrite nanoparticles obtained at 5, 50, 100, 200, and 300 K in the field range from -5 to +5 kOe. The inset of the figure is a magnification of curves at very low fields

This behavior is understood since thermal energy will enhance thermal fluctuations of the pinned magnetic moments and hence minimizing the effect of anisotropy barriers. We have fitted the $H_c(T)$ data with the Kneller's law (equation 1). H_o is estimated by extrapolating the $H_c(T)$ curve towards the magnetic field axis (at T = 0 K). The blocking temperature for our nanoparticles is predicted to be larger than 200 K. This prediction is based on earlier measurements of blocking temperature (determined from the peak of zero field cooled magnetization) of other smaller $Mn_{0.5}Zn_{0.5}Gd_{0.02}Fe_{1.98}O_4$ nanoparticle samples that were similarly synthesized [43, 44]. The solid line in figure 4(b) shows the fitting of our $H_c(T)$ data with the Kneller's law. We can see the good fitting of the data to the above relation but with $\alpha = 0.57$, T_B = 223 K and H_o = 274 Oe. Due to fluctuations that occurred at T = 300 K near H = 0 Oe, it was not possible to include H_c value for T = 300 K in the fitting.

Figure 5(a) displays the saturation magnetization, M_s as a function of temperature. We can see that $M_s(T)$ decreases monastically with temperature. However, at temperature below 100 K, M_s decreased slowly with increasing temperature after which, M_s dropped sharply. We have fitted the $M_s(T)$ data for our nanoparticles with the Bloch's law (equation 2). The solid line in figure 5(b) shows the fitted curve.

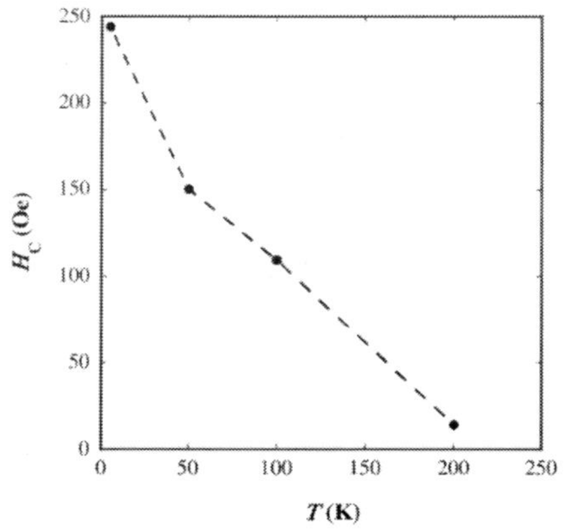

(a)

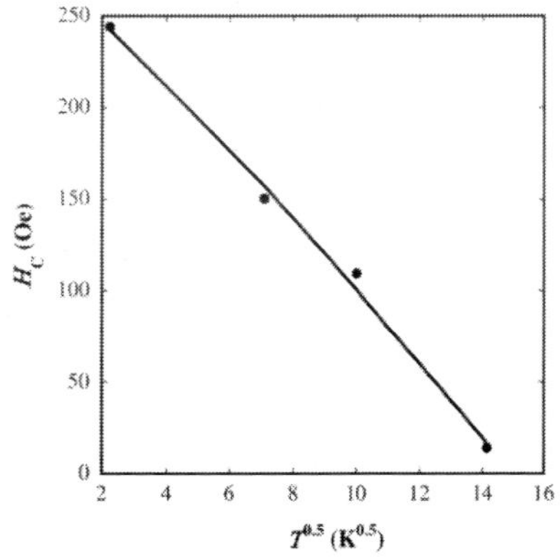

(b)

Figure 4. (a) Coercivity as function of temperature, $H_c(T)$ for the ferrite nanoparticles. The dashed line is a guide for the eye. (b) Coercivity as function of temperature $H_c(T)$ for the ferrite nanoparticles. The solid line represents the fitting to the modified Kneller's law.

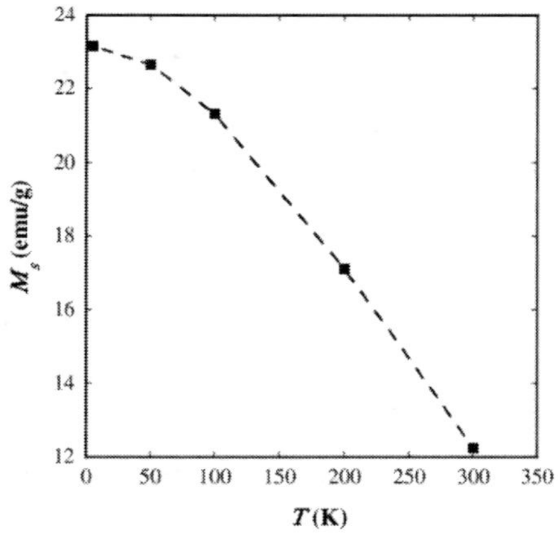

(a)

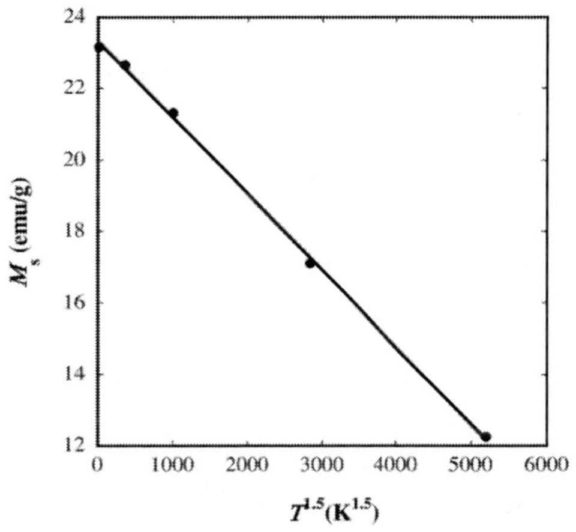

(b)

Figure 5. (a) Saturation magnetization as function of temperature, $M_S(T)$ for the ferrite nanoparticles. The dashed line is a guide for the eye. (b) Saturation magnetization as function of temperature, $M_S(T)$. The solid line represents the fitting to the modified Bloch's law.

We can see the excellent fitting of the data with the above relation with $\alpha = 1.514$. $M(0)$ was determined from the fit to be 23.3 emu/g and T_o is nearly 490 K. The dependence of the saturation magnetization, M_s of these ferrite particles on temperature is well described by Bloch's law over the whole temperature range.

We suggest that the excellent fitting of our data to Bloch's law and the good fitting to Kneller's and are due to the large volume of our nanoparticles where the bulk properties are nearly recovered. The slight deviation from Kneller's law could be attributed to several factors such as the size distribution of the particles, significant inter-particle interactions [41, 42] and the disordered surface spins which influence the surface anisotropy and hence the effective anisotropy in the particles. We believe that the smaller the particles are the larger is the influence of surface effects and the larger is the deviation from Kneller's and Bloch's laws. The size at which a crossover between the nanoparticle properties and bulk properties occurs is currently under investigation.

4. Conclusions

The coercivity, H_c as function of temperature was found to have good agreement with the classical Kneller's with small deviations. The temperature dependence of saturation magnetization, $M_s(T)$ was found to have excellent agreement with Bloch's laws. The good fitting of our data with these laws is suggested to be due to the large size of the particles where bulk properties might be significant. The small deviation from Kneller's law is suggested to be due the volume distribution, strong inter-particle interactions and enhancement of the effective anisotropy of these nanoparticles due to the disordered surface spins.

ACKNOWLEDGMENT

This work was financially supported by the UAEU Program for Advanced Research (UPAR) under the Grant no. 31S162.

References

[1] V. Skumryev, S. Stoyanov, Y. Zhang, G. Hadjipanayis, D. Givord, and J. Nogues, Nature **423**, 850 (2003).
[2] B. Issa, I. M. Obaidat, B. A. Albiss, and Y. Haik, Int. J. Mol. Sci. 14, 21266 (2013).
[3] I. M. Obaidat, B. Issa, and Y. Haik, Nanomaterials **5**, 63 (2015).
[4] R. H. Kodama, A.E. Berkowitz, E.J. McNiff Jr., S. Foner, J. Appl. Phys. **81**, 5552 (1997).
[5] R. H. Kodama, A.E. Berkowitz, Phys. Rev. B **59**, 6321 (1999).
[6] F. Gazeau, E. Dubois, M. Hennion, R. Perzynski, Yu.Raikher, Europhys. Lett. **40**, 575 (1997).
[7] M. Hennion, C. Bellouard, I. Mirebeau, J.L. Dormann, M.Nogues, Europhys. Lett. **25**, 43 (1994).
[8] C. Bellouard, I. Mirebeau, M. Hennion, Phys. Rev. B **53**, 5570 (1996).
[9] L. Del Bianco, A. Hernando, M. Multigner, C. Prados, J.C. Sanchez-Lopez, A. Fernandez, C.F. Conde, A. Conde, J. Appl. Phys. **84**, 2189 (1998).
[10] A. Millan, A. Urtizberea, N. J. O. silva, F. Palacio, V. S. Amaral, E. Snoeck, V. Serin, J. Magn. Magn. Mater. **312**, L5 (2007).
[11] A. K. Gaharwar, J. E. Wong, D. Müller-Schulte, D. Bahadur, and W. Richtering, J. Nanosci. Nanotechnol. **9**, 5355 (2009).
[12] P. V. Hendriksen, F. Bodker, S. Lenderoth, C. A. Oxborrow, S. Wells, and S. Morup, J. Phys: Condens. Matter **6**, 3081 (1994).
[13] D. Lin, A. C. Nunes, C. F. Majkrzak, A. E. Berkowitz, J. Magn. Magn. Mater. **145**, 343 (1995).
[14] S. Lenderoth, P. V. Hendriksen, F. Bodker, S. Wells, K. Davies, S. W. Charles, S. Mørup, J. Appl. Phys. **75**, 6583 (2004).
[15] Y. Xuan, Q. Li, and G. Yang, J. Magn. Magn. Mater. **312**, 464 (2007).
[16] J. Wang, C. Zeng, Z. Peng, and Q. Chen, Physica B **349**, 124 (2004).
[17] P. Mathur, A. Thakur, M. Singh, J. Magn. Magn. Mater. **320**, 1364 (2008).
[18] C. Caizer, M. Stefanescu, C. Muntean, I. Hrianca, J. Adv. Mater. **3**, 919 (2001).
[19] C. Caizer, Appl. Phys. A **80**, 1745 (2005).
[20] A. E. Berkowitz, R. H. Kodama, S. A. Makhlouf, F. T. Parker, F. E. Spada, E. J. McNiff Jr., S. Foner, J. Magn. Mater. **196–197**, 591 (1999).

[21] L. Zhang, G. C. Papaefthymiou, R. F. Ziolo, J. Y. Ying, Nano Structur. Mater. **9**, 185 (1997).
[22] B. Martinez, A. Roig, X. Obradors, E. Molins, P. Claret, C. Monty, J. Appl. Phys. **79**, 2580 (1996).
[23] C. Caizer, M. Stefanescu, J. Phys. D: Appl. Phys. **35**, 3035 (2002).
[24] K. Maaz, A. Mumtaz, S. K. Hasanain, M. F. Bertino, J. Magn. Magn. Mater. **322**, 2199 (2010).
[25] E. F. Kneller, F. E. Luborsky, J. Appl. Phys. **34**, 656 (1963).
[26] X. Batlle, M. Garcia del Muro, J. Tejada, H. Pfeiffer, P. Goand, E. Sinn, J. Appl. Phys. **74**, 3333 (1993).
[27] F. Bloch, Z. Phys. **61**, 206 (1930).
[28] E. Della Torre, L. H. Bennett and R. E. Watson, Phys. Rev. Lett. **94**, 147210 (2005).
[29] V. Senz, R. Röhlsberger, J. Bansmann, O. Leupold, K. –H. Meiwes-Broer, New J. Phys. **5**, 47 (2003).
[30] P. V. Hendriksen, S. Linderoth, P. A. Lindgard, J. Magn. Magn. Mater. **104–107**, 1577 (1992).
[31] P. V. Hendriksen, S. Linderoth, P. A. Lindgard, Phys. Rev. B **48**, 7259 (1993).
[32] S. Linderoth, L. Balcells, A. Labarta, J. Tejada, P. V. Hendriksen, S. A. Sethi, J. Magn. Magn. Mater. **124**, 269 (1993).
[33] A. S. Eggeman, A. K. Petford-Long, P. J. Dobson, J. Wiggins, T. Bromwich, R. Dunin-Borkowski and T. Kasam, J. Magn. Magn. Mater. **301,** 336 (2006).
[34] D. Ortega, E. Vélez-Fort, D. A. García, R. García, R. Litrán, C. Barrera-Solano, M. Ramírez-del-Solar and M. Domínguez, Phil. Trans. R. Soc. A **368**, 4407 (2010).
[35] Y. A. Koksharov, In "Magnetic Nanoparticles", S. P. Gubin (Ed.), Wiley-VCH, Berlin, 2009.
[36] C. R. Alves, R. Aquino, M. H. Sousa, H. R. Rechenbrg, G. F. Goya, F. A. Tourinho, J. Depeyrot, J. Met. Nanocrys. Mater. **20–21**, 694 (2004).
[37] K. Mandal, S. Mitra, P. A. Kumar, Europhys. Lett. **75**, 618 (2006).
[38] L. D. Tung, V. Kolesnichenko, D. Caruntu, N.H. Chou, C.T. O Connor, L. Spinu, J. Appl. Phys. **93**, 7486 (2003).
[39] S. Morup, Europhys. Lett. **77**, 27003 (2007).
[40] C. R. Alves, R. Aquino, J. Depeyrot, T. A. P Cotta, M. H. Sousa, F. A. Tourinho, H. R. Rechenberg, G. F. Goya, J. Appl. Phys. **99**, 08M905 (2006).
[41] S. P. Gubin, Yu. A. Koksharov, G. B. Khomutov, G. Yu. Yurkov, Russian Chemical Reviews **74**, 489 (2005).
[42] A. K. Giri, J. Appl. Phys. **81**, 1348 (1997).
[43] I. M. Obaidat, B. Issa, and Y. Haik, J. Nanosci. Nanotech. **11**, 3882 (2011).
[44] I. M. Obaidat, B. Issa, B. A. Albiss, G. Qabaja, N. E. Al Khaili, Z. A. Karam, A. M. Hefaity, S. Qadri, A. Al-Okour, and Y. Haik, J. Nanoeng. Nanomanuf. **2**, 325 (2012).

IOP Conf. Series: Materials Science and Engineering **92** (2015) 012013 doi:10.1088/1757-899X/92/1/012013

Resistance to moist conditions of whey protein isolate and pea starch biodegradable films and low density polyethylene non-degradable films: a comparative study

G F Mehyar[1] and A Al Bawab[2,3]
[1]Department of Nutrition and Technology, Faculty of Agriculture, The University of Jordan, Amman, Jordan
[2]Department of Chemistry, Faculty of Science, The University of Jordan, Amman, Jordan
[3]Hamdi Mango Center for Scientific Research (HMCSR), The University of Jordan, Amman, Jordan
E-mail: g.mehyar@ju.edu.jo

Abstract. Biodegradable packaging materials are degraded under the natural environmental conditions. Therefore using them could alleviate the problem of plastics accumulation in nature. For effective replacement of plastics, with biodegradable materials, biodegradable packages should keep their properties under the high relative humidity (RH) conditions. Therefore the objectives of the study were to develop biodegradable packaging material based on whey protein isolate (WPI) and pea starch (PS). To study their mechanical, oxygen barrier and solubility properties under different RHs compared with those of low density polyethylene (LDPE), the most used plastic in packaging. Films of WPI and PS were prepared separately and conditioned at different RH (30-90%) then their properties were studied. At low RHs ($\leq$50%), WPI films had 2-3 times lower elongation at break (E or stretchability) than PS and LDPE. Increasing RH to 90% significantly ($P<0.01$) increased the elongation of PS but not WPI and LDPE films. LDPE and WPI films kept significantly ($P<0.01$) higher tensile strength (TS) than PS films at high RH (90 %). Oxygen permeability (OP) of all films was very low (<0.5 cm^3 μm m^{-2} d^{-1} kPa^{-1}) below 40 % RH but increased for PS films and became significantly ($P<0.01$) different than that of LDPE and WPI at $\geq$ 40 % RH. Oxygen permeability of WPI and LDPE did not adversely affected by increasing RH to 65 %. Furthermore, WPI and LDPE films had lower degree of hydration at 50 % and 90 % RH and total soluble matter than PS films. These results suggest that WPI could be successfully replacing LDPE in packaging of moist products.

1. Introduction

The interest in biodegradable packaging such as edible films and coatings is increasing because they consist of natural materials, therefore control environmental pollution [1,2]. The problem of accumulation large amounts of synthetic plastics, in particularly LDPE, is increasing. Furthermore, the increase usage of LDPE in packaging accompanied with their slow degradation rate, leaded to ecological problems [3,4]. Biodegradable materials may be applied on a food or between heterogeneous food components to prevent mass transfer phenomena within foods that could deteriorate the quality [5]. For successive replacement of LDPE in moist products such as foods and cosmetics, biodegradable materials should retain their properties in moist conditions [1,5,6].

Biodegradable materials should also be tasteless, free of toxic substances, barrier to water, aroma, and gases in particularly oxygen; the most stimulating gas for spoilage reactions. They could also be used as a carrier or encapsulating material for functional ingredients such as antioxidants, antimicrobial agents, pigments or nutritional substances contributing to microbial and chemical stability [7,8,9].

Protein based biodegradable films and coatings such as those based on pea protein, soy protein and whey protein have received increased interest due to their high barrier properties, superior mechanical properties and nutritional quality. This allows them to either substitute the synthetic packaging materials or simplify the total packaging requirements [6,10]. Being amphophilic in nature, proteins in films and coatings showed enhanced functionality in food preservation when being mixed with variety of hydrophobic and hydrophilic additives such as antimicrobial and antioxidant compounds [11,12]. Strong peptide and covalent bonds in protein polypeptide structure could enhance superior mechanical properties of proteins. Among proteins, WPI could be the most appropriate for large scale production of biodegradable packages because it is a by-product of cheese industry, and readily available at reasonable price.

Starch is complex carbohydrate consists of repeated glucose molecules and represents the major constituent of legumes and cereals. It contains of about 25% amylose (regions of linear compact arrangement) and 75% amylopectin (regions of branched arrangement) [13,14]. Pea starch is attractive raw material for biodegradable packaging because of its low cost, renewability, biodegradability, high-amylose content (40%) and has low gelation temperature [14,15]. Films made of high-amylose starches, showed excellent oxygen barrier properties, lower water solubility and retrogradation temperature and more stable mechanical properties at high RHs compared to those made of normal starches [16,17,18]. High-amylose starch is a very useful film-forming material due to its strong gelation properties and helical linear polymer structure [14]. For appropriate application of these packaging materials in moist products, they should keep their properties when exposed to high humidity conditions (RH > 50%). Although biodegradable materials appeared useful alternative to the conventional synthetic packaging materials, the amount of studies on resistance WPI and PS to high RH comparing to LDEP is still at limited level nevertheless the potential of huge market [19]. The objective of the current study is to test mechanical, oxygen barrier and solubility properties of WPI and PS based biodegradable film material at different RH levels comparing that with LDPE. Therefore determine the possibility to replace LDPE by these biodegradable materials in particularly in moist products.

2. Methodology

2.1. Materials

High-amylose pea starch (PS; 40% amylose) and whey protein isolate (WPI) were supplied from Parrheim Foods Co. (Portage-la-Prairie, MB, Canada), and Davisco Foods Int., (Mn., USA.), respectively. Glycerol from Sigma Chemical Co. (St. Louis, MO) was used as a plasticizer. LDPE films were donated from Al-Hadaf Int. Co. RZ (Amman, Jordan).

2.2. Film preparation

Aqueous dispersions of 3% and 12 % (w/w) PS and WPI, respectively and separately were prepared then glycerol was added at the ratio of 1:2 (w/w) of glycerol:polymer. The dispersions were heated with continuous mixing until boiling and kept at the boiling temperature for 3 min to allow complete gelatinization of the polymers. The film-forming solutions were cooled down to ~ 60 °C then casted into poly styrene petri-dishes (10 cm dia) by pouring 12 g per dish resting on leveled smooth surface. The petri-dishes (six for each composition) were kept at room temperature for 24 hr to allow the film-forming solution to be dried before the films being peeled off. The thickness of the films was measured with a caliper micrometer (B.C. Ames Co., Waltham, MA) at 5 random positions of the film. The average thickness was used for the proceeding mechanical test and oxygen permeability determination. The films were transferred into large plastic bags (LDPE sandwich bags) and arranged side by side into one layer thickness. The filled bags were then gently pressed on the both sides of the films which allowed both sides of the films to stick onto the bag to minimize the access of air or moisture to the films. Then the bags were sealed and stored at room temperature maximum for one week, till the testing day.

2.3. Tensile test

Film specimens (1 cm × 8 cm) were cut and conditioned in a desicator for 3 days at 30, 50 or 90% RH over the super-saturated salt solutions of magnesium chloride, magnesium nitrate and potassium nitrate, respectively and separately at room temperature. Tensile strength (TS) and elongation-at-break (E) were determined from strain-stress curve using a texture analyzing instrument (MuliTest, 1-D, Mecmesin, West Sussex, UK) following the procedure outlined in ASTM method D882-91 (ASTM 1991). Initial grip distance and crosshead speed were 5 cm and 100 mm/min, respectively. Tensile strength was calculated by dividing the peak load by the cross-section area (thickness × 1 cm) of the initial specimen. E was expressed as the percentage of change in the length of the specimen to the original length between the grips (5 cm). TS and E measurements were done from three replications.

2.4. Oxygen permeability (OP)

The oxygen transmission rate (OTR) was measured using OX-TRAN 2/61 (Mocon, Minneapolis, MN). The measurements (in triplicates for each film) were done at 25 °C and the films were equilibrated into RH of 30, 40, 50 and 65% during the experiment using appropriate saturated salt solutions. The OP was calculated by the following equation:

$$OP = OTR \times Thickness/\ \Delta P_{O2}$$

The ΔP_{O2} is the partial pressure difference of oxygen between the two sides of the film, which was 1 atm during the experiment.

2.5. Soluble matter (SM) and degree of hydration (DH)

Film specimens (1.5 cm × 1.5 cm, in triplicates) were dried for 3 days in a desiccator over anhydrous calcium sulphate. The specimens were accurately weighed to obtain the initial dry weight (W_i). The specimens were immersed in test tubes containing 5 mL distilled water. The tubes were covered with aluminum foil and incubated at ambient temperature for 24 hr with occasional gentle agitation. The films then were dried in the oven at 70 °C for 24 hr and weighed to obtain the final dry weight (W_f). The percentage of soluble matter (% SM) of the films was calculated using the following equation:

$$\%\ SM\ of\ the\ films = [(W_i - W_f)/\ W_i] \times 100$$

The degree of hydration was calculated by conditioning the prepared films at 50% and 90% RH. The films were weighted before (W_b), after (W_a) conditioning and after drying (W_f) in an oven at 70 °C for 24hr.

The amount of water absorbed by the films was determined using the following equations:
Water absorbed by conditioning ($W_a - W_b$)
Total water content of conditioned films ($W_a - W_f$)
Proportion of water in the conditioned films ($W_a - W_f / W_a$)

3. Results and discussion

3.1. Films appearance

WPI and PS films were easily handled, and flexible (figure 1). WPI films were transparent and had mild yellowish color whereas PS films were opaque to translucent and have off-white color.

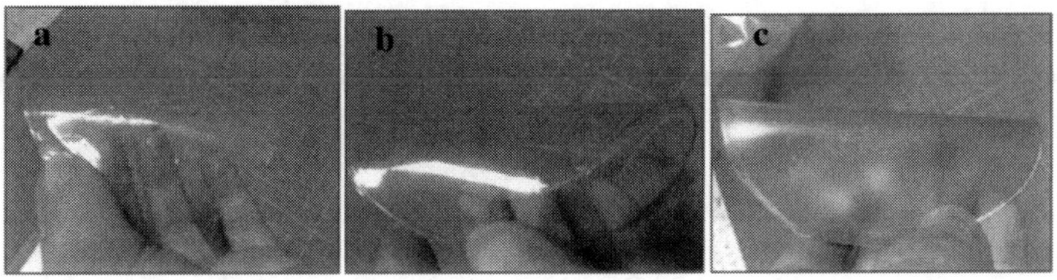

Figure 1. Photographic images of: (a) LPDE film, (b) prepared WPI film and (c) prepared PS films.

3.2. Mechanical properties

Figure 2 shows representative strain-stress curves for LDPE, WPI and PS films at different RHs. Tensile strength and E of LDPE were not affected as RH increases from 30% to 90% (Table 1). Their values are comparable to those reported by Salih et al., [18] under similar conditions. Amin et al., [21] reported that LDPE films at high RH conditions (75%) underwent dielectric relaxation and increased molecular movement that reduced TS and increased E of these films. In the current results, no such effect of high RH on LDPE films was noticed, which could be due to shorter conditioning period (3hr) applied on films compared to that (12hr) by Amin et al., [21]. This longer conditioning period may allow better plasticization effect of water on polymer molecules than the shorter period. Tensile strength of WPI significantly reduced by increasing RH to 90 % whereas E was not significantly ($P \geq 0.01$) affected (Figure 2 and Table 1). Furthermore, WPI films had higher TS and lower E than PS films at 90 % RH indicating that WPI films are more resistant to high RH conditions than PS films. Pea starch films were the most affected ones because TS significantly reduced and E significantly increased (Table1) at 90% RH. Films made of whey protein concentrate (WPC) was reported by Tulamandi and Rizvi [22] to have lower TS and higher E at high water activity (0.8-1) of films. Bertuzzi et al [17] found out that stress at break decreases and strain at break increased for high amylose corn starch films plasticized with glycerol (30% of starch weight) by increasing water activity of films from 0.1 to 0.9. The microstructure of these films changes from dense and homogeneous to a more open heterogeneous and more motile structure by increasing water activity [21].

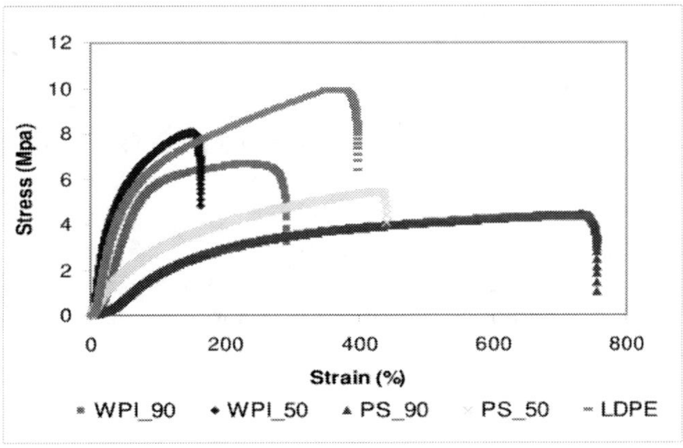

Figure 2. Representative strain-stress curves for LDPE, WPI and PS films at 50% and 90% RH.

Table 1. Tensile strength (TS) and elongation at break (E) of the starch films at different relative humidity

Film	RH (%)	TS (MPa)	E (%)
LDPE	30	9.903 ± 0.014^a	436.3 ± 1.2^b
	50	9.443 ± 0.224^a	421.0 ± 9.2^b
	90	9.645 ± 0.153^a	419.2 ± 2.6^b
WPI	30	8.561 ± 0.014^b	181.3 ± 1.2^c
	50	8.493 ± 0.224^b	182.9 ± 9.2^c
	90	6.932 ± 0.153^c	271.4 ± 10.6^c
PS	30	5.773 ± 0.061^c	491.3 ± 32.6^b
	50	5.261 ± 0.454^c	483.7 ± 12.0^b
	90	4.326 ± 0.370^d	765.6 ± 13.6^a

$^{a\text{-}c}$Means within the same column with different letters are significantly ($p<0.01$) different.

Although TS and E of PS films were the most affected by high RHs conditions compared to LDPE and WPI films (Figure 2 and table1). Our previous study [24] showed that PS films are more resistant to high RHs than rice starch (RS) films. This was related to the higher amylose content in PS than RS films; 40% and 30%, respectively. Amylose is the helical linear polymer of the films and responsible for keeping the tensile strength properties of high-amylose starch films. In another study [18], it was reported that stress at break of cassava-starch films increased with amylose content and decreased with glycerol and water contents. The high-amylose films were stiffer, more resistant to fracture and less stretchable than high amylopectin films and have more stable microstructure at higher RHs (70%) with strong molecular orientation, crystalinity and heat resistance [25,26].

3.3. Oxygen permeability (OP)

LDPE, WPI and PS films are excellent oxygen barriers (OP < 0.5 cm^3 μm m^{-2} d^{-1} KPa^{-1}) at RH below 40% (figure 3). This indicates good oxygen barriers properties for all films at low RH conditions. Increasing RH from 40% to 65% increases OP of all films in particularly in PS films; OP was 1.7 cm^3 μm m^{-2} d^{-1} KPa^{-1} at 65% RH. At intermediate humidity (~50% RH), LDPE and WPI films kept low OP ($\leq$0.6 cm^3 μm m^{-2} d^{-1} KPa^{-1}), whereas PS films lost their oxygen barrier properties (OP was 1.45 cm^3 μm m^{-2} d^{-1} KPa^{-1}) (figure 3). It was reported that LDPE films needs longer exposure period ($\geq$12h) to high RH to lose its barrier properties [21]. Oxygen permeability of WPI and PS films was significantly different ($p < 0.01$) from each other at high RH (50%-60%) whereas no significant differences in OP between WPI and LDPE films at the same RHs. Therefore WPI had oxygen barrier properties similar to LDPE than PS films at high RH condition. Forssell and others [27] found out that amylose films plasticized with glycerol have good oxygen barrier at RHs below 50% and glycerol content up to 30% but this barrier properties lost at higher RHs (50-90%). In another study [26], it was reported that increasing the surrounding RH from 50 to 90% resulted in increasing of oxygen permeability of both amylose and amylopectin films. Increasing the RH above 50% produced more loosen structure in more hydrophilic than hydrophobic polymers that increased mobility of oxygen molecules through these films [22,24]. WPI films kept their properties at high RHs could be due to their higher hydrophobicity than PS films [12].

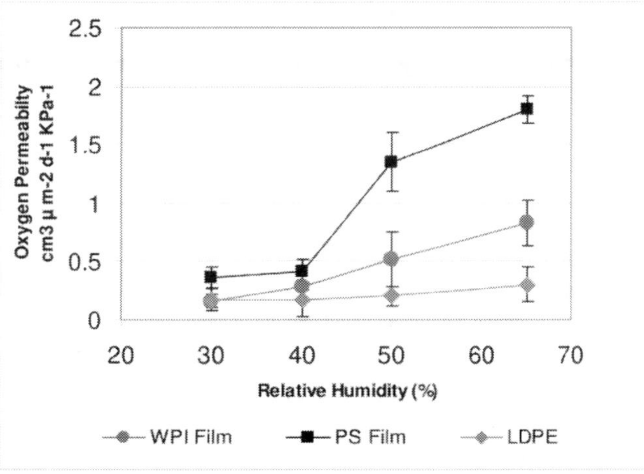

Figure 3. Oxygen permeability of WPI, PS and LDPE films at different relative humidity.

3.4. Degree of hydration (DH) and soluble matter (SM)

Table 2 shows degree of hydration of LDPE, WPI and PS films when conditioned at 50% and 90%. LDPE absorbed very low amounts of moisture; this could be due to the high hydrophobicity of this polymer or due to the short conditioning period [21]. At any of the tested RHs, the amount of moisture absorbed by WPI is lower than that by PS films. This is expected as WPI is more hydrophobic therefore could have lower capability to bind water [12]. The higher degree of hydration in PS films is likely to be responsible for inferior mechanical and barrier properties of this film at high RH (≥50%) conditions than the less hydrated films (e.g. LDPE and WPI). WPI and PS films have similar soluble matter (SM) of 32% and 34.3%, respectively (Table 3). This may be caused by the incorporated soluble component that is glycerol. The glycerol:polymer ratio in these films is 1:2. Therefore, total loss of glycerol from the films may result in 33% of solubility that is very close to glycerol content. Although PS has higher hydrophilicity than WPI, both polymers form gelatinized network via electrostatic interactions in WPI and starch granular swelling in PS by heating [28]. These more integrated microstructure could be more resistance to dissolution by water, therefore no statistical difference in SM% between these two films was detected.

Table 2. Degree of dehydration (DH) as affected by RH and total soluble matter (SM) of LDPE, WPI and PS films.

	Weight of film (g) at RH					
	LDPE		WPI		PS	
	90 %	50 %	90 %	50 %	90 %	50 %
Weight before conditioning (W_b)	1.290	1.330	1.160	1.190	0.935	1.048
Weight after conditioning (W_a)	1.295	1.331	1.580	1.110	1.760	1.250
weight after drying(W_f)	1.281	1.329	0.860	0.920	0.917	0.916
Water absorbed by conditioning ($W_a - W_b$)	0.005	0.001	0.42	0.08	0.725	0.202
Total water content of conditioned films ($W_a - W_f$)	0.014	0.002	0.52	0.19	0.943	0.334
DH ($W_a - W_f / W_a$)	0.011	0.004	0.38	0.17	0.44	0.267
SM (%)	0.001		32.0		34.3	

5. Conclusion

LDPE films showed no detectable differences in mechanical properties, oxygen barrier and degree of hydration at different RH levels. Tensile strength of WPI films was reduced by increasing RH to 90 %, but it was significantly higher than that of PS films. WPI and LDPE films kept reasonable oxygen barrier up to 65% RH whereas PS film quickly lost its oxygen barrier. LDPE and WPI films had lower degree of hydration and soluble matter at all the tested RHs than PS films. The more hydrophilic polymer (PS) is likely to lose its mechanical and barrier properties by hydration at higher RH than less the hydrophilic polymers (LDPE and WPI). WPI could better replace LDPE in packaging of moist products such as foods and cosmetics.

6. References

[1] Siracusa V, Rocculi P, Romani S and Dalla Rosa 2008 *Tends Food Sci Technol* **19** 634-43
[2] Brody A L 2008 *Food Technol* **62** 121-126
[3] Brody A L 2009 *Food Technol* **63** 103-104
[4] Marsh K and Bugusu B 2007 *J Food Sci* **72** R39-R55
[5] Bourtoom T 2009 *Int Food Res* J **15** 237-48
[6] Olivas G I, Barbosa-Cánovas G V 2005 *Crit Rev Food Sci Nutr* **45** 657-70
[7] Kulkarni V K, Butte K D and Rathod S S 2012 *Int J Res Pharm Biomed Sci* **3** 1597-613
[8] Mehyar G F, Al-Ismail K, Han J H, Chee G W 2012 J Food Sci **77** E52-E59
[9] Mehyar G F Al-Ismail K M, Al-Ghizzawi H M, Holley R A 2014 J Food Sci **79** C1939-49
[10] Choi W S, Han J H 2001 *J Food Sci* **66** 319-22
[11] Choi W S, Han J H 2002 *J Food Sci* **67** 1399-406
[12] Embuscado M E, Huber K C 2009 *Edible Films and Coatings for Food Application* vol 1 (NY: Springer Science +Business Media LLC) p25-133
[13] Aggarwal V, Singh N, Kamboj S S and Brar P S 2004 *Food Chem* **85** 585-90
[14] Ratnayake W S, Hoover R, Warkentin 2002 *Starch* **54** 217-34
[15] Polesi L F, Sarmento S B S, Anjos C B P 2011 *Braz J Food Technol* **14** 74-81
[16] Stading M, Rindlav-Westling A, Gatenholm P 2001 *Carbohydr Polym* **45** 209-17
[17] Bertuzzi M A, Gottifredi J C Armada M 2012 *Braz J Food Technol* **15** 219-27
[18] Alves V D, Mali S, Beléia A, Grossmann M V E 2007 *J Food Eng* **78** 941-46
[19] Janjarasskul T and Krochta J M 2010 *Annu Rev Food Sci Technol* **1** 415-48
[20] Salih S, Hammod A F and Abd Alsalam A H 2013 *Mod Appl Sci* **7** 33-44
[21] Amin A, Dantuluri A K R, Basal A K 2012 *Int J Pharm* **422** 302-9
[22] Tulamandi S and Rizvi S 2015 *Schol Res J* **1** 39-47
[23] Ferreira F A B, Grossmann M V E, Mali S, Yamashita F and Cardoso L P 2009 *Braz Arch Biol Technol* **52** 1505-12
[24] Mehyar G F and Han J H 2004 *J Food Sci* **69** E449-54
[25] Cano A, Jiménez A, Cháfer M, Gónzalez C, Chiralt A. 2014 *Carbohydr Polym* **13** 543-55
[26] Romero-Batida C A, Bello-Pérez L A, Velazquez G, Alvarez-Ramirez J 2015 *Carbohydr Polym* **127** 195-201
[27] Forssell P, Lahtinen R, Lahelin M, Myllarinen P 2002 *Carbohydr Polym* **47** 125-129
[28] Wang H, Sun C and Huang L 2010 *World Conference on Packaging* **17th** IAPRI 259

Raman imaging to study structural and chemical features of the dentin enamel junction

M. Anwar Alebrahim [1,2], C. Krafft [1], and J. Popp [1,3]

[1] Leibniz Institute of Photonic Technology, Jena, Germany
[2] Jordan University of Science and Technology, Irbid, Jordan
[3] Institute of Physical Chemistry and Abbe Center of Photonics, University Jena, Jena, Germany.

E-mail: nmarneh@yahoo.com

Abstract. The structure and chemical features of the human dentin enamel junction (DEJ) were characterized using Raman spectroscopic imaging. Slices were prepared from 10 German, and 10 Turkish teeth. Raman images were collected at 785 nm excitation and the average Raman spectra were calculated for analysis. Univariate and multivariate spectral analysis were applied for investigation. Raman images were obtained based on the intensity ratios of CH at 1450 cm^{-1} (matrix) to phosphate at 960 cm^{-1} (mineral), and carbonate to phosphate (1070/960) ratios. Different algorithms (HCA, K-means cluster and VCA) also used to study the DEJ. The obtained results showed that the width of DEJ is about 5 μm related to univariate method while it varies from 6 to 12 μm based on multivariate spectral technique. Both spectral analyses showed increasing in carbonate content inside the DEJ compared to the dentin, and the amide I (collagen) peak in dentin spectra is higher than DEJ spectra peak.

1. Introduction

Enamel, dentin and cementum are the major tissues which make up the tooth in humans. Enamel is the hardest and most highly mineralized substance in the human body with a mineral content of 96%. The mineral content of dentin is reduced to approximately 70% whereas the organic content is increased to 20%. Enamel's and dentin's most abundant mineral is hydroxyapatite, which is a crystalline calcium phosphate. The dentin-enamel junction, abbreviated DEJ, is the interface between the enamel and the underlying dentin. DEJ is known for its unique biomechanical properties that provide a crack-arrest barrier for flaws formed in the brittle enamel [1]. The DEJ structure is generally described as a series of 25–100 μm diameter scallops with their convexities directed toward the dentin [2, 3]. The DEJ is also investigated by parallel 80–120 nm diameter collagen fibrils that directly insert into the enamel mineral and also merge with the interwoven fibrillar network of the dentin collagen matrix [4]. It is thought that the DEJ is greatly influenced by enamel knots that form before the

mineralization of the tissue secret [5]. Lucas (2004) describes that the first enamel formed, at the DEJ, has not been well studied, and appears to have a random orientation under light microscope [6]. The structure and chemical composition of the human dentin-enamel junction (DEJ) was studied using confocal Raman microscopy and the authors reported that the Phosphate to organic ratio showed an accumulation of organic material under the enamel surface and high phosphate content compared to enamel in the vicinity of the DEJ [7]. During recent years, vibrational spectroscopic imaging has matured into an important tool for the study of musculoskeletal tissues. Both IR and Raman spectroscopies contain signatures for the stages of development, for tissue disease and damage, and even for mechanical properties. Raman spectroscopy is non-destructive and label-free vibrational spectroscopic technique that enables structural and chemical analysis of mineralized tissue such as bone and teeth. Two dimensional Raman mapping/imaging of the specimen enables us to obtain spectra at points on the specimen separated by 1 μm. Thus the data consists of three dimensions – the x and y coordinates at every pixel and the intensity value at each Raman shift of the spectrum at that pixel. The data thus obtained is known as hyperspectral data, in view of its large size in each dimension. Therefore, analysis of the hyperspectral data is very important to give a spatial distribution of the chemical makeup of the specimen. However, analysts apply different techniques such as univariate and multivariate analysis for well understand the data. Univariate methods consider one wave number, of the Raman shift, presenting results about only characteristic functional group based on a particular wave number not on all pixels as a whole. Multivariate analysis methods help to analyze the hyperspectral data by manipulating each spectrum as a whole, rather than corresponding individual peaks in each spectrum [8].
The aim of this computing was to apply univariate and multivariate analysis to characterize the structural and chemical properties of dentin enamel junction by using Raman imaging.

2. Materials and methods

2.1. Specimen preparation
Six primary and four permanent non-carious, human teeth were provided from the Department of Conservative Dentistry, University Hospital Jena, Germany. Six primary and four permanent non-carious, human teeth were collected under informal consent from children and adults in Hacettepe University, Ankara, Turkey. Soft tissue remnants of contemporary samples were quickly removed from the root surface. The samples were stored in an aqueous solution of 4% formalin to inhibit the microbial growth. German specimens were embedded in Kallocryl-CPGH red and Turkish specimens in autopolymerizing polymethylmethacrylate (PMMA). A stainless steel saw with galvanic diamond coating, grain size 120 micron and 4 cm diameter (Komet, Germany), was used to prepare 1–2 mm thick slices. The specimens were manually polished before the measurements were performed.

2.2. Raman spectroscopic imaging
A single mode diode laser (model Xtra, Toptica, Germany) with 785 nm emission was coupled to a microscope and the microscope was coupled to the Raman spectrometer (RXN1 microprobe, Kaiser Optical System, USA). The laser light with an intensity of 100 mW was focused on the sample with a 100x/NA 0.9 objective (Nikon, Japan). The Raman signal was detected on a Peltier-cooled (−60°C), back-illuminated, deep-depletion CCD chip (Andor, Ireland). Images were obtained over the spectral region of 200 to 3450 cm^{-1} at spectral resolution of 4 cm^{-1} with a step size of 1 μm and 6 s exposure time plus 3 s dwell time per Raman spectrum. The Holograms software (Kaiser) automatically performed corrections of dark counts, cosmic spikes, and calibrations of the wavenumber and intensity axis.

3. Result and discussion
Figure1A represents the tooth slice illustrating where spectral maps were obtained. Regions of interest were selected from normal appearing sites. Figure1B shows Raman spectra (800 - 1800 cm^{-1}) of

human dentin and enamel. The difference spectrum of dentin minus enamel was calculated to better visualize the organic bands. The bands associated with inorganic component (mineral) appeared in the spectral range from 900 to 1100 cm^{-1}, including ν_1 and ν_3 vibration bands. The organic components (collagen) appeared in the spectral region of 1200 to 1720 cm^{-1}. Moreover, the most intense peak on Raman spectra is the symmetric stretch vibration of PO_4^{2-} at 962 cm^{-1}. The Raman assignments are included in table 1.

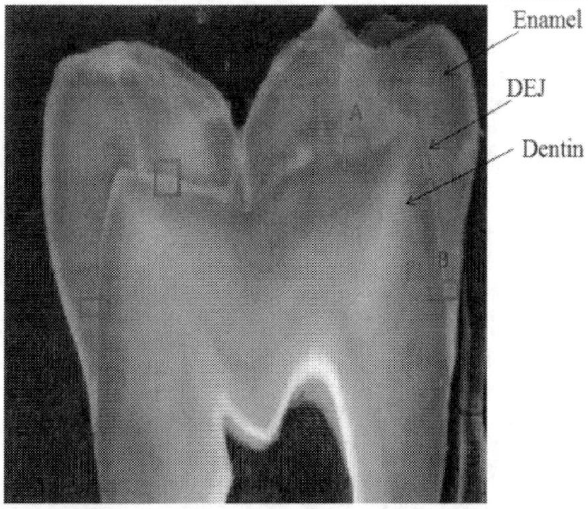

Figure1A: Tooth slice illustrating where spectral maps were obtained. Where A is an occlusal position and B is the cervical position.

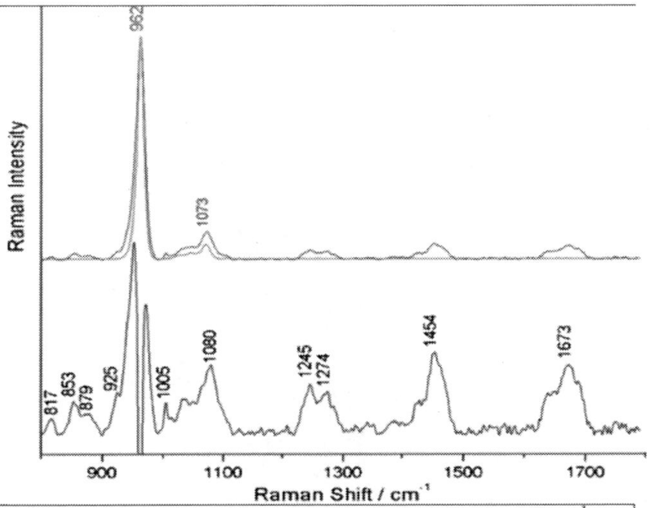

Figure 1 B: Raman spectra from 800 to 1800 cm^{-1} of permanent dentin (blue) and enamel (red). Difference spectra (dentin–enamel) visualize spectral contributions of organic matrix and collagen (black).

Table 1: Band assignment in Raman spectra of enamel and dentin [9]

Raman band	Assignment
817, 853, 920	Collagen
960	ν_1 of PO_4^{3-}
1005	Collagen (Phe)
1070	ν_1 of CO_3^{2-}
1248, 1273	Collagen (amide III)
1450	Collagen (CH_2)
1670	Collagen (amide I)

3.1. Univariate spectral analysis

Representative Raman mapping results of the DEJ at occlusal and cervical sites of a human tooth are shown in figure 2 A and B, respectively. Raman images were obtained based on the intensity ratios of CH at 1450 cm^{-1} (matrix) to phosphate at 960 cm^{-1} (mineral). The relative matrix/mineral (1450/960) ratios are represented by the color differences. Red/pink color represents the lower relative ratio (Enamel), while green/yellow/orange represents the higher relative ratio (dentin). A region where the color changes from blue is related to the DEJ transition zone. Comparison of the two images indicates that the width of DEJ transition zone at cervical position is the same of at occlusal position, and it equals to 4 μm based on the univariate method spectral analysis.

The representative relationships of the carbonate to phosphate (1070/960) ratios as a function of positions across the DEJ transition zone are shown in figure 3A and B. The matrix/mineral ratios of the DEJ were higher than those of enamel and lower than those of dentin. These ratios of the DEJ gradually increased during the transition from the enamel to dentin zone. The width of transition zone equals to 5 μm at both positions (occlusal & cervical). On the other hand, average Raman spectra from the dentin and DEJ area in the protein- derived spectral region of 1200- 1750 cm^{-1} are shown in figure 4A, in which the vibration peak of CH at 1450 cm^{-1} was selected as internal standard for the normalization adjustment. As compared to dentin spectra, the amide I (collagen) peak increased, but the amide III peak decreased in the DEJ spectra. Also the average spectra from the dentin and DEJ zone in the spectral region of 900- 1200 cm^{-1} are shown in figure 4B. The spectra were normalized based on phosphate peak at 960 cm^{-1}. The spectra showed that the peak at 1070 cm^{-1} which is related to the carbonate type-B from the DEJ spectra is higher than the carbonate peak from the dentin average spectra.

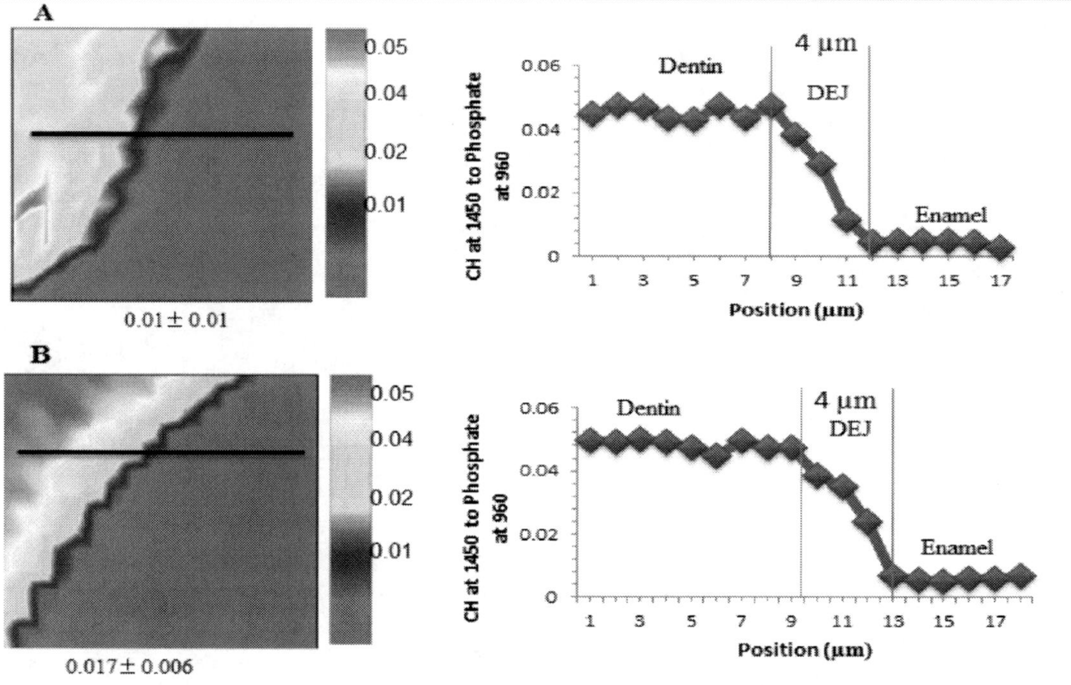

Figure 2: Representative Raman images and relative intensity ratios of the DEJ based on intensity ratios of 1450/960 at occlusal (A) and cervical (B) positions for permanent teeth. The bold black line represents the position where the DEJ determined.

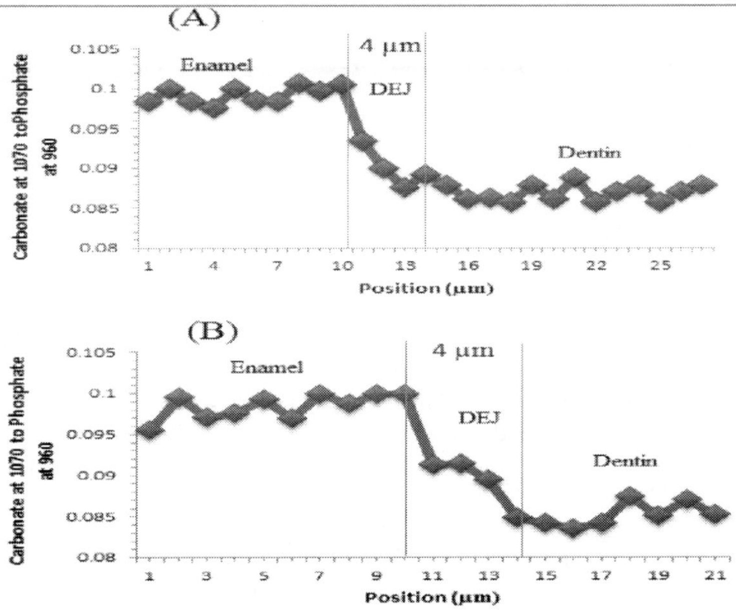

Figure 3: relative intensity ratios of 1070/960 across the enamel/DEJ/dentin zone at occlusal (A) positions and at cervical (B) positions.

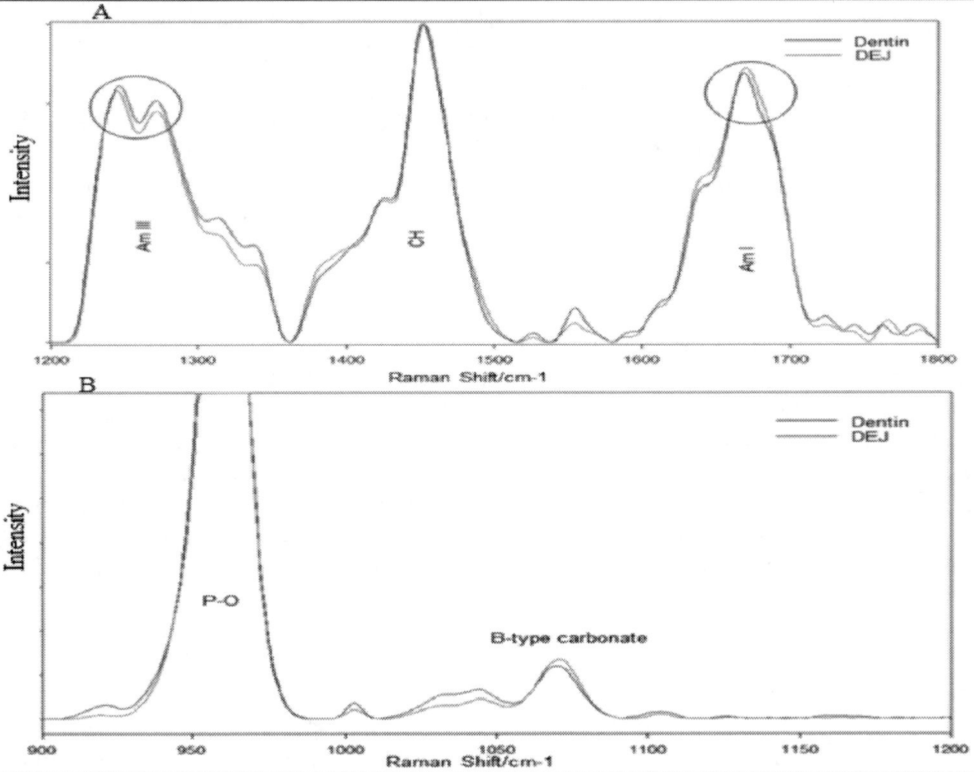

Figure 4: Raman spectra in protein-derived (A) and in mineral-derived (B) spectral regions from dentin and the DEJ.

3.2. Multivariate spectral analysis

Raman spectra collected are often contaminated with various interfering signals. The sources of interference include fluorescence and stretching of spectral intensity amplitudes [8]. The amplitude stretching (intensity variations) could be due to specimen inclination, variation in specimen thickness and absolute intensity of light. For instance, mineral and protein contents of dentin show variations compared to the enamel. In order to eliminate interfering signals and bring the spectra to a common platform of comparison, some preprocessing were performed. Data was carried out by using the Cytospec software and MatLab procedures written in-lab. A baseline correction by polynomial fitting of a second degree was used to remove of the florescence background with no smoothing.

3.2.1. K-Means Cluster & hierarchical cluster analysis result

Cluster analysis is an unsupervised statistical method of grouping the data, without a priori knowledge of its nature. The initial centroids of K-means clustering are randomly chosen. The distances are calculated between a spectrum and the centroid. Then each spectrum is assigned to the cluster whose centroid is nearest and the data set is grouped into a preselected number (K) of clusters. Figure 5 A, B, C & D shows the HCA & k-means cluster analysis maps of 70 μm X 70 μm of Raman image A& B and Raman averaged spectra of protein and mineral region C & D. HCA and K-means cluster analysis showed identical result to distinguish the transition zone of DEJ across the enamel and dentin. The cluster analyses do not give a fixed width for the DEJ. The width was varied from 6 μm in narrower width to the wider width of 12 μm. In HCA we were used three clusters (enamel, DEJ, and dentin). While in K-means analysis, five clusters were used (figure 5 B). The first two clusters were related to the enamel and DEJ and the rest were attributed to dentin region. Dentin contains around 75% in mineral (HAP) and 25% proteins and water. On the hand the enamel contains 99% in mineral considering a homogeneous layer. Based on that different content the K-means cluster analysis carried out by using more than three clusters to give agreement result with the HCA cluster analysis. On the other hand the molecular features of DEJ in comparison to the dentin and enamel were the same results from HCA and K- means cluster analysis. In protein spectral Raman region (figure 5 C) after normalization based on CH band, the amide I and amide III peaks in dentin were higher than DEJ spectra peaks. The mineral Raman spectral region (figure 5 D) normalized to phosphate peak at 960 cm^{-1}. Moreover the intensity of carbonate at 1070 cm^{-1} increased from dentin to enamel across the DEJ area indicating the lower carbonate contents in dentin.

3.2.2. Vertex component analysis (VCA)

VCA offers an unsupervised solution to the blind source separation problem. VCA represents the image raw data in a space of smaller dimensionality aiming to retain all relevant information and it calculates endmembers instead of loadings. Scope of VCA is that endmembers represent spectra of most dissimilar chemical constituents. Then, scores denote the concentration of the endmember spectra. The VCA algorithm iteratively projects data onto a direction orthogonal to the subspace spanned by the endmembers already determined. Figure 6 shows the Raman image of vertex component analysis, the composite image, dentin in red, DEJ in blue and the enamel in green (A) and three components with its related spectra (B). It showed that the DEJ width varies from 12 microns in some positions to 6 microns on another positions. On the hand, the organic components in DEJ were not clearly observable compared to the dentin, while the mineral in DEJ clearly appeared to be higher than dentin in comparison to the intensity of the phosphate band at 960 cm^{-1} as shown in figure 6 B. Usually the interface between two materials with different mechanical properties like elasticity is a weak link in structure. The DEJ zone acts to transfer applied loads from the enamel to dentin like during the masticatory or impact and also to inhibit the crack propagation. However, many different instruments have been used to identify the DEJ, such as SEM, Raman spectroscopy and optical spectroscopy. The researchers were reported different width to the transition area, for example it was reported that the DEJ width varies from 50 to 100 μm [4], 10 μm [10] and around 2-3 μm [11]. DEJ width has been investigated later to be dependent on the intratooth location and the authors reported that the mean width is 12.9 ± 3.2 μm at occlusal position and 6.3 ± 1.3 μm at cervical position. To date, there has been minimal chemical information related to the DEJ transition zone due to its small size [12]. In this study, it is the first time that we used the multivariate analysis in detection of transition zone. The obtained results from different algorithms (HCA, K-means cluster and VCA) were indicated that the width of transition zone varies as a function of position in microscale. However, the chemical structure features based on multivariate analysis suggested that the amide I and amide III peaks decreased in DEJ in comparison to the dentin Raman peaks. The decrease in these peaks could be related to the increasing disorder of the functional groups in organic matrix [13]. The spectral differences indicate that the structure of organic matrix in the dentin and DEJ was diverse. Formation of the DEJ begins at the early stages of tooth morphogenesis and is thought to be linked to a mixture of dentin proteins secreted by odontoblasts and enamel proteins from ameloblasts [13-15]. The proteins associated with DEJ formation and collagen fibril bundles that cross the transition zone and insert into enamel [4] might be responsible for the variations and distribution deviations in composition and structure of the organic matrix within the DEJ. The width and shape of phosphate peak at 960 cm^{-1} of dentin and DEJ spectra were similar and no significant change. This indicating that the mineral crystallinity within the DEJ might be not much different compared to the mineral of dentin, (figure 6.B).

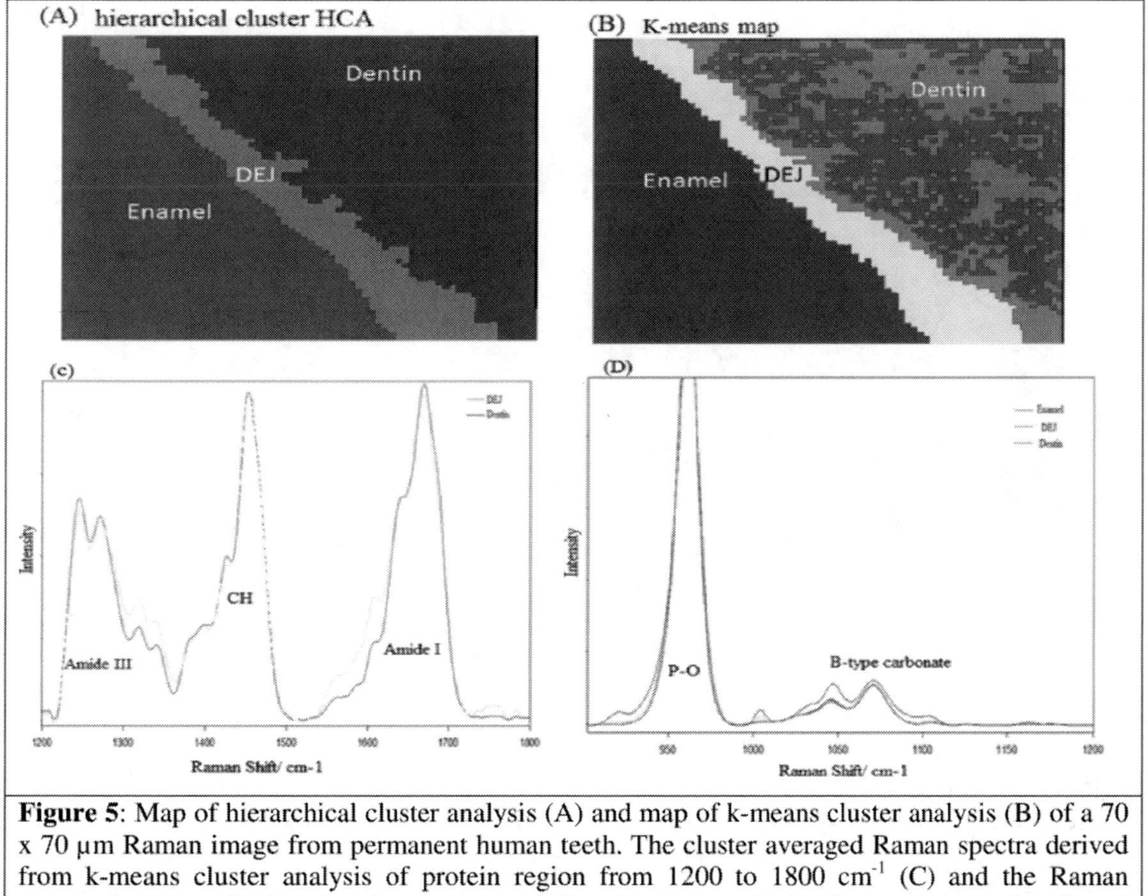

Figure 5: Map of hierarchical cluster analysis (A) and map of k-means cluster analysis (B) of a 70 x 70 μm Raman image from permanent human teeth. The cluster averaged Raman spectra derived from k-means cluster analysis of protein region from 1200 to 1800 cm^{-1} (C) and the Raman averaged spectra of mineral region constructed from k-means analysis of mineral spectral region of 900 – 1200 cm^{-1} (D).

IOP Conf. Series: Materials Science and Engineering **92** (2015) 012014 doi:10.1088/1757-899X/92/1/012014

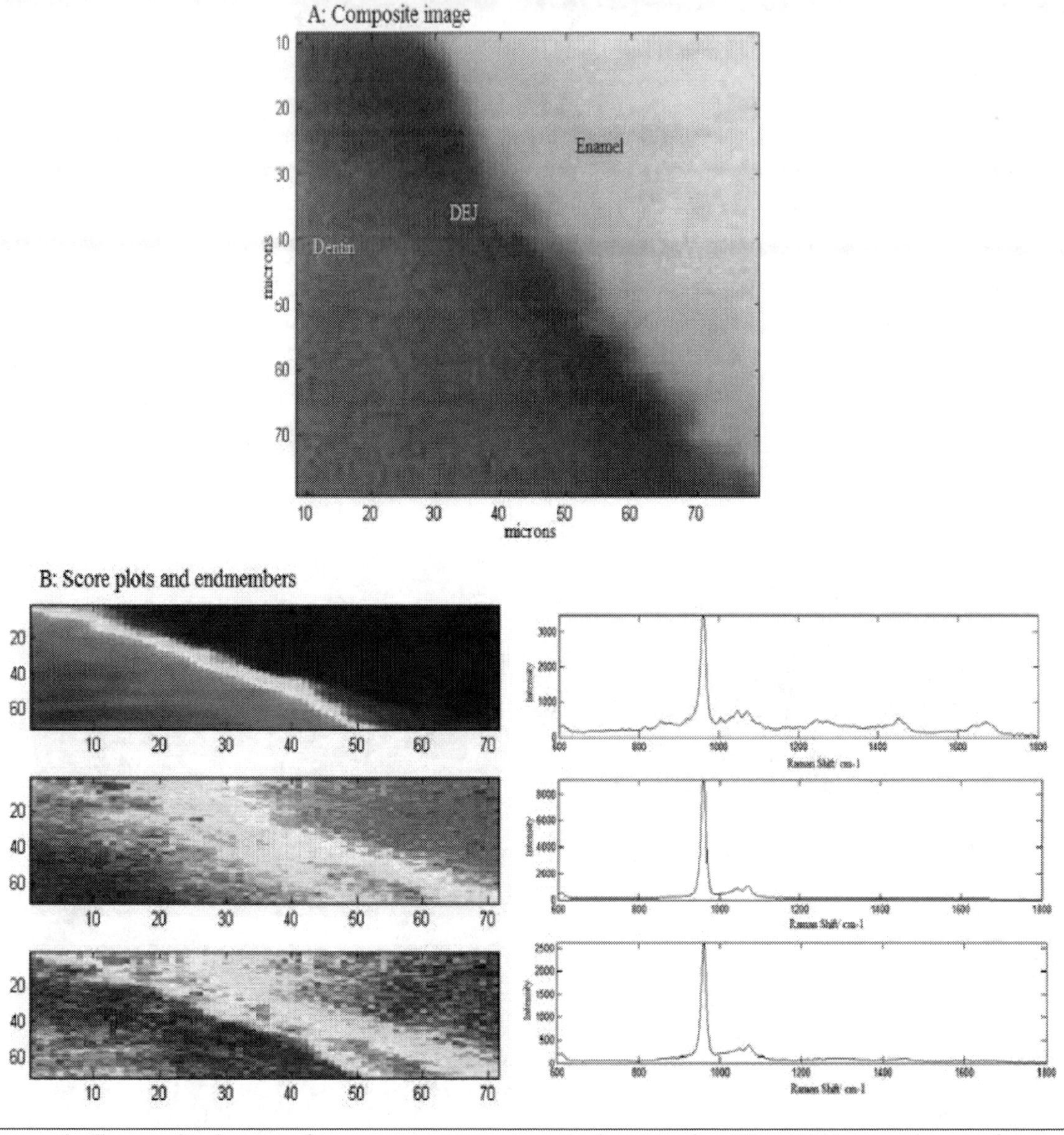

Figure 6: Composite image of vertex component analysis (A) of a 70 x 70 Raman image from a permanent human tooth, of the first three components (B) are shown: dentin, enamel and DEJ, respectively.

4) Conclusion

Limited work has been done to study the DEJ by Raman imaging, the data were analyzed by applying univariate and multivariate methods analysis. Univariate spectral analysis might be not suitable spectral technique to characterize the DEJ, especially for the DEJ width which was not more than 5 µm in comparison to the multivariate spectral analysis. . The obtained results from different algorithms (HCA, K-means cluster and VCA) confirm each other, were indicated that the width of transition zone (DEJ) varies as a function of position in microscale. The molecular features derived from univariate and multivariate spectral analysis showed increasing in carbonate content inside the DEJ compared to the dentin. VCA showed similarity of the width and shape of phosphate peak at 960 cm^{-1} of dentin and DEJ spectra, indicating the mineral crystallinity within the DEJ is not much different compared of the mineral crystallinity of dentin. Both methods showed that the amide I peak in dentin was higher than DEJ spectra peak. These results probably give more information beside the published scientific articles in the literature for understanding of this critical zone.

References

[1] Imbeni V, Kruzic J J, Marshall G W, Marshall S J, and Ritchie R O 2005. *Nature materials* **4**, 229 - 232

[2] Avery J K, SteeleP F, editors 2002 Oral Development and History. New York p. 74

[3] Ten Gate, AR. 1994 Oral Histology. Mosby; St Louis

[4] Lin C P, Douglas W H, Erlandsen S L 1993. *Journal of Histochemistry and Cytochemistry* **41**:381. [PubMed: 8429200]

[5] Tafforeau P and Smith T M *Journal of Human Evolution*. **54**, P. 272- 278

[6] Lucase P W 2004. *Cambridge University Press*, Cambridge, UK

[7] Desoutter A, Salehi H, Slimani A, Marquet P, Jacquot B, Tassery H, Cuisinier F J G 2014 *Lasers in Dentistry* SPIE 8929

[8] Ranganathan Parthasarathy, Ganesh Thiagarajan1, Xiaomei Yao, Yu-Ping Wang, Paulette Spencer, and Yong Wang 2008. *J.Biomed Opt* **13** (1)

[9] M. Anwar Alebrahim, C. krafft, W. Sechaneh, B. Sibusch, and J. Popp 2014. Biomedical Spectroscopy and Imaging. **3** P. 15-27

[10] Marshall Jr G W, Balooch M, Gallagher R R, Gansky S A, Marshall S J 2001. *J. biomed Mater Res* **54**, P. 87–95.

[11] Balooch G, Marshall G W, Marshall S J, Warren O L, Asif S A S, Balooch M 2004. *J Biomech.* **37** P. 1223-32

[12] Xu C, Yao X, Walker M P, and Wang Y 2009. *Calcif Tissue Int.* **84** (3) P. 221- 228

[13] Nagano J T, Oida S, Ando H, Gomi K, Arai T, and Fukae M 2003 *J Dent Res.* **82**, P.982–986.

[14] Begue-Kir C, Krebsbach P H, Bartlett J D, Butler W T 1998 *Eur J Oral Sci.* **106**, P. 963–970.

[15] Nanci A, Zalzal S, Lavoie P, Kunikata M, Chen W, Krebsbach P H, et al 1998 *J Histochem Cytochem.* **46,** P.911–934.

IOP Conf. Series: Materials Science and Engineering **92** (2015) 012015 doi:10.1088/1757-899X/92/1/012015

Exergy efficiency analysis of a flat plate solar collector using graphene based nanofluid

Z.Said[1], M.A. Alim[2] and Isam Janajreh[1*]

[1]Department of Mechanical and Materials Engineering (MME), Masdar Institute of Sicence and Technology, 54224 Abu Dhabi, United Arab Emirates
[2]Solar Energy Technologies, School of Computing, Engineering and Mathematics, University of Western Sydney, Penrith, NSW 2751, Australia

*e-mail: ijanajreh@masdar.ac.ae

Abstract: The thermal efficiency of a flat plate solar thermal collector is largely affected by the thermal conductivity of the fluid used. In this paper, we theoretically analyzed the heat transfer performance, the entropy generation rate, and the exergy efficiency of the two different graphene based nanofluids (graphene/Acetone and graphene/water). From the analyses, it is revealed that by inserting a small amount of graphene nanoparticles in water, exergy efficiency could be enhanced by 21%, comparing to conventional fluids and entropy generation is decreased by 4 %. However, the graphene/water nanofluid shows a lower entropy generation. This characteristic suggests that graphene/water nanofluid is a better candidate for flat solar thermal application.

1. Introduction

The solar radiation intensity in Malaysia varies due to high humidity and unpredictable weather, especially during the period of monsoon. Othman et al. [1] observed that the immediate solar radiation intensity or insolation can reach a maximum value of 1400 W/m^2 in Malaysia. Various parts of Malaysia have a short duration of sunshine hours. Therefore, an efficient solar collector system for Malaysian conditions should be considered to absorb the maximum heat at minimum convective losses. Solar insolation differs throughout the day, a minimum in the morning, reaching its peak at 2 p.m. and then reducing after that.

Remarkable consideration has been given to solar energy in the recent years. The use of fossil fuels will be limited due to their shortage in reserve as well as for environmental concerns. As a result, the circumstances are inspiring researchers to find other potential sources of energy. Heat transfer improvement in solar thermal collectors is one of the key issues of energy saving and compact design. Solar thermal energy is broadly used for numerous applications, such as electricity generation, chemical processing, and thermal heating, due to its renewable and nonpolluting nature [2]. Reports on using nanofluids as heat transfer fluids in solar collectors while continue to emerge are limited in number up to date [3-11]. Aravind et al. [11] investigated graphene and graphene–multiwalled carbon nanotube (MWNT) based nanofluids. They reported an enhancement in thermal conductivity of 9.2% and 10.5% for graphene and graphene–MWNT nanofluids in de-ionized water at room temperature for 0.04% volume fraction. Further, an enhancement in heat-transfer coefficient of 193% at Reynolds number 2000 for 0.02% volume fraction of aqueous graphene–MWNT nanofluids suggested the potential application of the present hybrid material-based nanofluids in cooling circuits. Eswaraiah et al. [12] investigated the frictional characteristics (FC), antiwear (AW), and extreme pressure (EP) properties of graphene based engine oil nanofluids. They explained that the improvement in FC, AW, and EP properties of nanofluids was respectively by 80, 33, and 40% compared with base oil.

The present objective of this paper is to numerically investigate the efficiency of flat plate solar collector using graphene based nanofluids as the working fluid compared to the base fluid.

2. Scope and Theoretical Approach

Energy and exergy analysis were carried out to evaluate the system efficiency of the solar collector with graphene based nanofluids. Hence, the purposes of this article are to (i) perform the elaborated exergy, entropy generation analysis, the exergy destruction and the pressure drop analyses of flat plate solar collectors using two different nanofluids as the working medium to assess the thermal and exegetic performance, and (ii) discover the effect of the volume flow rate and nanoparticles volume concentration on exergy efficiency, pressure drop, pumping power and convective heat transfer coefficient under particular operating situation. The two considered nanofluids are based on graphene infused nanoparticles in water and in acetone as the two based fluids.

2.1 Energy Analysis: The possible heat gain (Q_u) by the absorbing medium can be inscribed as,

$$Q_u = \dot{m} C_p \left(T_{f,out} - T_{f,in} \right) \quad (1)$$

The heat capacity and density of nanofluid are calculated according to the simple mixing law as follows [8, 9]:

$$C_{p,nf} = \varphi C_{p,np} + C_{p,bf} (1 - \varphi) \quad (2)$$

$$\rho_{nf} = (1 - \varphi) \rho_{bf} + \varphi \rho_{np} \quad (3)$$

IOP Conf. Series: Materials Science and Engineering **92** (2015) 012015 doi:10.1088/1757-899X/92/1/012015

The Absorbed irradiation per unit area of solar collector absorber plate (S) is determined by,

$$S = I_T(\tau\alpha) \tag{4}$$

The instantaneous collector efficiency relates the beneficial energy to the total radiation incident on the collector surface, and is written as:

$$\eta_{En} = \frac{\dot{Q}_u}{A_p I_T} = \frac{\dot{m}C_p\left(T_{f,out} - T_{f,in}\right)}{I_T} \tag{5}$$

Present analysis was performed by considering the normal incidence condition, hence, the $F_R(\tau\alpha)$, heat removal factor (F_R), and overall loss coefficient of solar collector (U_l) were set to be constants within the range of tested temperatures.

2.2 Exergy analysis: The irreversibility can then be quantified as the difference in exergy measured at the inlet and outlet sections of the control volume or the deficit. The simplest exergy balance equation per unit interception area of a solar collector in steady state can be expressed as shown below:

$$\dot{E}_g = \eta_o \dot{E}_{sun} - \dot{E}_{loss} \tag{6}$$

The exergy collection rate in steady state is exergy gained by heat transfer fluid while the fluid temperature increases from $T_{f,in}$ at the inlet to $T_{f,out}$ at the outlet. The expression of the exergy collection rate, assuming that the fluid is incompressible, can be obtained by use of the following equation [13], without considering the mechanical exergy,

$$\dot{E}_g = \dot{m}C_p\left(T_{f,out} - T_{f,in} - T_a \ln\frac{T_{f,out}}{T_{f,in}}\right) \tag{7}$$

The exergy flux from the sun is defined here as:

$$\dot{E}_{sun} = I_T\left(1 - \frac{T_a}{T_s}\right) \tag{8}$$

The exergetic efficiency (η_{Ex}) is defined here and is expressed as follows:

$\eta_{Ex} = \dfrac{\dot{E}_g}{\dot{E}_{sun}}$, and after rearranging the equations, it is reduced to:

$$\eta_{Ex} = \frac{\dot{m}C_p\left[\left(T_{f,in} - T_a - \frac{S}{U_1}\right)\left(\exp\left(-\frac{U_1 A_p F}{\dot{m}C_p} - 1\right)\right)\right] - \dot{m}C_p\left[T_a\left(\frac{\left(\exp\left(-\frac{U_1 A_p F}{\dot{m}C_p}\right) - 1\right)}{T_{f,in}}\left(T_{f,in} - T_a - \frac{S}{U_1}\right) + 1\right)\right]}{A_p I_T\left[1 - \left(\frac{T_a}{T_s}\right)\right]} \tag{9}$$

IOP Conf. Series: Materials Science and Engineering **92** (2015) 012015 doi:10.1088/1757-899X/92/1/012015

where, T_a is the ambient temperature (K), $\dot{m}$ *is the mass flow rate (kg/s)*, A_p is the collector area (m^2) and C_p is the specific heat (*J/kg·K*). The required system data including the thermophysical properties are given in Table 1 below.

Table 1. Environmental and analysis conditions for the flat plate solar collector

Physical Properties	Molecular Formula	Density, at 20°C (kg/m^3)	Viscosity, at 25°C ($N{\cdot}s/m^2$)	Thermal Conductivity, at 25°C (W/m·K)	Specific Heat Capacity at 25°C (J/kg·°C)
Acetone	$(CH_3)_2CO$	0.791 E^3	3.075E^{-4}	0.16	2.160 E^3
Water	H_2O	1E^3	7.8E^{-4}	0.58	4.182 E^3
Graphene [14,15]		2.25E^3		3E^3-5E^3[16]	6.50 E^3[14]

3. Results and Discussion

Results of the exergy efficiency with respect to nanoparticle loading and volume flow rate are depicted in Fig. 1 & Fig. 2. Increasing the volume flow rates resulted in enhanced collector efficiency. Efficiency was calculated using Eq. (5) and data of Table 1. This behaviour may be described by the longer residence times of nanofluid with the hot surfaces inside the collector and is in agreement with Benli's work [17]. Insignificant difference is observed with the increase in volume concentration of nanoparticles in the base fluid. On the other hand, it is observed that the exergy efficiency increases substantially with the increase in the volume flow rate. The analysis demonstrates that the lowest efficiency belonged to the collector, operated by Graphene/acetone while the maximum was for Graphene/water.

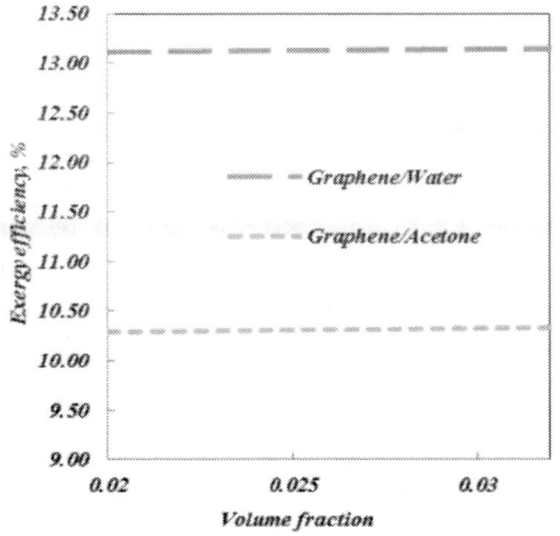

Fig. 1. Effect of nanofluid volume fraction on the exergy efficiency.

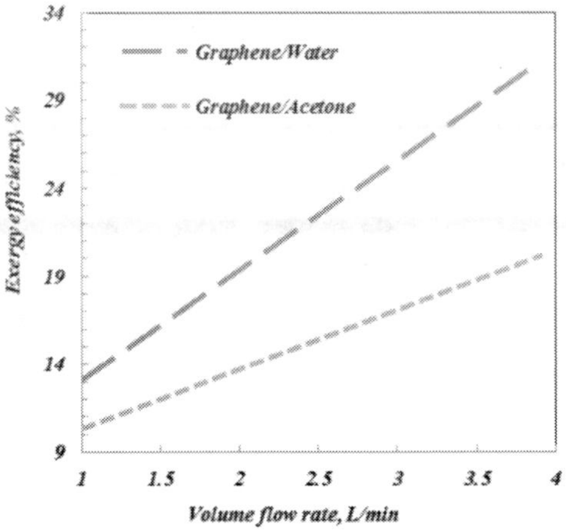

Fig. 2. Effect of the flow rate on the exergy efficiency.

Results of the variations of entropy generation with volume flow rate and nanoparticles volume fraction are shown in Fig. 3(a) and Fig. 3(b). It is observed that the entropy generation drops fast with the increase of the volume fraction of nanoparticles and less obvious with respect to volume flow rate. This reduction takes place due to increase of the heat flux on the absorber plate; the irreversibility turns out to be dominant. As a result, rising the volume fraction results in enhanced thermal conductivity of the nanofluids, which happens to improve the thermal conductance. As a result of heat transfer, the irreversibility has a developed an effect which is more pronounced than that of the viscous influence to the entropy generation. With increasing the volume fraction of nanoparticles, the functional viscosity of the nanofluids grows and subsequently the nanofluid friction contributes to the entropy generation. However as a whole, the entropy generation (the summation of fluid friction contributions and the heat) reduces in the gap. Graphene/water mixture shows lower entropy generation rate compared to Graphene/acetone mixture.

The analysis presented here is based on Bejan's work [18, 19]. The analysis, however, is adapted to flat plate collectors because entropy generation minimization is more important for high temperature systems. The minimization of the entropy generation rate implies the maximization of the power output.

In order to quantify the exergy of a system, we must specify both the system and the surroundings. The exergy reference environment is used to standardize the quantification of exergy. The exergy reference environment or simply the environment is assumed to be a large, simple compressible system. In addition, it is assumed that any process does not significantly change the intensive properties of the environment. Some energy and exergy values are dependent on the intensive properties of the dead state. Dead state definitions of exergy analysis have been studied by Krakow [20]. He reported that an exergy analysis was an implicit comparison of the performance of real thermal systems with the performance of ideal, reversible thermal systems. Gogus et al. [21] studied the influence of the variation of environmental condition in the process. In their study, the general exergy balance of a system was given based on the variation of the temperature of the surroundings. Fig. 4 demonstrates the dead state temperature.

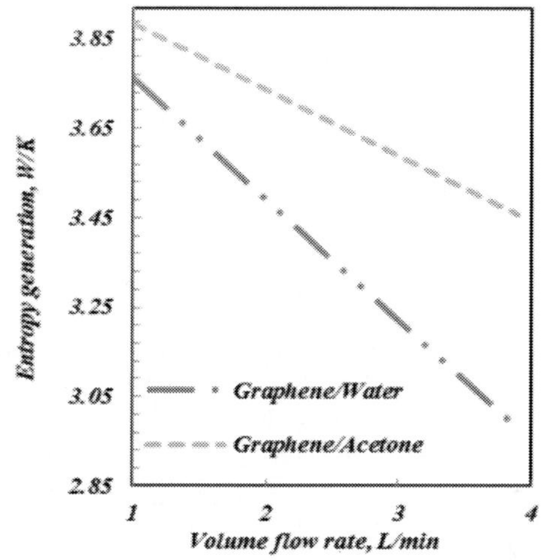

Fig. 3(a) Effect of flow rate on entropy generation

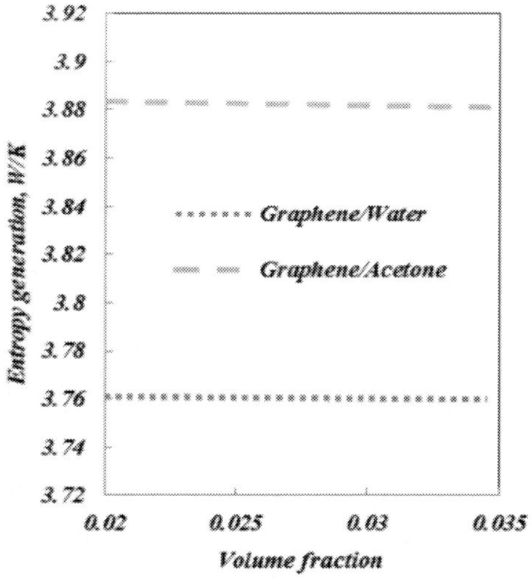

Fig. 3(b) Effect of volume fraction on entropy generation.

In this paper, the conducted theoretical studies to investigate how varying dead-state temperatures (ranging from 0 to 25 °C) would affect the second law efficiency of the flat plate solar collector. Many authors used a deadstate temperature of 20 or 25 °C as a single value. The deadstate is normally selected to simulate the accessible/actual natural. Fig. 4 shows the variation of the second law efficiencies of various dead state temperatures with the fixed volume fraction and the volume flow rate. It was calculated using Eq. (9) and input data tables. The study proposed that, when the reference deadstate temperature is increased, the second law efficiency value is decreased. This means it increases the value of exergy destruction. Exhaust exergy rates changed at different deadstate temperatures. Exergy loss rates increased with the increase in the deadstate temperatures.

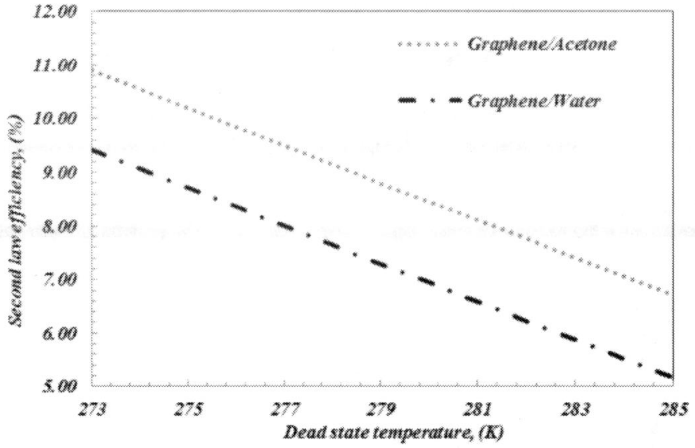

Fig. 4 Effect of dead state temperature on second law efficiency.

4. Conclusion

A comprehensive theoretical study was carried out to evaluate the thermal performance of flat plate solar collector using altered nanofluid as absorbing medium. The assumptions can be drawn that the efficiency of the collector advances with growing nanoparticles volume concentration due to an improved heat transfer to the nanofluid flow. The exergy loss of the system reduces subject to the increase of the collector efficiency. With increasing nanoparticles volume concentration, exergy can be enhanced with reducing exergy loss. From the analyses, it is revealed that by inserting a small amount of graphene nanoparticles in water, exergy efficiency could be enhanced by 21%, comparing to conventional fluids the entropy generation, however, is decreased by 4%.

Acknowledgements

The authors are grateful for the financial support from MASDAR Institute of Science & Technology and to members of the Waste to Energy Laboratory for their support throughout this research.

References

[1] Othman, M.Y.H., K. Sopian, B. Yatim, and M.N. Dalimin, Diurnal pattern of global solar radiation in the tropics: a case study in Malaysia. Renew Energ, 3(6): p. 741-745, 1993.

[2] Said, Z., M. Sajid, R. Saidur, M. Kamalisarvestani, and N. Rahim, Radiative properties of nanofluids. Int Commun Heat Mass, 2013.

[3] Natarajan, E. and R. Sathish, Role of nanofluids in solar water heater. Int J Adv Manuf Tech: p. 1-5, 2009.
[4] Tyagi, H., P. Phelan, and R. Prasher, Predicted efficiency of a low-temperature nanofluid-based direct absorption solar collector. J Sol Energ, 131(4), 2009.
[5] Otanicar, T.P. and J.S. Golden, Comparative environmental and economic analysis of conventional and nanofluid solar hot water technologies. Environ Sci Technol, 43(15): p. 6082-6087, 2009.
[6] Otanicar, T.P., P.E. Phelan, R.S. Prasher, G. Rosengarten, and R.A. Taylor, Nanofluid-based direct absorption solar collector. Renew Sust Energ Rev, 2: p. 033102, 2010.
[7] Yousefi, T., F. Veysi, E. Shojaeizadeh, and S. Zinadini, An experimental investigation on the effect of Al_2O_3–H_2O nanofluid on the efficiency of flat-plate solar collectors. Renew Energ, 39(1): p. 293-298, 2012.
[8] Saidur, R., T. Meng, Z. Said, M. Hasanuzzaman, and A. Kamyar, Evaluation of the effect of nanofluid-based absorbers on direct solar collector. Int J Heat Mass Tran, 55(21): p. 5899-5907, 2012.
[9] Said, Z., M.H. Sajid, M.A. Alim, R. Saidur, and N.A. Rahim, Experimental investigation of the thermophysical properties of Al_2O_3-nanofluid and its effect on a flat plate solar collector. Int Commun Heat Mass, 48(0): p. 99-107, 2013.
[10] Pak, B.C. and Y.I. Cho, Hydrodynamic and heat transfer study of dispersed fluids with submicron metallic oxide particles. Exp Heat Tran An Int J, 11(2): p. 151-170, 1998.
[11] Aravind, S.S.J. and S. Ramaprabhu, Graphene-multiwalled carbon nanotube-based nanofluids for improved heat dissipation. Rsc Adv, 3(13): p. 4199-4206, 2013.
[12] Eswaraiah, V., V. Sankaranarayanan, and S. Ramaprabhu, *Graphene-based engine oil nanofluids for tribological applications.* ACS applied materials & interfaces, 2011. **3**(11): p. 4221-4227.
[13] Jafarkazemi, F. and E. Ahmadifard, Energetic and exergetic evaluation of flat plate solar collectors. Renewable Energy, 56: p. 55-63, 2013.
[14] Xia, M., Y. Song, and S. Zhang, Specific heat of graphene nanoribbons. Phys. Lett. A, 375(42): p. 3726-3730, 2011.
[15] Wang X., X. Xu, and S.U. S. Choi, *Thermal conductivity of nanoparticle-fluid mixture.* Journal of thermophysics and heat transfer. 1999. **13**(4): p. 474-480.
[16] Balandin A.A., *Thermal properties of graphene and nanostructured carbon materials.* Nat Mater. 2011. **10**(8): p. 569-581.
[17] Benli, H., Experimentally derived efficiency and exergy analysis of a new solar air heater having different surface shapes. Renew Energ, 50: p. 58-67, 2013.
[18] Bejan, A., Entropy generation minimization: The new thermodynamics of finite-size devices and finite-time processes. J. Appl. Phys. , 79(3): p. 1191-1218, 1996.

[19] Bejan, A., D. Kearney, and F. Kreith, Second law analysis and synthesis of solar collector systems. ATJSE 103: p. 23-28, 1981.
[20] KI, K., 3478 - Exergy Analysis: Dead-State Definition. ASHRAE Trans, 97(1), 1991.
[21] Göğüş, Y.A., Ü. Çamdalı, and M.Ş. Kavsaoğlu, Exergy balance of a general system with variation of environmental conditions and some applications. Energy, 27(7): p. 625-646, 2002.

Assisted Sonication vs Conventional Transesterification Numerical Simulation and Sensitivity Study

Isam Janajreh*, Mohammed Noorul Hussain, Tala ElSamad

Mechanical and Materials Engineering Department
Masdar Institute of Science and Technology
Masdar City, Abu Dhabi, UAE

*e-mail:ijanajreh@masdar.ac.ae

Abstract— Transeterification is known as slow reaction that can take over several hours to complete as the two immiscible liquid reactants combine to form biodiesel and the less favorable glycerol. The quest of finding the perfect catalyst, optimal operational conditions, and reactor configuration to accelerate the reaction in mere few minutes that ensures high quality biodiesel, in economically viable way is coming along with sonication. This drastic reduction is a key enabler for the development of a continuous processing that otherwise is fairly costly and low throughput using conventional method. The reaction kinetics of sonication assisted as inferred by several authors is several time faster and this work implements these rates in a high fidelity numerical simulation model. This flow model is based on Navier-Stokes equations coupled with energy equation for non-isothermal flow and the transport equations of the multiple reactive species. The model is initially validated against experimental data from previous work of the authors using an annular reactor configuration. Following the validation, comparison of the reaction rate is shown to gain more insight to the distribution of the reaction and its attained rates. The two models (conventional and sonication) then compared on the basis of their sensitivity to the methane to oil molar ratio as the most pronounced process parameter. Both the exit reactor yield and the distribution of the species are evaluated with favorable yield under sonication process. These results pave the way to build a more robust process intensified reactor having an integrated selective heterogeneous catalyst to steer the reaction. This can avoid the downstream cleaning processes, cutting reaction time, and render economic benefit to the process.

1. Introduction:

Whether conventional or sonication assisted the transesterification process undergoes the following main and reversible reactions:

$$\text{Triglyceride} + \text{Alcohol} \ \frac{K1}{K2} \ \leftrightarrow \ \text{FAME} + \text{Diglyceride} \quad (1)$$

$$\text{Diglyceride} + \text{Alcohol} \ \frac{K3}{K4} \ \leftrightarrow \ \text{FAME} + \text{Monoglyceride} \quad (2)$$

$$\text{Monoglyceride} + \text{Alcohol} \ \frac{K5}{K6} \ \leftrightarrow \ \text{FAME} + \text{Glycerol} \quad (3)$$

$$\text{Triglyceride} + 3\text{Alcohol} \ \frac{K7}{K8} \ \leftrightarrow \ 3\text{FAME} + \text{Glycerol} \quad (4)$$

Summing the first three reactions provide the overall fourth reaction in which one mole of oil reacts with three moles of alcohol leading to the production of three moles of Fatty Acid Methyl Ester (FAME) and the less desired one mole of glycerol. Remains of Triglycerides (TG), Diglyceride (DG), or Monoglyceride (MG) in the reaction are considered impurities and thus one needs find the conditions to speed up the overall forward reaction to reach to the favorable reaction equilibrium state. Contrary to hydrocarbon fuel in which their thermodynamic properties is well documented and are known as priori, the Triglyceride, Diglyceride, and Monoglyceride is not a fixed formula and their properties including their standard enthalpy, specific heat, and even their physical

IOP Conf. Series: Materials Science and Engineering **92** (2015) 012016 doi:10.1088/1757-899X/92/1/012016

properties including density, viscosity, etc. are not surprisingly inconsistent in literature. This adds some uncertainty in modeling of their reactivity as far as establishing robust energy minimization approach such as Gibbs free energy based modeling. Alternatively, intrinsic modeling that based on experimental evaluated chemical kinetic data comes very handy. In this modeling quest the most recurring and cited properties is been used in this work. Conventional transesterification method is slow to be considered as a viable continuous method at high throughput. Ultrasound assisted method, however, reduces the immiscibility of the two reacting fluids as the sonication energy creates cavitation bubbles which continue to grow rapidly, collapsing violently, generating energy and mechanical effect leading to drastic mass transfer rate enhancement at the boundary. The pressure within the bubbles can be as high as 5000 atm which can cause a very high local temperature rise (about 7200°C) [2]. As this takes place at the phase boundary of the two immiscible species the reactions between the two species intensify [3].

Apart from the authors work on numerical simulation of the transesterfication there no other reported literature treating this subject. There are, however, several works that focused on kinetics of these reactions. Noureddini et al. [4] calculated the activation energies and rate constants for forward and backward overall reaction and subsidiary reactions for soybean oil transesterification at different conditions of a well-stirred batch reactor. Jose et al. [3] compared kinetics of sonicated and conventional transesterification and of soybean oil and reported that the rate constant for the sonication reactions were three times to half order of magnitude higher than the conventional process. This suggests that for similar activation energies the reaction constant must be compensated so that the reaction rates need to be much higher in the case of sonication. Vishwanath et al. [5] demonstrated the dependency of rate constants on molar ratio, catalyst percentage and temperature for forward and backward reaction of palm oil fatty distillate ultrasound assisted transesterification with isopropanol. The highest rate constants for forward reaction were reported at five to one of alcohol to oil molar ratio. At catalyst concentration 7%, the highest forward rate constants was achieved. The forward rate constant also depicted a direct proportionality with temperature. The highest attempted experimental temperature was 600C which resulted in the highest forward reaction constant. Other parameters showed direct proportionality with the reverse reaction constant, however molar ratio increase resulted in a lower backward reaction constant. Investigation of the kinetics of transesterification reaction of waste cooking oil with a heterogeneous catalyst (K3PO4) and methanol carried out by Dipak et al. [6]. A 375 W at 22 kHz ultrasonic sonication device is used. The reaction was carried out at six to one molar ratio. Similar to the Vishwanath and coworkers they reported the increase in the forward rate constant with the increase in temperature. The evaluated activation energy was 64,241 KJ/Mol and the rate constant was in the range 0.02 to 0.18 when the temperature swept from 300C to 600C. Earlier conducted works of the authors demonstrate the feasibility of CFD in evaluating species concentrations as far as their distributions and output product. Less emphasis was given to the influence of key reaction parameters including molar ratio and flow speed. Another work by the same author shows the evaluation of the reaction kinetics of the waste cooking oil transesterification followed by implementing these values in a reactive flow model [7]. In this work sensitivity study is carried out considering molar ratio, flow condition, and different geometrical configuration. The details of transesterification process inside the three dimensional cylindrical single tubular reactor is considered. The influence of the molar ratio for both conventional and sonicated transesterification reaction will be carried out. The results aim to gain deeper insight of the difference in the two reaction methods as far as the yield and their distributions.

2. Model Development

The none-isothermal, viscous, turbulent flow is governed by the Navier-Stokes equations associated with temporal, advective, viscous, and source terms and in general form is written as:

$$\frac{\partial}{\partial t}(\phi) \quad + \quad \frac{\partial}{\partial x_i}(u_i\phi) = -\frac{\partial}{\partial x_i}\left(\Gamma_\phi \frac{\partial \phi}{\partial x_i}\right) + S_\phi \qquad (5)$$

Time rate advective diffusion source Where u is the velocity and Sφ is the source term due to the interaction, destruction or creation of other species. Φ is the dependent variable which is density (ρ), the velocity density multiple (ρu), or the temperature (T) to constitute the continuity, momentum, and the energy, respectively. Equation (5) also can represent the turbulence scalars, the kinetic energy and the turbulence dissipation rate, and these are written as:

$$\rho u_i \frac{\partial k}{\partial x_i} = \mu_t\left(\frac{\partial u_j}{\partial x_i} + \frac{\partial u_i}{\partial x_j}\right)\frac{\partial u_j}{\partial x_i} + -\frac{\partial}{\partial x_i}\left(\frac{\mu_t}{\sigma_k}\frac{\partial k}{\partial x_i}\right) - \rho\varepsilon \qquad (6)$$

$$\rho u_i \frac{\partial \varepsilon}{\partial x_i} = C_{1\varepsilon}\frac{\mu_t \varepsilon}{k}\left(\frac{\partial u_j}{\partial x_i} + \frac{\partial u_i}{\partial x_j}\right)\frac{\partial u_j}{\partial x_i} + \frac{\partial}{\partial x_i}\left(\frac{\mu_t}{\sigma_\varepsilon}\frac{\partial \varepsilon}{\partial x_i}\right) - C_{2\varepsilon}\frac{\rho \varepsilon^2}{k} \qquad (7)$$

The three terms on the right hand are the diffusion, the generation, and destruction/dissipation of the turbulent quantities, respectively. They are coupled through the eddy viscosity (μ t) coefficient which is expressed as:

$$\mu_t = f_\mu C_\mu k^2 / \varepsilon \qquad (8)$$

Where f and C are pre-defined constants additional to the empirical constants C1ε, C2ε, □□□ and□□that are tuned for a specific flow categories, e.g. channel, swirl, separating, and jet flows.

For system closer, additional transportation specie equations of equal numbers to the numbers of reacting and product species are used. Each equation incorporates a reaction source term and is written in the following general form [10]:

$$\frac{\partial}{\partial t}(\rho m_i) + \frac{\partial}{\partial x_i}(\rho u_i m_i) = \frac{\partial}{\partial x_i}(\rho D_{i,m} + \mu_t / Sc_t)\frac{\partial m_i}{\partial x_i} + R_i + S_i \qquad (9)$$

Where Di,m is the diffusion coefficient of mi species, Sct is the turbulent Schmidt number and is defined as the ratio of the eddy viscosity μt to the eddy diffusivity Di,m. The reaction source term Ri accounts for formation or destruction of species according to the stoichiometric chemical reaction which is written as:

$$\sum_{i=1}^{N} v'_{i,r} S_i \underset{k_{b,r}}{\overset{k_{f,r}}{\Leftrightarrow}} \sum_{i=1}^{N} v''_{i,r} S_i \qquad (10)$$

Where v are the stoichiometric coefficients of reactant (prime) and product (double prime) kf and kb are the forward and backward reaction constants respectively and follows the Arrhenius for as:

$$k_{f\,or\,b} = AT^{\alpha} e^{-E/RT} \qquad (11)$$

Where A is the reaction pre-constant, T and □ are the temperature and its exponential dependency constant. E is the activation energy, and R is the universal gas constant.

The reaction constants (kf and kb) are coupled with the Ri source term in the transport equation according to the species concentrations and following the stoichiometry of equation (10) and is written as:

$$R_{i,r} = M_{i,r}(v''_{i,r} - v'_{i,r})\left(k_f \prod_{j=1}^{N} C_{j,r}^{\eta'_{j,r}} - k_b \prod_{j=1}^{N} C_{j,r}^{v''_{j,r}} \right) \qquad (12)$$

Where Cj is the molar concentration of jth specie raised to stoichiometric coefficients v and reaction order η, and Mi is the molecular weight of species i. The analyses and reactions are achieved within the framework of Ansys Fluent [8]. Janajreh and Alshrah [9] have carried out both non-reactive and reactive flow for tubular reactor to model the transesterification of the two reactants and their products. The properties of the species are listed in Table 1 and the reaction kinetics of each reaction are given elsewhere and are summarized in Table 2 [4].

Table 1. Summary of species properties and molecular weight

Species	Chem. formula	Mol. weight	Viscosity (kg/m.s)	Cp (J/kg.0C)	Density Kg/m3
Methanol	CH4O	32	3.96e-4	1.470e3	791.8
Waste oil or Triglyceride	C54H104O6	848	1.61e-2	2.2e3	883.3
Diglyceride	C37H72O5	596	-	-	880
Monoglyceride	C20H40O4	344	-	-	875
Biodiesel	C18H36O6	284	1.12e-3	1.187e3	870
Glycerol	C3H9O3	93	1.412e0	0238.6	1261

Table 2. Chemical Kinetic Constants of Reactions [4]

Reaction	K at 50 0C and Reynold's Number=6200	Activation Energy (J/mol)	Pre-exponential factor (A)-calculated
1. Triglyceride ► Diglyceride	0.05	56,902.4	77,983,755.8
2. Diglyceride ►Triglyceride	0.11	40,919.52	449,020.085
3. Diglyceride ►Monoglyceride	0.215	78,659.2	1.0976E+12
4. Monoglyceride ► Diglyceride	1.228	46,860.8	4,570,1701.3
5. Monoglyceride ► Glycerol	0.242	21,673.12	767.86
6. Glycerol ► Monoglyceride	0.007	41,296.08	32,870.58

The overall reactions for the conventional and sonication transesterification process are reported in the literature and are summarized in Table 3.

Table 3. Overall chemical kinetic constants as reported [3] [4]

Reaction	K	Activation Energy (J/mol)	Pre-exponential factor (A)-calculated
Forward Conventional.	7.8E-05	164958.38	3.4868E+22
Forward with sonication.	2.34E-04	164958.38	1.046E+23

The main assumption in these analysis is the state of well mixing between the two immiscible reactants represented by the methanol and triglyceride. Therefore, the two fluid enters the reactor as premixed fluid implying significant agitation incurs to the fluid prior its introduction to the reactor. This assumption simplified the physical aspect of the problem by pursuing the Euler multiple species numerical approach. This modeling departs from the multiple phase approach where bubbly, sludge, and intermittent flow can take place in the case of none premixed and slowly moving flow. Nevertheless, it is well know that transeterfication requires intense agitations to avoid any mass transfer limitation of the two reactants and thus additional to premixing flow needs to be configured to preserve the well mixing state of the reactants. Experimentally, the increase in molar ratio of methanol to triglyceride would push the reaction forward [11], numerically this considered indirectly as the increase in the molar ratio will deplete the triglyceride faster forming the FAME and the Glycerol. Because the FAME and Glycerol now at lower concentration stage there revers reaction proceeds at slower rate.

Simulation Setup and model validation

The model setup is in the line of the author's previous work [9]. It consists of a single cylindrical reactor (30cm length) flowing axially as depicted in Fig. 1.

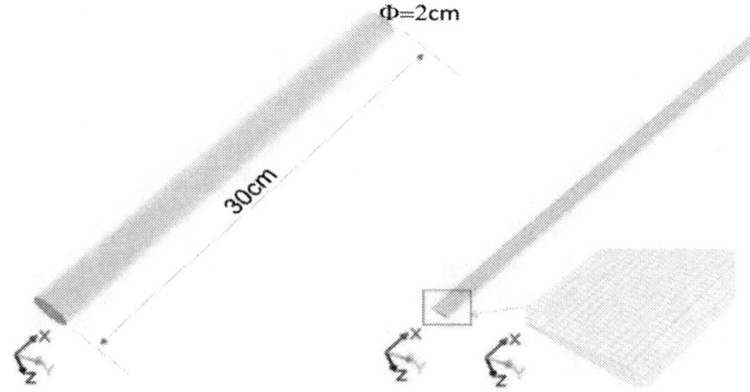

Fig. 1: The topography of the reactor, the considered wedge and its discretized hybrid hexagonal and prismatic mesh

A 3D cylindrical wedge of 5deg is considered where two periodic faces coincided at the centerline. It is subjected to prescribed dirichlet inlet (velocity, species concentration, and temperatures), and Neumann of zero axial gradient at the outlet, and no-slip and no-penetration at the outer cylindrical reactor wall. A hybrid hexagonal and prismatic mesh is considered at three refined levels to assess mesh independency comprises of a course, baseline, and refined. Table 4 shows the independency of the mesh as far as pressure drop near the turbulence Reynolds values. It demonstrates the suitability of the baseline mesh as a compromise between the accuracy and computational time.

Table 4. Summary of the mesh sensitivity results

Mesh level	Number of cells	Pressure Drop (Pas) and Rel. Error
Coarse	6,300	2.503; 28.5%
Baseline	50,400	2.007; 3.2%
Fine Mesh	403,200	1.948;---

The validation of the model is carried out against the experimental measurement obtained elsewhere [4] under conventional transesterfication. These measurements of the triglyceride (TG), FAME, Glycerol, and Alcohol (AL) are conducted using Thermoscientific GC/MS that equipped with FID detector. FAME column is calibrated against several standard biodiesel samples for qualitative and quantitative species measurements. Results of the model and experimental data are plotted in Fig. 2 in function of the residence time. The model residence time corresponds to the length of the reactor for the specific downstream location divided by the inlet flow velocity. These measurements are obtained at 6:1 AL to TG ratio, in 3,600rpm stirred batch reactor operated at atmospheric pressure and 50oC temperature. The trend of species depletion and formation is well-produced despite the small deviation shown in the initial reaction stage (20 minutes) in the biodiesel, glycerol and alcohol. A fairly decent match between the model and experimental measurements prevails to over the rest of the time for all species.

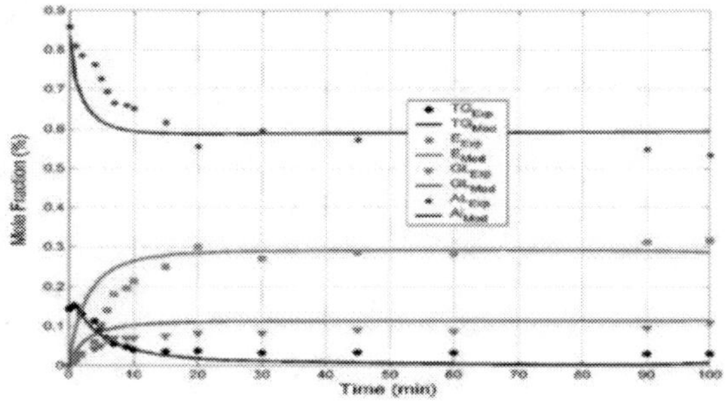

Fig. 2: Validation of model and experimental measurements

3. Results of Simulation and Discussion

Reaction rate

As kinetic reaction rate of the sonication transesterfication process is nearly three times higher than those of conventional process as noted in Table 3, simulation results also reflect this high reaction rate trend as depicted in Fig. 3. The results clearly demonstrate the earlier higher value as well as the occurrence of the peak in sonication transesterification which is nearly three times higher than the conventional process. The reaction rates distribution is more clearly shown in the contour plot in Fig. 4 reaching nearly 54 kmol/m3.s for the sonication versus 18 kmol/m3.s for the conventional transesterification.

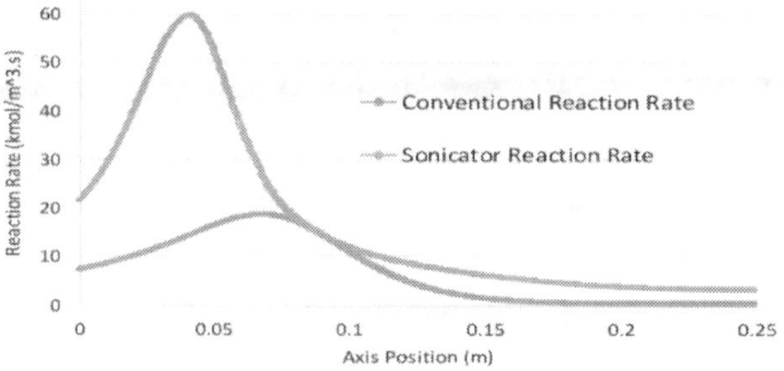

Fig. 3: Rate of reaction along the reactor axis for conventional and sonicated transesterfication process.

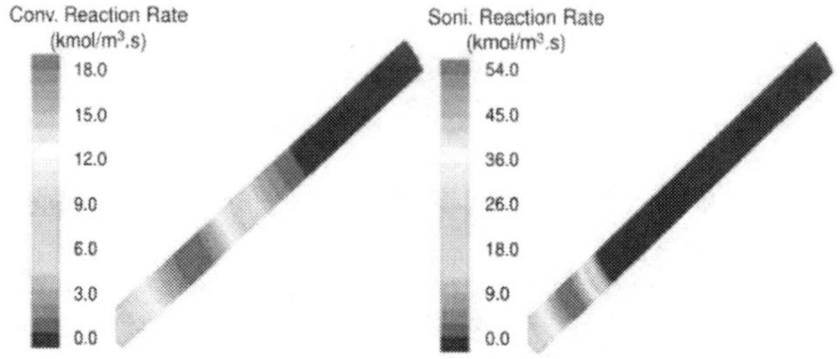

Fig. 4: Distribution of the rate of reaction for conventional and sonicated transesterfication process.

Species distribution

As for the resulted species, the spatial variation of the Ester and Glycerol is depicted in Fig. 5 which also depict a faster and higher concentrations are formed in the favor of the sonication transeterification.

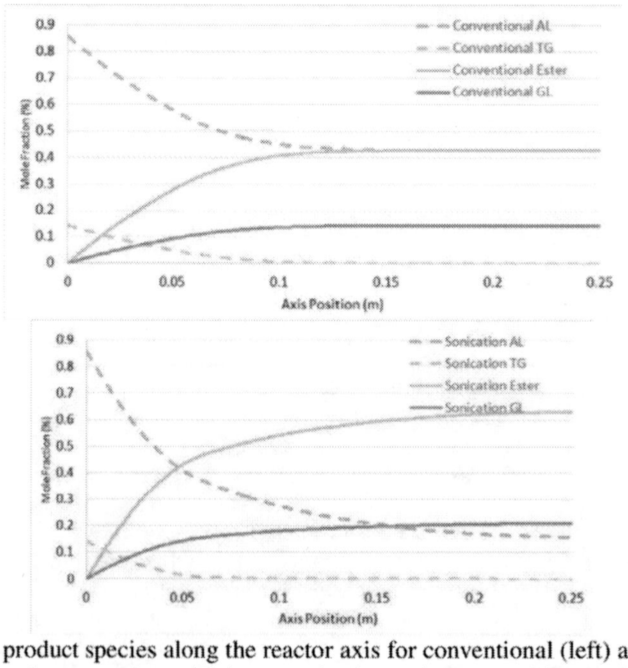

Fig. 5: Distribution of the product species along the reactor axis for conventional (left) and sonication (right)

The results at downstream shows a substantial increase in the sonicator reaching molar ratio that goes beyond theoretical stoichiometry values (3:3:1 of Biodiesel:AL:GL). The final sonication yield in molar fraction are 0.63, 0.21, and 0.16 for Biodiesel, GL and AL, respectively. It however attained near the exact theoretical stoichiometry values (0.43 FAME, 0.14 GL, and 0.43 AL) in the case of conventional process. The results shows the anticipated trend however and demonstrates the faster and higher molar fraction formation during the sonication method. Table 5 summarizes the molar fraction at the exit of the reactor.

TABLE 5. Molar Fraction at the exit (yield)

Species	Conventional Input & Output		Sonication Input & Output	
AL	0.856	0.428	0.856	0.159*
TG	0.144	6.10E-06	0.144	8.80E-04
Biodiesel	0	0.429	0	0.629**
GL	0	0.143	0	0.21

*lower than the anticipated value; **higher than the anticipated value

As the results reveal in Table 5, a higher molar concentration than what the stoichiometry suggests is generated in the sonication method. This may suggest the inaccuracy of the implemented sonication activation energy and rate constant values with no restriction to up-hold the reaction to the stoichiometric values. Never the less the current simulations bring more insight to the trend rather than the absolute measure for these fractions. The subsequent analysis is based on implementing these values in an attempt to verify the influence of the most pronounced transesterification parameter, i.e. molar ratio.

Molar ratio sensitivity

Fig.7 depicts the influence of the change of the molar ratio of the alcohol to the TG for both conventional and sonication transesterification. At 3:1 AL to TG molar ratio some of the TG stayed unconverted in the conventional process while is totally consumed during the sonication process. At higher AL to TG molar ratio of 6:1 the total TG conversion is achieved at the middle stream and albeit faster for the sonication transesterification with an asymptotic conversion starting at nearly 1/5th upstream and nearly completed at 2/5th upstream.

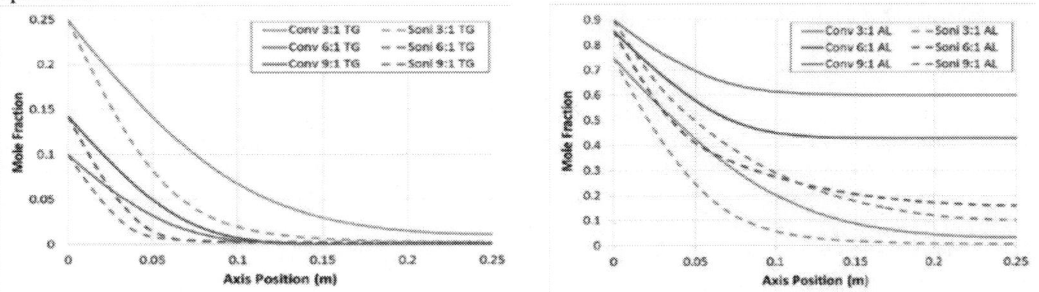

Fig. 7: TG (left) and AL (right) results of the conventional and Sonication transesterification at different AL to TG molar ratio

Further increase in the AL to TG molar ratio to 9:1 resulted in a noticeable earlier start of the asymptotic conversion of the sonication process, but would also result in near completion at the same upstream 2/5th station. As for the AL conversion, the alcohol residual for the conventional case at the 3:1 molar ratio signifies incomplete conversion contrary to the sonication process in which all the alcohol and TG are utilized. These results show clearly the considerable advantage for the conventional process in using larger molar rate to achieve completed conversion, this additional AL value, however this is not required in the case of sonication assisted transesterification.

Fig. 8 shows the product yield of the process captured by the molar fraction of Ester and GL. At the stoichiometric molar ratio (3:1) the values for the sonication process are at 0.75 and 0.25 for the Easter and GL, respectively. It signifies a complete conversion which is contrary to the yield of the conventional molar fraction values for Easter and TG which are at 0.71 and 0.24, respectively. The unconverted AL and TG was at 0.04 and 0.01 respectively.

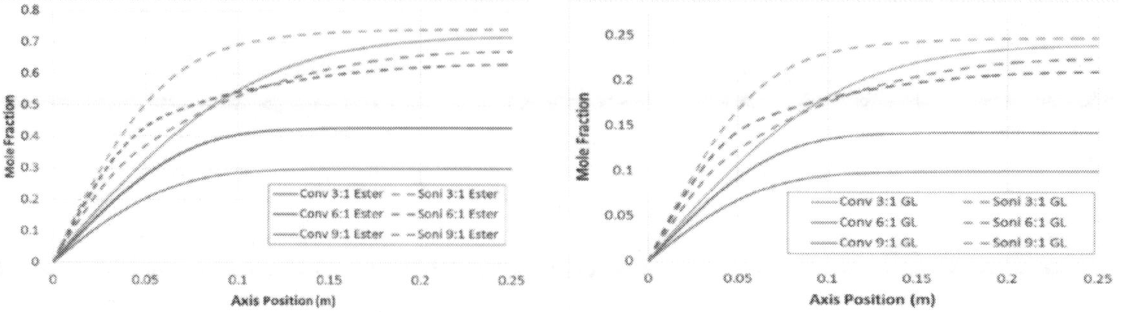

Fig. 8: Ester (left) and GL (right) results of conventional and Sonication transesterification at different AL to TG molar ratio

At higher molar ratio, and with complete conversion of the AL and TG, the Ester and GL reach to the theoretical stoichiometric values for the conventional transesterification, i.e. at 6 AL to one TG molar ratio, the Ester reach the 0.43 and the GL is 0.14 signifying the attainment of complete conversion. The sonication assisted, however, emerges at faster conversion and higher production of Ester and TG with values beyond what the stoichiometry would provide. It appears that AL continue to deplete also to value that is lower than the theoretical stoichiometry. The glycerol values accordingly is also increased beyond the stoichiometry suggesting that the used kinetics of the sonication assisted transesterification are loose and not that accurately measured. Therefore, this aspect needs to be taken further in the lab to capture these kinetics in a process similar to the work of Noureddini and Zhu [4]. At nine AL to one TG molar ratio similar and favorable trend is produced in the conventional transesterification (i.e. The molar fractions are 0.6 AL, 0.3 Ester, and 0.1 GL), but not so for the sonication as the AL continues to deplete to lesser value than the limited stoichiometry.

Conclusion

In this work, the numerical simulations of the transesterfication in a three dimensional reactor is achieved. A non-isothermal Navier stokes model coupled with species transport of reactive flow of the overall reversible reactions was carried out. The results clearly demonstrate the effectiveness of the sonication assisted transesterification over the conventional process. At low molar ratio whereas conventional transesterification fall short in completing the reaction, the sonication assisted transesterfication lead to complete stoichiometry conversion. These results shows clearly the considerable advantage for the conventional process in using larger molar rate to achieve completed conversion. This additional amount, however, can be avoided in the case of sonication assisted transesterification which eliminates the downstream AL separation and offset any associated cost. The results show the feasibility of reactive flow dynamics in capturing and numerically simulating the transesterification process. However, as at higher AL and TG molar ratio (6:1 and 9:1) the sonication reaction resulted in yield beyond the theoretical stoichiometric values. This is a "miss-behaved" system which is contrary to the "well behaved" conventional reaction. This suggest to seek more accurate sonication kinetic data. Hoping to continue construct a precise and insightful picture of the distribution of the reaction rates and their associated species for the development and design of more efficient reactors.

Nomenclature

Symbol	Meaning
K1	Forward reaction constant for first reaction
K2	Backward reaction constant for first reaction
K3	Forward reaction constant for second reaction
K4	Backward reaction constant for second reaction
K5	Forward reaction constant for third reaction
K6	Backward reaction constant for third reaction
K7	Forward reaction constant for overall reaction
K8	Backward reaction constant for overall reaction
E1	Activation energy for first reaction in forward direction
E2	Activation energy for first reaction in backward direction
E3	Activation energy for second reaction in forward direction

E4	Activation energy for second reaction in backward direction
E5	Activation energy for third reaction in forward direction
E6	Activation energy for third reaction in backward direction
E7	Activation energy for overall reaction in forward direction
E8	Activation energy for overall reaction in backward direction
u	Velocity component
S_Φ	Dependent variable- turbulent scalars.
T	Temperature
ρ	Density
μ_t	Eddy viscosity
Di,m	Diffusion coefficient
f,C	User-defined constants
σ_k , σ_ε	Empirical constants
mi	Species
Sct	Turbulent Schmidt number
kf and kb	Forward and backward reaction constants
Cj	Molar concentration of jth species
η	Reaction order
ν	Stoichiometric coefficient
Mi	Molar weight
Ri	Reaction source term
mf	Mass fraction

References

[1] Diasakou M, Louloudi A.: Papayannakos N. Kinetics of the non-catalytic transesterification of soybean oil. Fuel. 77, 1297-1302 (1998).
[2] Mason TJ.: Sonochemistry: Current uses and future prospects in the chemical and processing industries: Phil. Trans. R. Soc. London (1999).
[3] Colucci JA, Borrero EE, Alape F.: Biodiesel from an Alkaline Transesterification Reaction of Soybean Oil Using Ultrasonic Mixing. Journal of American Oil Chemists 82. (2005).
[4] Noureddini H, Zhu D.: Kinetics of Transesterification of Soybean Oil. Journal of the American Oil Chemists society. 74,1457 - 1463 (1997).
[5] Deshmane VG, Gogate PR, Pandit AB.: Ultrasound assisted synthesis of isopropyl esters from palm fatty acid distillate. Journal of Ultrasonic Sonochemistry. 16, 345-350 (2009).
[6] Pulakale DD, Maddikeri GL, Gogate PR, Pandit AB, Pratap AP.: Ultrasound assisted transesterification of waste cooking oil using heterogeneous solid catalyst. Journal of Ultrasonics sonochemistry. 22, 278 - 286 (2015).
[7] Janajreh I, ElSamad T, AlJaberi A, Diouri M.: Transesterification of waste cooking oil: kinetic study and reactive flow analysis. International conference of Applied Energy. (2015).
[8] Fluent 6.3.26 www.ansys.com: Fluent Inc; (2007).
[9] Janajreh I, Shrah MA. Numerical Simulation of Multiple Step Transesterification of Waste Oil in Tubular Reactor. Journal of Infrastructure Systems. (2013).
[10] Laidler K..: The development of the Arrhenius equation. Joural of Chemical Education. 61, 494, (1984).
[11] Isssariyakul T, Kulkarni MG, Meher LC, Dalai AK, Bakhshi NN.: Biodiesel production from mixtures of canola oil and used cooking oil. Journal of Chemical Engineering. 140, (2008).

IOP Conf. Series: Materials Science and Engineering **92** (2015) 012017 doi:10.1088/1757-899X/92/1/012017

Structural and electronic properties of Diisopropylammonium bromide molecular ferroelectric crystal

A Alsaad*[1], I A Qattan[2], A A Ahmad[1], N Al-Aqtash[3] and R F Sabirianov[3]

[1]Department of Physics, Jordan University of science and technology, P. O. Box 3030, Irbid 22110, Jordan.
[2]Department of App. Math. and Sci., Khalifa University of Sci., Tech. & Res., P.O. Box 127788, Abu Dhabi, UAE.

[3]Department of Physics, University of Nebraska at Omaha, Omaha, NE 68182, USA.

*email: alsaad11@just.edu.jo

Abstract: We report the results of *ab-initio* calculations based on Generalized Gradient Approximation (GGA) and hybrid functional (HSE06) of electronic band structure, density of states and partial density of states to get a deep insight into structural and electronic properties of $P2_1$ ferroelectric phase of Diisopropylammonium Bromide molecular crystal (DIPAB). We found that the optical band gap of the polar phase of DIPAB is ≈ 5 eV confirming it as a good dielectric. Examination of the density of states and partial density of states reveal that the valence band maximum is mainly composed of bromine *4p* orbitals and the conduction band minimum is dominated by carbon *2p*, carbon *2s*, and nitrogen *2s* orbitals. A unique aspect of $P2_1$ferroelectric phase is the permanent dipole within the material. We found that $P2_1$ DIPAB has a spontaneous polarization of 22.64 $\frac{\mu C}{cm^2}$ consistent with recent findings which make it good candidate for the creation of ferroelectric tunneling junctions (FTJs) which have the potential to be used as memory devices.

1. INTRODUCTION

Investigation of molecule-based ferroelectrics [1, 2, 3, 4, 5 ,6 ,7, 8, 9] revealed that a simple organic salt, Diisopropylammonium Chloride (DIPAC) displays the highest T_c among the known molecule-based ferroelectrics (T_c = 440 K). It undergoes a reversible phase transition from a high-temperature paraelectric phase (2/m) to a low-temperature ferroelectric one (2m). Its ferroelectric activity occurs below the T_c with a large spontaneous polarization $P_s \approx 8.9 \frac{\mu C}{cm^2}$, comparable with those of organic polymers such as Poly (Vinylidene Difluoride) (PVDF, ≈ 8 $\frac{\mu C}{cm^2}$) and Nylon-11 (≈ 5 $\frac{\mu C}{cm^2}$). This molecule is quite remarkable, with some of the key properties that are comparable with the most common inorganic crystals.

Previously reported molecule-based ferroelectrics usually show low T_c and small P_s values that extremely limit their practical applications. Ferroelectricity, involves the relationship of dielectric and elastic behavior in highly nonlinear, anisotropic, and polarizable, deformable crystals. It was found that crystals of Seignette (Rochelle) salt behave like a ferromagnet (They show the hysteresis of polarization and undergo phase transitions from the polar to the nonpolar phase) [10, 11, 12, 13].

It has been found that some organic molecules show a capacity as an organic substitute to today's silicon-based semiconductors [14] demonstrate properties that make them well appropriate to a wide range of applications in memory, sensing and low-cost energy storage. Switchable large polarization of ferroelectrics has made them suitable for electronics applications such as capacitor, sensor, and data storage, as well as to electromechanical and optical devices [15, 16]. For ''order-disorder-type'' ferroelectrics, the sum of permanent dipole moments of polar ions determines the polarization, while the product of relative displacement and charge of the ions had been the classical picture of the polarization for the ''displacive-type'' ferroelectrics. The macroscopic polarization of ferroelectrics can be described using Berry phase quantum mechanical approach [17, 18]. This fact is especially important for the displacive-type ferroelectrics having some orbital hybridization between ions [19, 20].

In this perspective, *ab-initio* calculations based on Berry phase have been intensively used for the understanding of the dielectrics. For instance, the displacive-type ferroelectric oxides have an amplified macroscopic polarization via the orbital hybridization between the occupied *p*-orbitals of oxygen and the empty *d*-orbitals of transition metals [21]. Molecular ferroelectrics are highly desirable for their easy and environmentally friendly processing, light weight, and mechanical flexibility. It has been found that Diisopropylammonium Bromide (DIPAB) is an organic crystal that is made from Bromine, a natural element isolated from sea salt, mixed with carbon, hydrogen and nitrogen. Researchers dissolved the elements in water and evaporated the liquid to grow the crystal. Because the molecule contains carbon, it is organic, and its pivoting chemical bonds allow it to flex. It was found to be a ferroelectric with a spontaneous polarization of 23 $\frac{\mu C}{cm^2}$ [of the order of that of Barium Titanate (BTO)], high Curie temperature of 426 K (above that of BTO), large dielectric constant, and low dielectric loss.

DIPAB exhibits fairly good piezoelectric response and well-defined ferroelectric domains. These characteristics make it a molecular alternative to perovskite ferroelectrics and ferroelectric polymers in sensing, actuation, data storage, electro-optics, and molecular or flexible electronics.

2. COMPUTATIONAL METHODS

First-principles density functional theory [22, 23] simulations are used in our study. The electronic structure was calculated using the Projector Augmented Wave (PAW) method [24] as implemented in the Vienna *ab initio* simulation package VASP [25]. The Generalized Gradient Approximation (GGA) of Perdew-Burke-Ernzerhof (PBE) form [26] was used for the exchange-correlation functional. Hybrid Fock exchange/density functional theory functional have shown to be very successful in describing a wide range of molecular properties. A hybrid functional [27] that evaluate the Fock exchange in real space was introduced. We used the hybrid functional HSE06 to calculate the electronic and optical properties of DIPAB.

Our results indicate that the hybrid functional indeed often improve the description of these properties. We set the plane-wave-cut-off energy to 400 eV. We used a 4 x 4x 4 Monkhorst- Pack grid to minimize the computation time. Instead of the all-electron electron-ion potential norm-conserving pseudopotentials were used. The relaxed atomic positions were followed by minimizing the total energy (i.e., as small as 10^{-6} eV), and the Hellmann-Feynman forces. These forces were as small as 0.003eV/Å at convergence in the unit cell of all different electronic structure relaxations that have been performed.

IOP Conf. Series: Materials Science and Engineering **92** (2015) 012017 doi:10.1088/1757-899X/92/1/012017

3. RESLUTS AND DISCUSSIONS

3.1. Structural properties of the $P2_1(\alpha)$ ferroelectric phase of DIPAB crystal

Table 1. The structural properties and spontaneous polarization of polar monoclinic $P2_1(\alpha)$ phases of DIPAB ferroelectric molecular crystal.

Parameter	$P2_1(\alpha)$ phase
Empirical formula	$C_6\ H_{16}\ Br\ N$
	$P2_1$
Space group	$P2_1$
Lattice parameter *a* (Å)	7.799
Lattice parameter *b* (Å)	8.067
Lattice parameter *c* (Å)	7.584
α°	90
β°	116.231
γ°	90
Equilibrium volume *V* (Å^3)	443.30
Spont. polarization Ps $(\frac{\mu C}{cm^2})$	-22.64

As shown in table 1, we employed XRD-derived structural parameters as initial structure in our *ab-initio* calculations [28]. We performed gradient-descent optimization of the structure. The obtained lattice parameters are presented in table 1. The relaxed structures are very close to the experimentally observed ones.

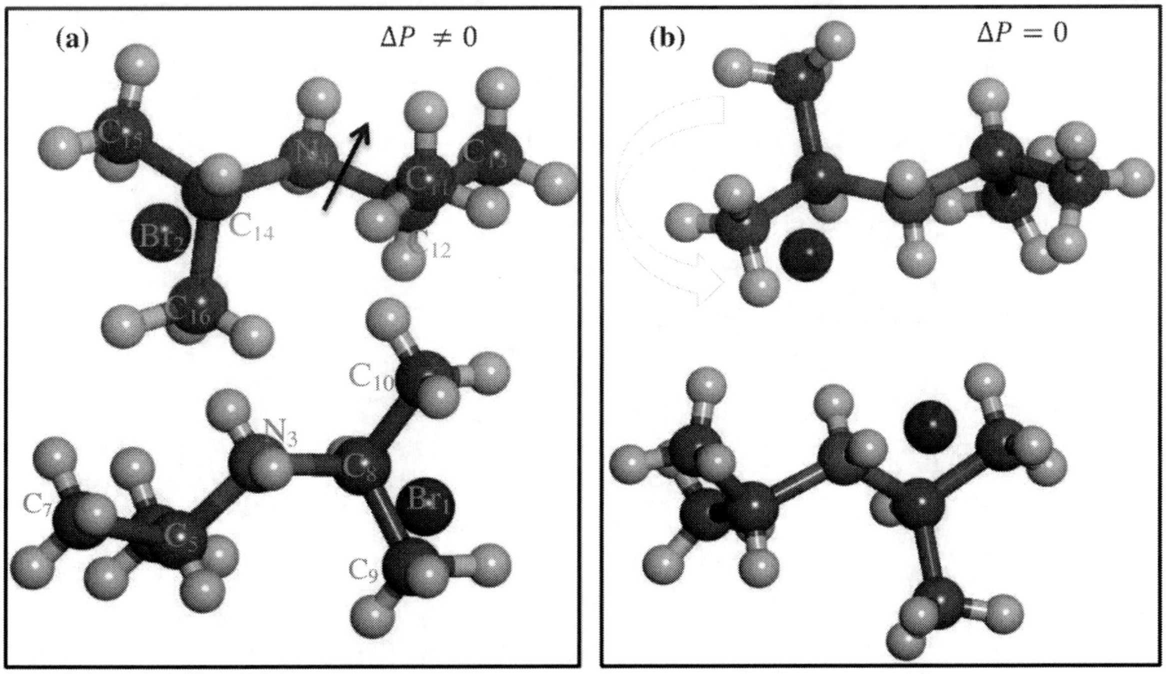

Figure 1. Schematic view of the molecular structure of the (a) $P2_1(\alpha)$ non cental-symmetric polar ferroelectric phase and (b) $P2_{1/m}(\beta)$ paraelectric centrosymmetric phase.

When a dielectric is placed in an electric field, it becomes polarized, producing a dipole with a dipole moment given by the product of the charges and their separation distance. The polarization is then defined as the dipole moment per unit volume. If the dielectric is non central-symmetric like the one shown in figure 1(a) (i.e., $P2_1(\alpha)$ polar phase of DIPAB), a relatively large piezoelectric strain proportional to electric field is possible. As can be seen from table 1, the magnitude of spontaneous polarization of $P2_1(\alpha)$ phase of DIPAB is 22.64 $(\frac{\mu C}{cm^2})$. This is quite remarkable and indicate that this molecular ferroelectric crystal could be used as an alternative for some perovskite materials in a wide range of application.

3.2. Band structure and density of states

For a better understanding of the electronic and optical properties of DIPAB, the investigation of electronic energy band structure can be useful. The calculated band structures of ferroelectric DIPAB using GGA method is shown in figure 2. The band structure has been computed and the high symmetry directions on the surface of the irreducible BZ are plotted. The Fermi level is set to zero. The obtained band gap of $P2_1$ polar phase is found ≈ 5 eV using GGA method (figure 2). To the best of our knowledge, we are not aware of any experimental or theoretical works that reported the electronic band structure of DIPAB. Furthermore, the density of states (DOS) and the partial density of states (PDOS) of monoclinic $P2_1$ DIPAB are shown in figures (3-9). The DOS and PDOS plots give a quick qualitative picture of the electronic structure of a material and sometimes they can be related directly to the electronic transitions and experimental spectroscopic results.

As can be clearly seen, the valence band maximum is mainly composed of Br *4p* orbitals and the conduction band minimum is dominated by C *2p*, C *2s*, and N *2s* orbitals. The figures indicate that the interband transitions are mainly from the valence bands of Bromine (Br) atoms to the conduction bands of Carbon (C) and Nitrogen (N) atoms. Examining the total density of states using GGA and HSE06 methods, we found that the optical band gap obtained using HSE06 is ≈ 6 eV, while that obtained using GGA method is ≈ 5 eV consistent with value of the optical band gap obtained from the electronic band structure plot using GGA method. Our results indicate that using HSE06 hybrid pseudopotentials rather than GGA method leads to a higher optical band gap. This is due to the fact that GGA underestimates the optical band gap.

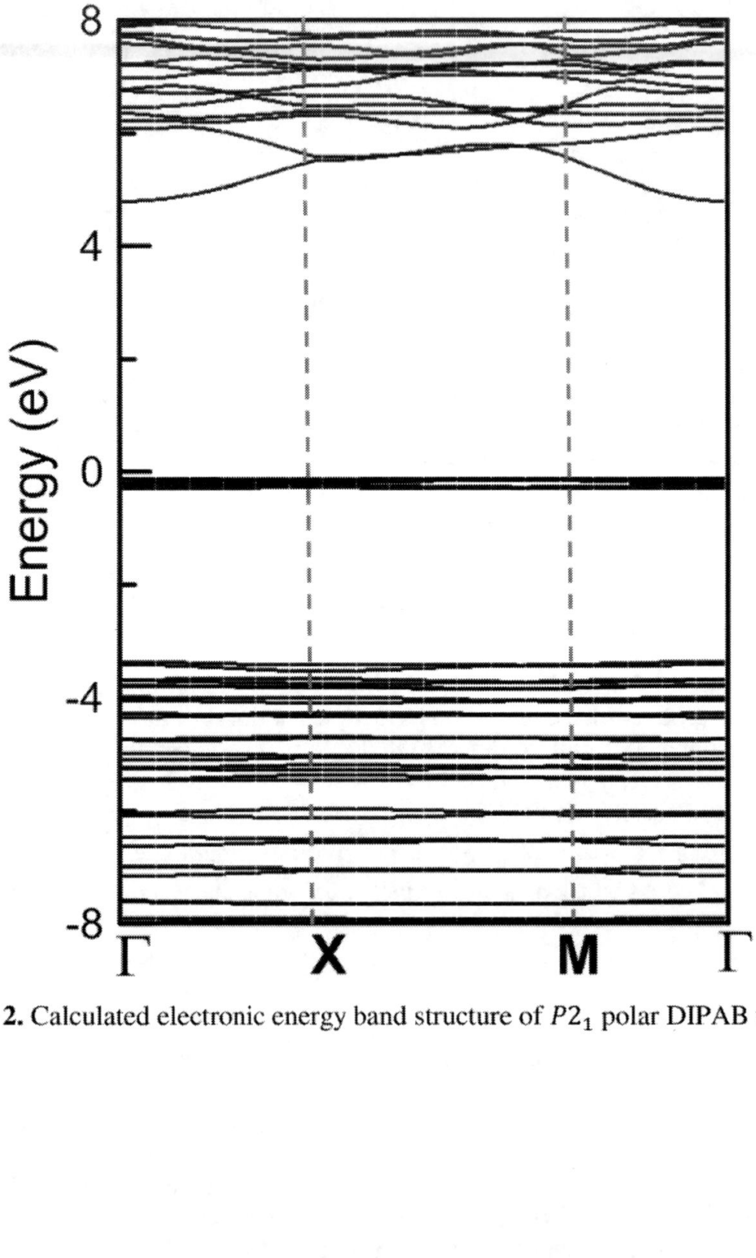

Figure 2. Calculated electronic energy band structure of $P2_1$ polar DIPAB using GGA method.

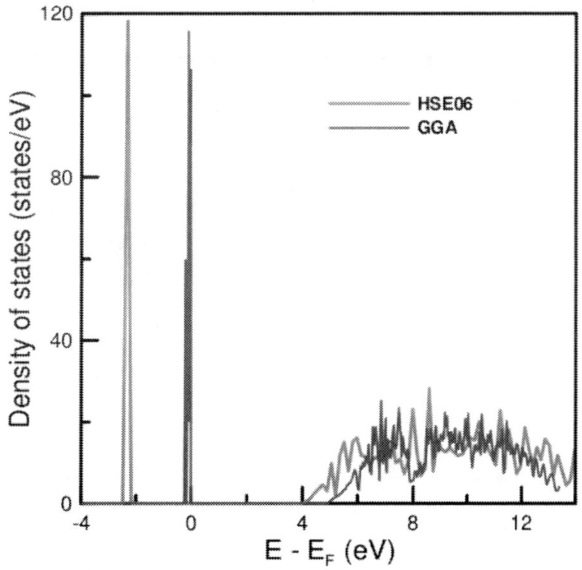

Figure 3. The total density of states (DOS) of $P2_1$ polar DIPAB calculated using GGA (Blue) and HSE06 Hybrid functional (Red).

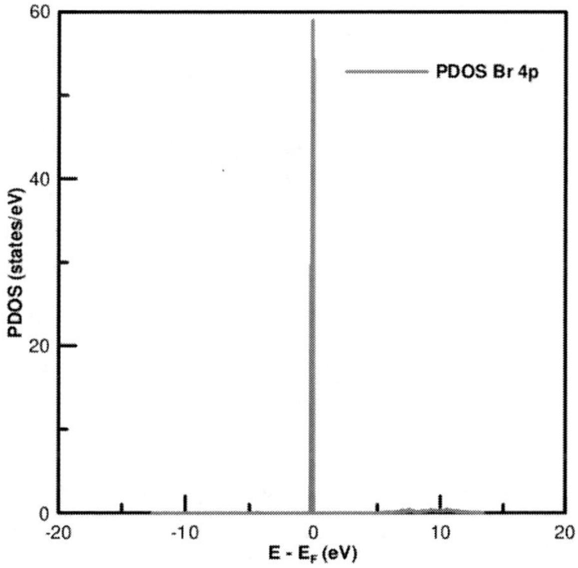

Figure 4.The partial density of states (PDOS) of Bromine *4p* orbitals of $P2_1$ polar DIPAB.

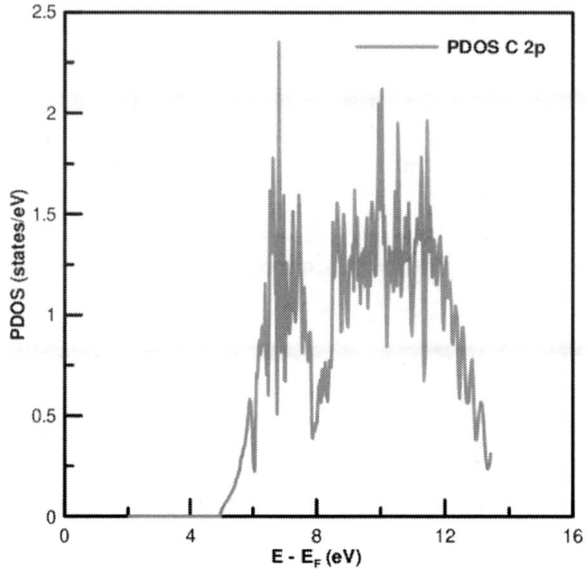

Figure 5.The partial density of states (PDOS) of Carbon *2p* orbitals of $P2_1$ polar DIPAB.

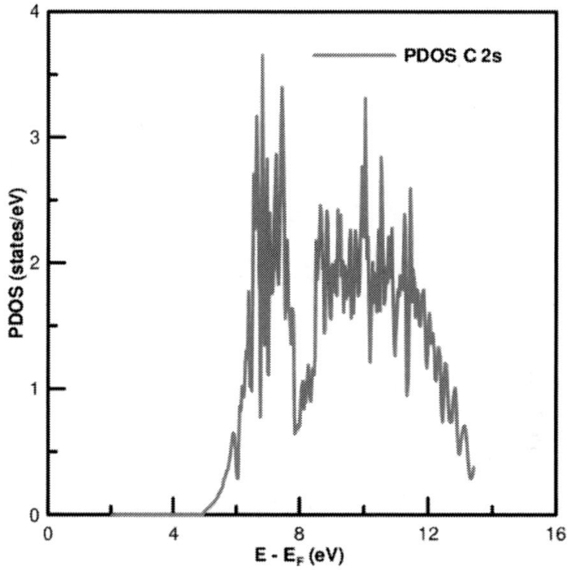

Figure 6.The partial density of states (PDOS) of Carbon *2s* orbitals of $P2_1$polar DIPAB.

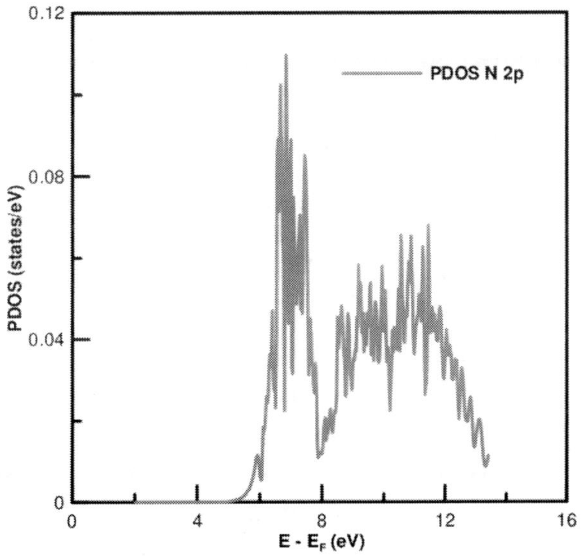

Figure 7. The partial density of states (PDOS) of Nitrogen *2p* orbitals of $P2_1$polar DIPAB.

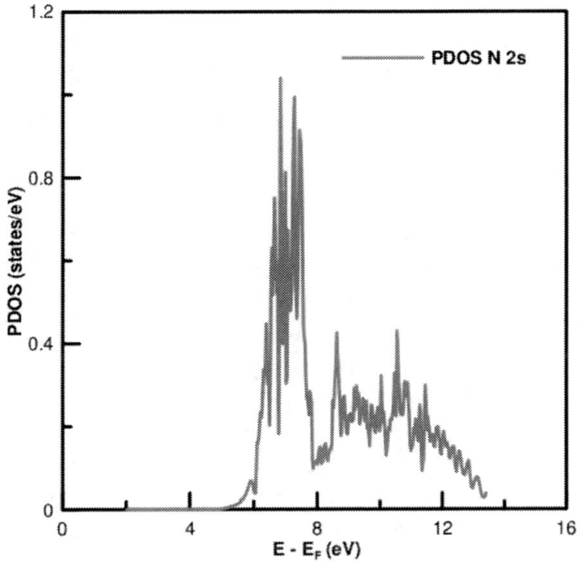

Figure 8. The partial density of states (PDOS) of Nitrogen *2s* orbitals of $P2_1$ polar DIPAB.

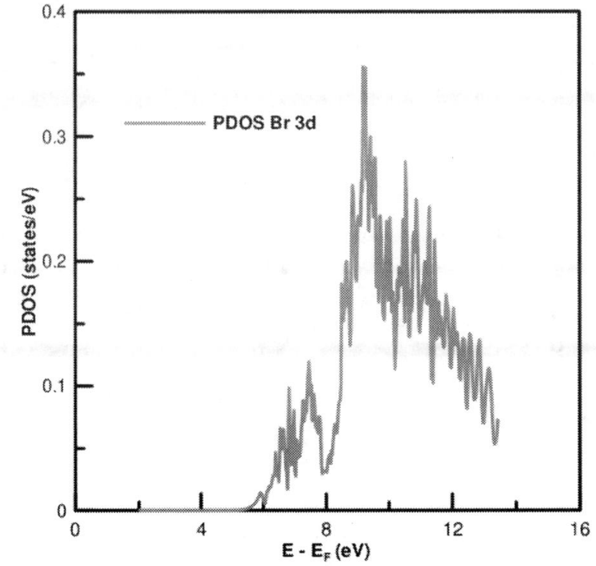

Figure 9. The partial density of states (PDOS) of Bromine *3d* orbitals of $P2_1$ polar DIPAB.

CONCLUSIONS

In summary, we present *ab initio* calculations of the electronic energy band structure, density of states, partial density of states and spontaneous polarization of the polar DIPAB molecular crystal. We found that the optical band gap of polar DIPAB is ≈ 5 eV indicating that the material is a dielectric. The density of states and partial density of states reveal that the major interband transitions in ferroelectric DIPAB occur from the valence band of bromine to the conduction bands of carbon and nitrogen atoms in polar DIPAB. Examining the total density of states using GGA and HSE06 methods, we found that the optical band gap obtained using HSE06 is ≈ 6 eV, while that obtained using GGA method is ≈ 5 eV. Furthermore, we report results on spontaneous polarization of non central-symmetric polar ferroelectric phase of DIPAB. We found that polar DIPAB exhibits a spontaneous polarization of 22.64 $\frac{\mu C}{cm^2}$ indicating that this remarkable ferroelectric molecule could be used as an alternative for some pervoskites in a wide range of piezoelectric applications, electronics applications such as capacitor, sensor, and data storage, as well as to electromechanical and optical devices. To the best of our knowledge, our results on electronic and ferroelectric properties of the polar phase of DIPAB crystal are the first to be reported. We hope that our findings can pave the way for further experimental investigations on electronic, piezoelectric and optical properties of this remarkable molecule.

ACKNOWLEDGMENTS

The authors would like to acknowledge the technical and financial support provided by the faculty of scientific research at Jordan University of Science and Technology (JUST), and by Khalifa University of Science and Research (KUSTAR), UAE.

References

[1] Q Ye , Y.-M Song, G.-X Wang, K Chen, D.-W Fu, P W H Chan, J.-S Zhu, S D Huang, R.-G Xiong 2006 *J. Am. Chem. Soc.* **128** 6554.
[2] H.-Y Ye, D.-W Fu, Y Zhang, W Zhang, R.-G Xiong, S D Huang 2009 *J. Am. Chem. Soc.* **131** 42.
[3] W Zhang, R.-G Xing, S D Huang 2008 *J. Am. Chem. Soc.* **130** 10468.
[4] D.-W Fu, Y.-M Song, G.-X Wang, Q Ye, R.-G Xiong , T Akutagawa , T Nakamura , P W H Chan , S D Huang 2007 *J. Am. Chem. Soc.* **129** 5346.
[5] W Zhang , L.-Z Chen , R.-G Xiong , T Nakamura , S D Huang 2009 *J. Am. Chem. Soc.* **131** 12544.

[6] W Zhang, H.-Y Ye, H.-L Cai, J.-Z Ge , R.-G Xing, S P D Huang 2010 *J. Am. Chem. Soc.* **132** 7300.

[7] T Hang, W Zhang, H.-Y Ye, R.-G Xiong 2011 *Chem. Soc. Rev.,* **40** 3577.

[8] H.-L Cai , W Zhang , J.-Z Ge , Y Zhang , K Awaga , T Nakamura , R.-G Xiong 2011 *Phys. Rev. Lett.* **107** 147601.

[9] G.-C Xu , W Zhang , X.-M Ma , Y.-H Chen , L. Zhang , H.-L Cai , Z.-M Wang , R.-G Xiong , S Gao 2011 *J. Am. Chem. Soc.* **133** 149.

[10] J Valasek, "Piezoelectricity and Allied Phenomena in Rochelle Salt" 1921 *Phys. Rev.* **17** [4] 475-81.
[11] J Valasek, "Piezoelectric Activity of Rochelle Salt Under Various Conditions" 1922 *Phys. Rev.* **19** 478-91.
[12] J Valasek, "Properties of Rochelle Salt Related to the Piezoelectric Effect" 1922 *Phys. Rev.* **20** 639-64.
[13] J Valasek, "Dielectric Anomalies in Rochelle Salt" 1924 *Phys. Rev.* **24** [5] 560-68.
[14] Da-Wei Fu, Hong-Ling Cai, Yuanming Liu, Qiong Ye, Wen Zhang, Yi Zhang, Xue-Yuan Chen, Gianluca Giovannetti, Massimo Capone, Jiangyu Li, Ren-Gen Xiong 2013 *SCIENCE* vol **339,** 425-428.
[15] K Uchino, *Ferroelectric Devices* 2000 (Marcel Dekker, NewYork).

[16] M E Lines and A M Glass, *Principles and Applications of Ferroelectrics and Related Materials* 1977 (Oxford University, New York).

[17] J F Scott 2007 *Science* **315** 954.

[18] R D King-Smith and D Vanderbilt 1993 *Phys. Rev. B* **47** 1651.

[19] R Resta 1994 *Rev. Mod. Phys.* **66** 899.

[20] R E Cohen 1992 *Nature* (London) **358** 136.

[21] T Egami, S Ishihara, and M Tachiki 1993 *Science* 261 1307.

[22] G Giovannetti *et al* 2011 Proximity of iron pnictide superconductors to a quantum tricritical point. *Nat. Commun.* **2** 398.

[23] P Hohenberg, W Kohn, Inhomogeneous electron gas 1964 *Phys. Rev.***136** (3B), B864.

[24] J P Perdew, J A Chevary, S H Vosko, K A Jackson, M R Pederson, D J Singh, and C Fiolhais 1992 *Phys. Rev. B* **46** 6671.

[25] G Kresse, D Joubert 1999 *Phys. Rev. B* **59** 1758.

[26] J P Perdew, K Burke, M Ernzerhof 1996 *Phys. Rev. Lett.* **77** 3865.

[27] HSE03, J Heyd, G E Scuseria, and M Ernzerhof 2003 *J. Chem. Phys.* **118** 8207.

[28] Anna Piecha, Anna Gągor, Ryszard Jakubas and Przemysław Szklarz 2013 "Room-temperature ferroelectricity in diisopropylammonium bromide" ***Cryst. Eng. Comm.*****15** 940-944.

Use of mesoporous silicate nanoparticles as drug carrier for mefenamic acid

F M Mustafa[1], H A Hodali[1]
[1]Department of Chemistry, Faculty of Science, University of Jordan, Amman, 11942, Jordan.

h.hodali@ju.edu.jo

Abstract. The aim of the study is to evaluate the use of mesoporous silicate nanoparticles for the loading and release of the non-steroidal anti-inflammatory drug, mefenamic acid. Nanoparticles of the mesoporous silicate materials, MCM-41 and SBA-16 were synthesized and characterized by XRD, SEM, FT-IR, TGA and BET surface area techniques. Both silicate systems were loaded with mefenamic acid with loading capacities of 18.6% and 11.6%, respectively. The *in vitro* release of mefenamic acid into simulated body fluid (pH = 7.4) at 37°C was investigated. The percent release was nonlinearly regressed against time (t) according to the first order kinetic release model; Higuchi's first burst model and Kopcha's empirical model. The overall %release was obtained for both silicate systems and was found to be about 60%. Analysis of results show the rate of drug release is more rapid from SBA-16 (the more interconnected porous network) than from MCM-41. It also show that drug release from either mesoporous silicate is a diffusion controlled process.

1. Introduction

The 2D hexagonal mesoporous material MCM-41 was first synthesized by Mobil Corporation scientists [1]. Following that, many other silicate mesoporous materials have been synthesized such as SBA-series. The SBA-16 system (Santa Barbara Amorphous No.16) exhibits a 3D cubic structure and characterized by thick walls [2]. Due to their ordered porous structure, thermal stability, tuneable pore sizes and ability to be functionalized, mesoporous silicate materials have been employed in several applications, including water treatment [3], gas sensing [4], catalysis [5] and drug delivery [6].

Recently, mesoporous silicate nanoparticles have been investigated extensively as drug carriers, due to their high permeability and good biocompatibility [7]. Other favourable features of nanoparticles of interest are: tuneable particle size, tuneable pore diameter and high surface area. Nanoparticles of MCM-41 were first prepared by two different research groups [8-9] and the first use of mesoporous silicate nanoparticles in drug delivery was by Lai *et al.* [10]. Herein, we report on the use of nanoparticles of MCM-41 and SBA-16 for loading and release of mefenamic acid (Figure 1), a non-steroidal anti-inflammatory drug (NSAID) [11]. It is a poorly soluble in aqueous media with a solubility in water = 4.18 mg/L at 37°C [12]. Enhancing the solubility of mefenamic acid have been investigated by many research groups via using solid dispersions [13-16], encapsulation by polymeric nanoparticles [17], inclusion into β-cyclodextrin [18-19], nanostructured lipid carrier [20] and microspheres [21-23].

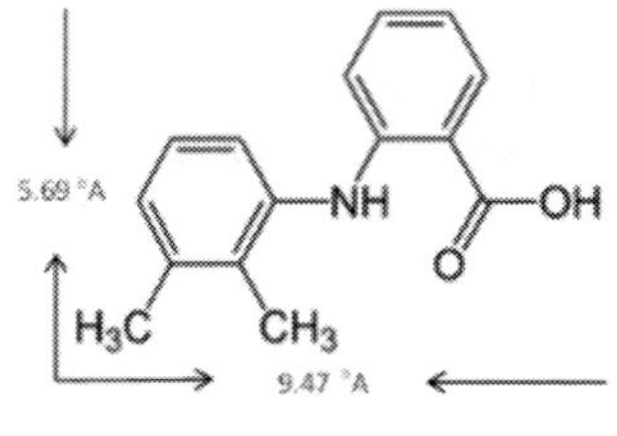

Figure 1. Schematic structure and dimensions of mefenamic acid

2. Experimental Details

2.1. Materials

Tetraethyl orthosilicate ($Si(OC_2H_5)_4$), cetyltrimethylammoniuom bromide $CH_3(CH_2)_{15}N\text{-}(CH_3)_3Br$ (CTAB) and pluronic 127 tirblock copolymer, poly(ethyleneoxide)-*block*-poly(propylene oxide)-

block-poly(ethylene oxide) ($EO_{106}PO_{70}EO_{106}$) with average molecular weight of ~12,500 Daltons were obtained from Sigma Aldrich. Mefenamic acid was kindly donated by Pharma International- Amman. The simulated body fluid was prepared, as described by Oyane *et al.* [25].

2.2. Instrumentation

The morphology and particle size estimates of synthesized mesoporous silicate materials were obtained by scanning electron microscopy (SEM) using Inspect F50 FEI SEM, Netherlands. XRD spectra were obtained using XRD-7000 Shimadzu diffractometer. XRD spectra of SBA-16 were obtained using the low angle instrument PW-1840 Philips diffractometer. Percent loading and release were estimated using UV-Visible spectrophotometry (Varian Cary 100 Bio UV-Visible spectrophotometer). BET surface areas, and pore diameters were evaluated through BET and BJH modeling using a nitrogen adsorption/desorption instrument (Nova 2200e). Thermal gravimetric analysis (TGA) was conducted using the NetzschSta409 PC instrument (NETZSCH-Ger) operating from room temperature up to 1000°C at rate of 20°C/min. FT-IR spectra were recorded using a Thermo Nicolet NEXUS 670 FT-IR spectrometer (4000 to 400 cm^{-1}, 4 cm^{-1} spectral resolution, KBr pellets).

2.3. Synthesis of mesoporous silicate nanoparticles

Nanoparticles of MCM-41 [9] and nanoparticles of SBA-16 [25] were prepared following literature methods with slight modifications.

2.4. Loading of mefenamic acid

A saturated solution of the drug was prepared by dissolving 1.6 g of mefenamic acid in 10 mL tetrahydrofuran (THF) and placed in a Schlenk flask with an O-ring cap. A 0.80 g sample of the mesoporous material was added to the solution and the mixture was stirred at room temperature for 24 hours. It was then centrifuged at 5000 rpm for 5 minutes and the solid material was separated, washed with propan-2-ol (2 mL) and dried at 50°C in vacuum for 3 hours. The percent loading was determined spectrophotometrically using the following procedure: Accurately weighed about 15 mg of the loaded mesoporous silicate were soaked in 100.0 ml of aqueous 0.1 M NaOH and the mixture was stirred for 24 hours. The mixture was filtered using a 0.2 μm nylon filter. A blank solution was prepared by dissolving 15 mg of the unloaded mesoporous material in 100.0 ml 0.1 M NaOH. With suitable dilution for calibration curve, the UV-absorbance was measured against the appropriately diluted blank at λ=285 nm.

2.5. In vitro release of mefenamic acid

An accurately weighed disc about 0.3 g of the drug-loaded nanoparticles were soaked in 250.0 mL of SBF at pH 7.4 and 37°C, in a thermostatic bath shaker. Out of the resulting solution, 5.00 mL samples were withdrawn at predetermined time intervals. The withdrawn samples were readily replaced with the same volume of fresh SBF. The withdrawn samples were filtered using a 0.2 μm nylon filter. The concentration of the drug released was estimated from the corresponding absorbance measured at 285 nm against a calibration curve of the drug in SBF at pH 7.4 and 37°C.

3. Result and Discussion

3.1. Characterization of mesoporous silicate nanoparticles

The synthesized samples of mesoporous silicates, MCM-41 and SBA-16, were characterized by SEM, XRD, FT-IR, N_2 adsorption/desorption isotherms and TGA. The SEM images for MCM-41 and SBA-

16 are shown in Figure 2. Both SEM micrographs of MCM-41 and SBA-16 show nanoparticles with spherical morphology [9,25] and with average diameter of about 31 nm. The XRD spectrum of MCM-41 shows a pattern with a strong peak at 2θ of about 2.7 degrees and low intensity peaks in the 2θ range of 4-6 degrees. The pattern is comparable with that reported by Cai *et al.* [9]. The XRD pattern of SBA-16 shows a peak at 2θ of 0.7 degrees in agreement with values reported in literature [25].

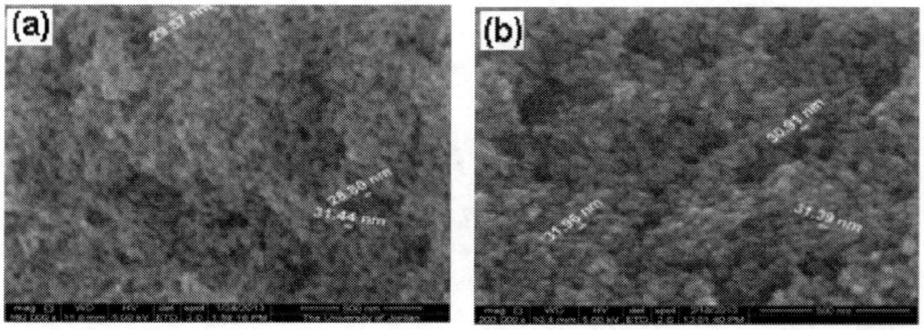

Figure 2. SEM images of (a) MCM- 41, and (b) SBA-16.

The FT-IR spectra of both systems show characteristic bands that are consistent with literature reported data [26]. The corresponding bands assignment were hydrogen-bonded Si-O-H and adsorbed water at 3437cm^{-1}, Si-O-Si asymmetric stretching (external) at 1250cm^{-1},Si-O stretching symmetric (external) at 965cm^{-1} and H-O-H bending at 1635cm^{-1}.

Nitrogen adsorption-desorption isotherms were used to estimate the specific surface area (m^2/g) and the average pore diameter (nm) for MCM-41 and SBA-16. The adsorption branch was used to estimate the specific surface area using the Brunauer-Emmett-Teller (BET) theory, while the desorption branch was used to estimate the average pore diameter using the Barrett-Joyner-Halenda (BJH) theory. Both systems show isotherms (Figure 3) characteristic of type IV. The increase in the adsorption curve at relative pressure P/P_o (0.3 for MCM-41 and 0.58 for SBA-16) corresponds to capillary condensation within a uniform mesoporous silicate material [25].The surface area and the pore size for MCM-41 and SBA-16 were estimated at 615 m^2/g, 3.05 nm and 957 m^2/g, 3.04nm, respectively.

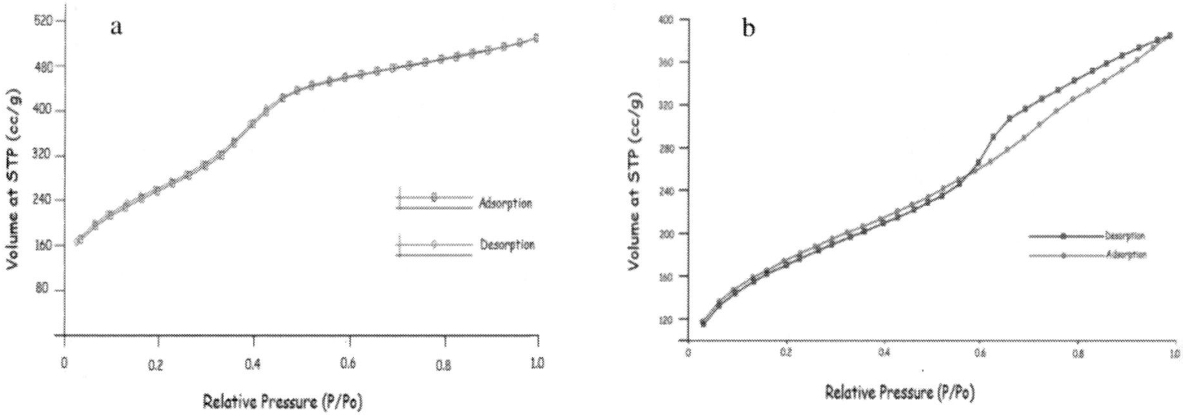

Figure 3. Nitrogen adsorption/desorption isotherms for (a) MCM-41 and (b) SBA-16.

The thermal gravimetric analysis for MCM-41were measured in the temperature range between room temperature to 1000°C. The mass loss in the temperature range, 25 -300°C is due to desorption of water molecules within the pores in addition to the water molecules that are hydrogen-bonded to the surface silanol groups, while the mass loss in the range (300-1000°C) is due to the condensation of silanol groups [9,27]. The densities of surface silianol groups that were extracted from TGA thermogravimetric curves of unloaded MCM-41 and SBA-16 are 1.25 and 1.28 N_{Si-OH}/nm^2 , respectively. The very close values of surface silanol group densities in both mesoporous silicate nanoparticles would suggest similar behaviour in loading and release.

3.2. Loading and release of mefenamic acid

As previously mentioned, the dimensions of mefenamic acid (Figure 1) are smaller than pore dimensions of MCM-41 or SBA-16 silicate systems; so it can be encapsulated into their pores. In the loading procedure, a saturated solution of mefenamic acid in THF was used. THF was chosen because of its high capacity to dissolve mefenamic acid, as compared with other solvents (acetone, ethanol, and aqueous ethanol). The loading capacity of mefenamic acid into the mesoporous materials was determined by UV absorption spectrophotometry. The results of %loading was 18.6 in MCM-41 and 11.6 in SBA-16. Loading of the drug into the silicate system was confirmed by IR spectrometry. The IR- spectra of the loaded samples show the presence of mefenamic acid by the appearance of the corresponding characteristic vibrational bands at 1651 cm^{-1}, 1576 cm^{-1}, 1507 cm^{-1}, 1447 cm^{-1}, which denote the carbonyl, and aromatic carbon-carbon stretching vibrations [28].

The *in vitro* release of mefenamic acid from the mesoporous silicates MCM-41 and SBA-16 were measured to explore possible factors affecting the release process. The mefenamic acid concentration in the release medium was analyzed using UV absorption spectrophotomerty, by measuring the absorbance at 285 nm using a blank sample for base line correction. The concentration of mefenamic acid was estimated with reference to a calibration curve of mefenamic acid in SBF at the same conditions according to the linear equation (1):

$$A_{285\,nm} = 36.697\ C\ (mg/ml) + 0.0194 \qquad (1)$$

The release experiments were conducted in triplicate, and the results for the three experimental runs were averaged and analyzed for %release kinetics and kinetic parameters.

The measured concentrations of mefenamic acid (C_t in mg/mL) released into the SBF buffer solution, at different time intervals t, were used to calculate the mass of drug (Q_t in mg) released into the medium of volume V in mL according to $Q_t = C_t \times V$. This was subsequently corrected for concentration depletion due sampling to obtain the cumulative drug release (equation 2)

$$Q_{tcorr} = Q_t + (v / V)\, Q_{t-1} = Q_t + (v / V) \Sigma_1^{t-1} Q_t \qquad (2)$$

where Q_t and Q_{tcorr} are both given in mg, the sample volume is typically 5.00 mL, V is the volume of the SBF release medium which was kept constant (typically 250 mL) withdrawn samples with same volumes from the SBF buffer. The corrected drug release Q_{tcorr} was divided by the drug load (W_D) to obtain the percent drug release, according to: % Release = $(Q_{tcorr}/ W_D)*100\%$. The percent release was nonlinearly regressed against time t according to the first order kinetic release model [29] according to equation (3):

$$Q_{tcorr} / W_D = 1 - e^{-k\,t} \qquad (3)$$

where the best fit estimate of the first order kinetic release rate constant (k) is obtained. Further modeling of drug release against time was carried out using Higuchi's first burst model [28],

$$Q_{tcorr}/W_D = k_H t^{1/2} + c \qquad (4)$$

and Kopcha's empirical model [30],

$$Q_{tcorr}/W_D = At^{1/2} + Bt \qquad (5)$$

In the Higuchi's first burst model, k_H ($h^{-1/2}$) is the Higuchi first burst constant, and c is a constant characteristic of the drug and host being formulated. It is based on Fickian diffusion, applied to the initial first burst of drug release. This model assumes that the % drug release (Q_t/W_D) varies with the square root of time ($t^{1/2}$), which typically occurs relatively fast, compared with subsequent release. In contrast, Kopcha's model yields estimates of the individual contributions of drug diffusion (A) and erosion (B) to %Release. The *first order* release profiles for mefenamic acid loaded into the mesoporous carriers MCM-41 and SBA-16 are shown in Figure 4, which depicts plots of cumulative drug quantity released against time (solid data points).

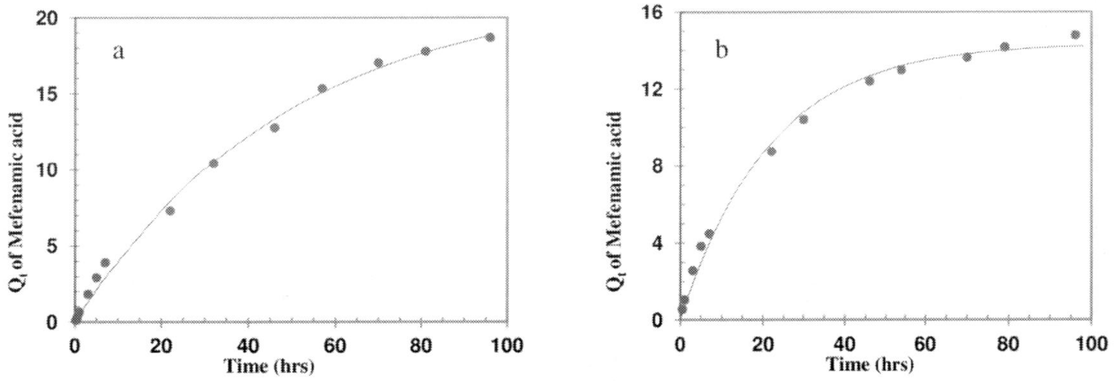

Figure 4. Release profile of (a) mefenamic acid from MCM-41 and (b) from SBA-16

A summary of the loading and release parameters for mefenamic acid are listed in Table 1, below.

Table 1. Mefenamic acid optimized loading and release, and kinetic release parameters

Carrier	% Loading	W_D [a] (mg)	% Release	k [c] (h^{-1})	k_H [d] ($h^{-1/2}$)	A/B [e]
MCM-41	18.6	37.8	59.6%	0.0202	0.044	0.036 / 0.002
SBA-16	11.6	23.9	61.6%	0.0457	0.090	0.075 / 0

[a] W_D = drug load in mg.
[b] % Release = 100 ×Q_{tcorr}/ W_D
[c] k = First order kinetic release rate constant (1st order model ⇒ % Release = $1 - e^{-kt}$)
[d] k_H =Higuchi first burst release model (Higuchi's model: % Release = $k_H \times t^{1/2}$ + constant)
[e] A & B are factors denoting the relative contributions of diffusion and erosion processes to drug release (Kopcha's empirical release model ⇒ % Release = A $\times t^{1/2}$ + B × t).

It is observed that %release of mefenamic acid from both MCM-41 and SBA-16 are close to each other. Further, the rate constants are approximately doubled in SBA-16. In addition, Kopcha's constant A is much higher than constant B, thus indicating that release of both drugs from either mesoporous silicate is a diffusion controlled process.

4. Conclusions

The results of the investigation of loading into, and %release, of mefenamic acid from the MCM-41 and SBA-16 reveals that: Mefenamic acid in the SBF does exist as a carboxylate anion (pK_a = 4.55), which interacts with silanol groups that are mostly neutral. Although both systems have the same %loading, MCM-41 exhibits half as much the rate for SBA-16. This could be due to the interconnected porous network of SBA-16. The pores of MCM-41 are one dimensional and thus imposing more restriction to free drug diffusion into the release buffer medium, and hence a lower rate of release.

References

[1] Kresge C, Leonowicz M, Roth W, Vartuli J and Beck J 1992 *Nature* **359** 710.
[2] Zhao D, Huo Q, Feng J, Chmelka B and Stucky G 1998 *JACS* **120** 6024.
[3] Deng Y and Zhao D 2013 Mesoporous Materials for Water Treatment *Functional Nanostructured Materials and Membranes for Water Treatment* (first edition) (Weinheim: Wiley-VCH) chapter 4 pp 67-84.
[4] Wagner T, Haffer S, Weinberger C, Klaus D and Tiemann M 2013 *Chem. Soc. Rev.* **42** 403.
[5] Perego C and Millini R 2013 *Chem. Soc. Rev.* **42** 3956.
[6] Vallet-Regi M, Ra´mila A, Del Real R and Pe´rez-Pariente 2001 *Chem. Mater.* **13** 308.
[7] Joglekar M, Roggers R, Zhao Yand and Trewyn B G 2013 *RSC. Adv.* **3** 2454.
[8] Fowler C E, Khushalani D, Lebeau B and Mann S 2001 *Adv. Mate.***13** 649.
[9] Cai Q, Luo Z-S, Pang W-Q, Fan Y-W, Chen X-H and Cui F-Z 2001 *Chem. Mater.* **13** 258.
[10] Lai C-Yu, Trewyn B G, Jeftinija D M, Jeftinija K Xu S, Jeftinija S and Lin V S-Y 2003 *J. Am. Chem. Soc.* **125** 4451.
[11] Adam A, Schrimp L and Schmidt P C 2000 *Drug Dev Ind Pharm* **26** 477.
[12] Patil1 P B, Gupta V R M, Udupi RH, Srikanth K and Prasad B S G 2010 *RJPBCS.* **1** 544
[13] Rao Ch V P and Nagabhushanam M V 2011 *IJRPC.* 1 1127.
[14] Rao C, Nagabhushanam M and Prabhakar C 2011 *RJPBCS.***2** 1025.
[15] Imai T, Nohdomi K, Acarturk F and Otagiri M 1991 *J. Pharm. Sci.* **80** 484.
[16] Kumar K, Parkash C, Kumar K and Singh S 2009 *IJPSDR.* **1** 164.
[17] Pandey M and Saraf S *JPR.* **3**146.
[18] Haldon T, Pawlaczyk J and Szafran B 1999 *J. Inclusion Phenom. Macrocyclic Chem.* **35** 497.
[19] Mudit D Keshavarao K Selvam P 2011 *IRJP.* **2** 207.
[20] Khurana S and Bedi P M S 2012 *IJNN.* **2** 232.
[21] Dixit M, Kini A G and Kilkarni P K 2012 *Turk. J. Pharm. Sci.* **9** 13.
[22] Etman M, Farid R, Nada A and Ebian 2010 *J. Microencapsulation* **27** 640.
[23] Nair R, Kumar A C K and Priya K V 2011 *JPST.* 3 563.
[24] Oyane A, Kim H-M, Furuya T, Kokubo T, Miyazaki T and Nakamura T 2003 *J. Biomed. Mater. Res. A.* **65** 188.
[25] Thomas M, Slipper I, Walunj A, Jain A, Favretto M E, Kallinteri P and Douroumis D 2010 *Int. J. Pharm* **387** 272.
[26] Marzouqa D M, Zughul M B, Taha M O and Hodali H A 2012 *J.Porous. Mat.***19** 825.
[27] Andrade G F, Soares D C F, Santos R G and Sousa E B 2013 *Microporous Mesoporous Mater.* **168** 102.
[28] Cunha V R R, Izumi C M S, Petersen P A D, Magalhães A, Temperini M L A, Petrilli H M and Constantino V R L 2014 *J Phys Chem B.* **118** 4333.
[28] Heikkila T, Salonen H J, Tuura J, Kumar N, Salmi T Y, Murzin D Y, Hamdy M S, Mul G,Laitinen L, Kaukonen A M, Hirvonen and Lehto V P 2007 *Drug Deliv* **14** 337.
[29] Higuchi T 1963 *J Pharm Sci* **52** 1145.
[30] Chevalier E, Viana M, Artaud A, Haddouchi S and Chulia D 2009 *Talanta* **77** 1545.

Synthesis and structural characterization of non-stoichiometric barium hexaferrite materials with Fe:Ba ratio of 11.5 – 16.16.

Yazan Maswadeh , Sami H. Mahmood, Ahmad Awadallah , Aynour N. Aloqaily

Physics Department, The University of Jordan, Amman 11942, Jordan

e-mail: s.mahmood@ju.edu.jo

Abstract: Synthesis of barium hexaferrites $BaFe_{12}O_{19}$ (BaM) is often accompanied by the presence of secondary nonmagnetic phases. The coexistence of these phases reduces the yield of the desired BaM magnetic phase and screens its intrinsic magnetic properties such as the saturation magnetization, and impacts the magnetic properties of the sample negatively. Therefore, assessment of the abundance of these phases and investigating their effect on the structural properties of the sample is of fundamental and practical importance. In this work, BaM hexaferrites were prepared by ball milling and sintering powder precursors with Fe:Ba molar ratios varying from 11.5 to 16.16. The structural properties of the phases in the samples were investigated by x-ray diffraction (XRD). The weight ratios of the different phases, as well as their refined structural parameters were determined using Rietveld analysis. XRD patterns revealed the development of α-Fe_2O_3 (hematite) phase with increasing relative diffracted intensity as the Fe:Ba molar ratio increased. The evolution of the intensity of this phase was used to monitor the weight ratio of the secondary hematite phase in the sample, and a relation between the its weight ratio and the Fe:Ba ratio was established. The optimal Fe:Ba ratio required to synthesis a pure barium hexaferrite phase was then determined, and found to be 11.7.

1. Introduction

Magnetic materials are widely used in our everyday life, covering the range from being employed in components of electronic devices and equipment to refrigerator stickers and parts of children toys [1-3]. Ferrites existing in different forms such as cubic ferrites, hexagonal ferrites, orthoferrites and garnets, are magnetic ceramics based on iron oxides, and form a class of important magnetic materials due to their desirable magnetic and electrical properties, chemical stability, and low cost of production [1 - 7] . These materials play an important role in the technological and industrial progress, and the evolution of new applications, where for example they were efficiently used in telecommunication and radar technologies, digital storage devices, and motor industry. The demand for low-cost novel magnetic materials with properties suitable for specific needs of the various technological and industrial applications had therefore generated a great interest in the synthesis and characterization of hexagonal ferrites due to their improved and tunable properties [1-3].

M-type barium hexaferrite $BaFe_{12}O_{19}$ is the most important type of hexaferrites in terms of the annual rate of production of magnetic materials [2]. In search of new magnetic materials with modified properties, M-type hexaferrites were synthesized with Ba replaced partially or fully by different cations such as Sr, Pb, and Ca [8-13], and Fe partially substituted by trivalent metals [12, 14-17], or combinations of divalent and tetravalent cations [13, 18-24]. In many cases, the product contained undesired secondary phases which can deteriorate the properties of the synthesized samples [25-32]. One of the commonly observed nonmagnetic phases in synthesized M-type hexaferrites is α-Fe_2O_3, which can be associated with the choice of the Fe:Ba molar ratio in the starting powder [10, 14, 29, 33-37]. Consequently, extensive research was carried out to synthesize M-type hexaferrites from starting powders with off-stoichiometric Fe:Ba ratio [25, 33, 37-43]. In an attempt to obtain a single BaM phase, different research groups used a ratio of Fe:Ba below the theoretical value of 12, and generally, a ratio between 11 and 12 was reported to yield almost single-phase samples [30, 31, 37, 41, 42, 44].

The presence of secondary phases together with the desired BaM phase poses difficulty in understanding the intrinsic properties of the synthesized samples. In addition, these phases influence the properties of the fabricated sample negatively, due to the reduction of the yield of the desired phase. Accordingly, accurate assessment of the fractions of secondary phases in the sample is of both fundamental and practical importance. In this paper the results of the structural refinement for the barium hexaferrites with Fe:Ba ratio ranging from 11.5 – 16.16 are discussed. The relative diffracted intensity of the evolving α-Fe_2O_3 (hematite) phase with increasing (Fe:Ba) ratio was used to determine the weight ratio of the secondary hematite phase in the sample. In addition, a relation between the weight ratio of the hematite phase and the starting Fe:Ba ratio was established, from which the optimal ratio required for the synthesis of a pure hexaferrite phase was determined. A summary of crystallographic information is also included, and the unit cell of each crystallographic phase is constructed from the refined crystallographic parameters.

2. Building Blocks of M-type Hexa Ferrites

The M-type barium hexaferrite (also known as Ferroxdure or BaM) is a ferrimagnetic material with chemical formula $BaFe_{12}O_{19}$, melting point of 1390° C, and theoretical density of 5.3 g/cm^3 [2]. The symmetry of the barium hexaferrite crystal structure is characterized by the space group ($P6_3/mmc$) with lattice parameter $a = b = 5.89$Å, c = 23.17 Å, $\beta = \alpha = 90°$, $\gamma = 120°$ [2].

The unit cell of M-type barium hexaferrite contains two $BaFe_{12}O_{19}$ molecules, and is built by stacking S and R structural blocks along the hexagonal *c*-axis in the sequence RSR*S* (the star indicates rotation by 180° about the *c*-axis) [2, 7]. The S block (Fig. 1) with chemical formula Fe_6O_8 consists of two close-packed oxygen layers arranged in an fcc stacking sequence, each layer consisting of four oxygen ions. The R block (Fig. 2) with chemical formula $BaFe_6O_{11}$ consists of three close-packed layers in an hcp stacking sequence, with Ba^{2+} ion replacing an O^{2-} ion substitutionally in the middle layer. The small metal cations occupy interstitial sites between the oxygen layers: one iron cation at (2a) octahedral site and two iron cations at ($4f_1$) tetrahedral sites within the S block, one iron cation at the (2b) trigonal bi-pyramidal site and two iron cations at the ($4f_2$) octahedral sites within the R-block, and six iron cations at (12k) octahedral sites in the R-S interfaces. (Table 1) summarizes the distribution of magnetic metal ions over the different sites of the M-type barium hexaferrite lattice, together with their coordinations, locations in the lattice, and spin orientations.

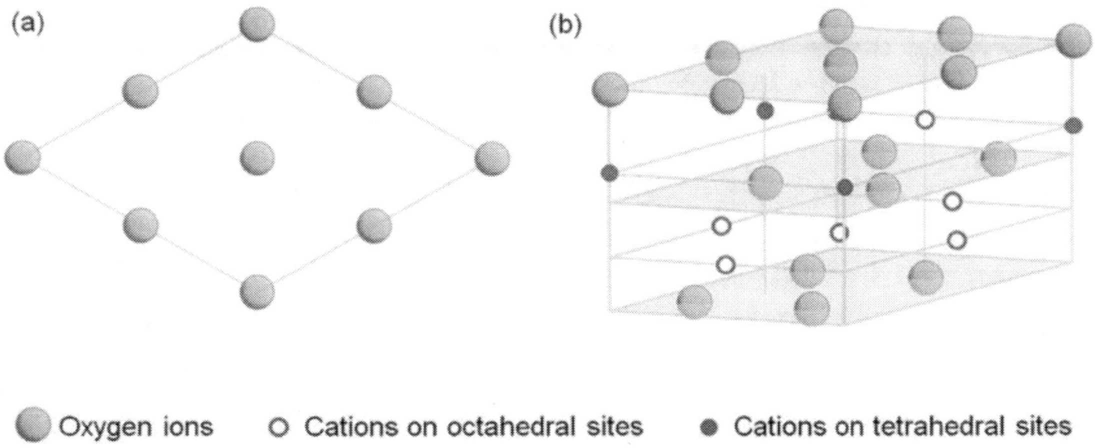

Figure 1. The S block in barium hexaferrite: (a) Top oxygen R layer viewed from above. (b) The two layers of the S-block with the oxygen ions and metal cations in the interstitial sites.

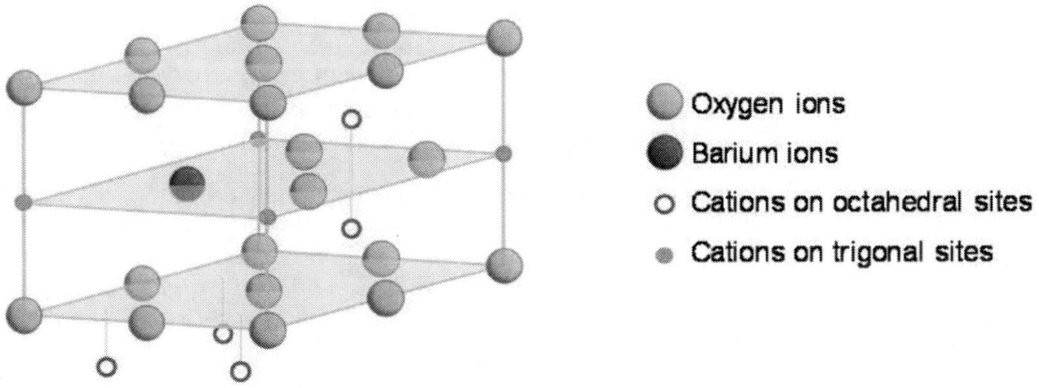

Figure 2. The R block structure in barium hexaferrite with the metal cations at the interstitial sites.

Table 1. Metallic sub-lattices of M-type hexaferrite

Block	Sublattice	Coordination	Cations	Spin
S	$4f_1$	Tetrahedral	2	↓
	2a	Octahedral	1	↑
R	$4f_2$	Octahedral	2	↓
	2b	Bi-pyramidal	1	↑
S-R	12k	Octahedral	6	↑

IOP Conf. Series: Materials Science and Engineering **92** (2015) 012019 doi:10.1088/1757-899X/92/1/012019

3. Experimental Technique

About 10 g of a mixture of the of high purity (≥ 99%) $BaCO_3$ and α-Fe_2O_3 powders with Fe:Ba molar ratio of 11.5 was prepared using high energy ball milling. Appropriate amounts of the powder precursors were weighed accurately (to five decimal places) and mixed in two zirconia cups and milled for 16 hours in an acetone bath using a (Fritsch Pulverissette-7) ball mill. Seven zirconia balls were used for the milling with a ball-to-powder mass ratio of 14. The rotational speed was 250 rpm. The milled powder was left to dry at room temperature. About one gram of the dry powder was pressed into 1.5 cm diameter disc under a 50 kN force, and sintered at 1100° C for 2 hours; this sample is labeled (0%). Then five different samples with different Fe:Ba ratios (between 11.5 and 16.16) were prepared by mixing portions of the original powder with different weight ratios of α-Fe_2O_3 (from 5% to 25%) for about an hour using an agate mortar and pestle, pressing into discs and sintering at 1100° C (Table 2).

Table 2. The iron to barium stoichiometry ratio for each sample.

Sample name	25%	20%	15%	10%	5%	0%
Fe:Ba Ratio	16.158	14.993	13.966	13.053	12.236	11.500

The structure of the samples was investigated by θ-2θ x-ray diffraction (XRD) using Shimadzu X-ray Diffraction Instrument, with Cu-K_α radiation ($\lambda_{\alpha1}$ = 1.54056 Å and $\lambda_{\alpha2}$ = 1.54439 Å). The XRD scan configuration was, (2θ = 20° – 70°) for the scan range, with step time of 1.2 s and 0.01° sampling pitch. A divergence slit of width DF = 2.4 mm, a scattering slits of width SS = 1.25 mm, and a receiving slit of width RS = 0.9 mm were used for the data collection.

XRD patterns of the samples were initially analyzed beginning with Panalytical X'Pert Highscore plus software based on PDF-2 ICDD library. This analysis revealed the phases in each sample by matching the observed pattern with those included in the library. Rietveld profile refinement was performed using the fitting parameters reported elsewhere [45-48] [50 - 53] as input parameters. A reliable fit was obtained, with a rather low Chi-Squared value, and the corresponding main structural output parameters were obtained.

4. Result and discussion

4.1 Fitting Result:

The diffraction patterns of the samples with different Fe:Ba ratios in Fig. 3 indicated the dominance of BaM phase, and the evolution of α-Fe_2O_3 phase with increasing Fe:Ba ratio. The patterns were refined by Rietveld fitting, and the residue (blue line) was a horizontal line with small fluctuations at the positions of the peaks, indicating good fitting. The goodness of fit was also deduced from the rather low values of χ^2. In what follows we present the main structural results for the two phase in the patterns of the samples. Table 3 shows the general structural characteristics of the two phases the samples, and Tables 4-9 show the main structural data including the cell parameters and cell volume, the wt.% of each phase, the ionic fractional positions in the lattice, and the symmetry of each ionic site, in addition to the indicators of the goodness of fit (χ^2, Bragg R-factor, and R_f-factor). The ionic positions obtained in the refinement process did not change significantly with the variations of Fe:Ba ratio, and our results are in good agreement with previously reported results [45, 46].

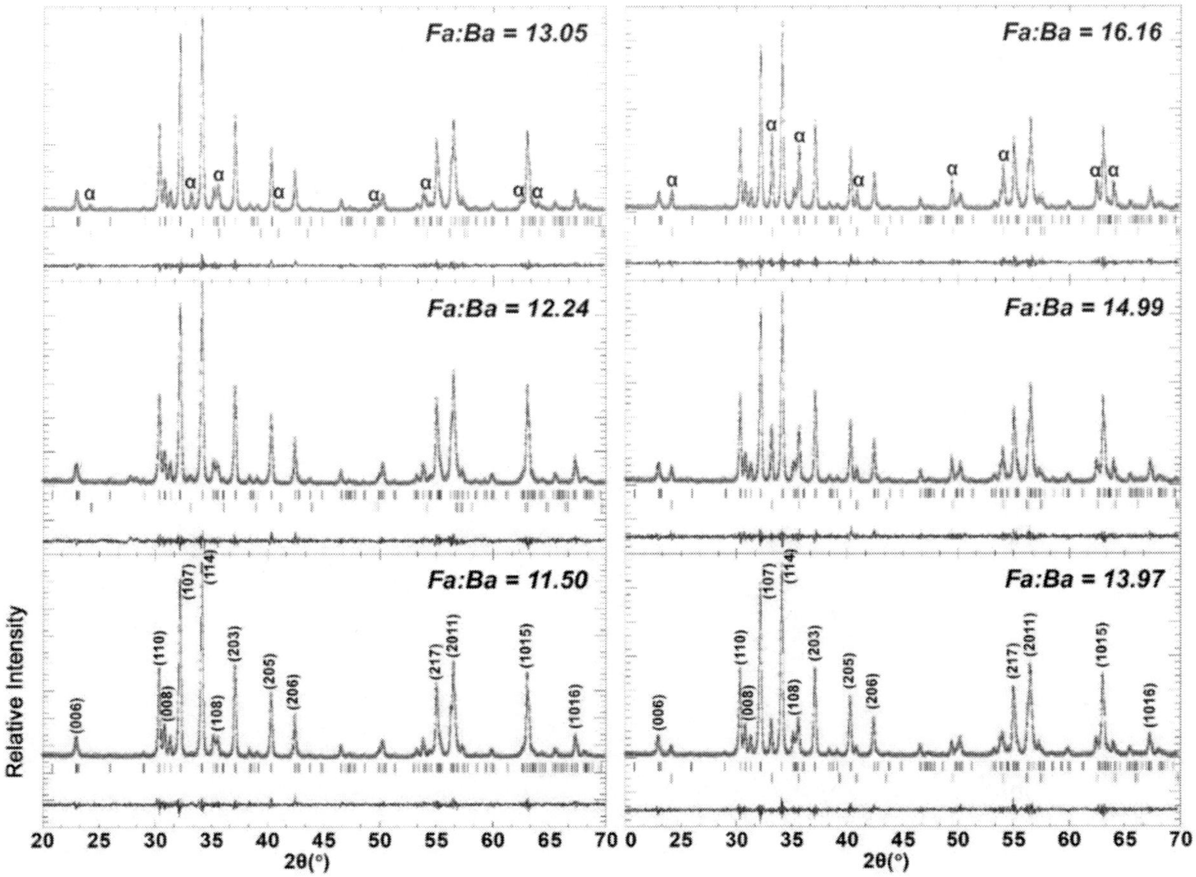

Figure 3. Fitted XRD pattern using Rietveld method for the samples with Fe:Ba ratio = 11.5 – 16.16.

Table 3. Barium hexaferrite and Hematite Phase data.

Phase data	$BaFe_{12}O_{19}$	Fe_2O_3
Formula sum	$Ba_2Fe_{24}O_{38}$	$Fe_{12}O_{18}$
Formula weight	2222.96 g/mol	958.153 g/mol
Crystal system	hexagonal	trigonal
Space-group	$P6_3/mmc$ (194)	*R*-3*c* (167)
Pearson code	hP64	hR10
Wyckoff sequence	k3hf3ecba	ec

Table 4. Fitting factors, fractional coordinates and cell parameters for barium hexaferrite in the sample with Fe:Ba ratio = 11.5 (Sample 0%)

χ^2:	**1.16**		
Phase 1	$BaFe_{12}O_{19}$		
Cell parameters	a = 5.8917 Å c = 23.213 Å	*B.R-factor:*	2.57
Cell ratio	c/a = 3.9399	*R_F-factor:*	2.83
Cell volume	697.81 Å^3	Wt. %	100.00

Atomic Parameters

Atom	Ox.	Wyck.	Site	S.O.F.	*x/a*	*y/b*	*z/c*
Ba	+2	2d	-6m2	1	0.66667	0.33333	0.25000
Fe 1	+3	2a	-3m.	1	0.00000	0.00000	0.00000
Fe 2	+3	2b	-6m2	1	0.00000	0.00000	0.25000
Fe 3	+3	$4f_1$	3m.	1	0.33333	0.66667	0.02668
Fe 4	+3	$4f_2$	3m.	1	0.33333	0.66667	0.18998
Fe 5	+3	12k	.m.	1	0.16883	0.33769	0.89142
O 1	-2	4e	3m.	1	0.00000	0.00000	0.15270
O 2	-2	4f	3m.	1	0.33333	0.66667	0.94377
O 3	-2	6h	mm2	1	0.17887	0.35795	0.25000
O 4	-2	12k	.m.	1	0.15612	0.31250	0.05217
O 5	-2	12k	.m.	1	0.50293	1.00585	0.15073

Table 5. Fitting factors, fractional coordinates and cell parameters for barium hexaferrite and hematite in the sample with Fe:Ba ratio of 12.24 (Sample 5%).

χ^2:	**1.24**		
Phase 1	$BaFe_{12}O_{19}$		
Cell parameters	a = 5.8932 Å c = 23.223 Å	*B.R-factor:*	2.94
Cell ratio	c/a = 3.9407	*R_F-factor:*	3.11
Cell volume	698.48 Å^3	Wt. %	96.3

Phase 2	Fe_2O_3		
Cell parameters	a = 4.9861 Å c = 13.864 Å	*B.R-factor:*	6.75
Cell ratio	c/a = 2.7805	*R_F-factor:*	6.21
Cell volume	298.50 Å^3	Wt. %	3.69

Atomic Parameters

$BaFe_{12}O_{19}$

Atom	Ox.	Wyck.	Site	S.O.F.	*x/a*	*y/b*	*z/c*
Ba	+2	2d	-6m2	1	0.66667	0.33333	0.25000
Fe 1	+3	2a	-3m.	1	0.00000	0.00000	0.00000
Fe 2	+3	2b	-6m2	1	0.00000	0.00000	0.25000

Fe 3	+3	$4f_1$	3m.	1	0.33333	0.66667	0.02663
Fe 4	+3	$4f_2$	3m.	1	0.33333	0.66667	0.19011
Fe 5	+3	12k	.m.	1	0.16839	0.33637	0.89169
O 1	-2	4e	3m.	1	0.00000	0.00000	0.15251
O 2	-2	4f	3m.	1	0.33333	0.66667	0.94459
O 3	-2	6h	mm2	1	0.17894	0.35785	0.25000
O 4	-2	12k	.m.	1	0.15878	0.31746	0.05133
O 5	-2	12k	.m.	1	0.50223	1.00441	0.15027

Fe_2O_3

Atom	Ox.	Wyck.	Site	S.O.F.	*x/a*	*y/b*	*z/c*
Fe	3	12c	3	1	0.00000	0.00000	0.37500
O	-2	18e	0.2	1	0.31775	0.00000	0.25000

Table 6. Fitting factors, fractional coordinates and cell parameters for barium hexaferrite and hematite in the sample with Fe:Ba ratio = 13.05 (Sample 10%).

χ^2:	**1.13**		
Phase 1	$BaFe_{12}O_{19}$		
Cell parameters	*a* = 5.8949 Å *c* = 23.2250 Å	*B.R-factor:*	2.32
Cell ratio	*c/a* = 3.9399	*R_F-factor:*	2.34
Cell volume	698.93 Å^3	Wt. %)	93.5

Phase 2	Fe_2O_3		
Cell parameters	*a* = 5.0364 Å *c* = 13.750 Å	*B.R-factor:*	4.29
Cell ratio	*c/a* = 2.7301	*R_F-factor:*	3.49
Cell volume	302.04 Å^3	Wt. %	6.50

Atomic Parameters

$BaFe_{12}O_{19}$

Atom	Ox.	Wyck.	Site	S.O.F.	*x/a*	*y/b*	*z/c*
Ba	+2	2d	-6m2	1	0.66667	0.33333	0.25000
Fe 1	+3	2a	-3m.	1	0.00000	0.00000	0.00000
Fe 2	+3	2b	-6m2	1	0.00000	0.00000	0.25000
Fe 3	+3	$4f_1$	3m.	1	0.33333	0.66667	0.02696
Fe 4	+3	$4f_2$	3m.	1	0.33333	0.66667	0.18981
Fe 5	+3	12k	.m.	1	0.16920	0.33834	0.89179
O 1	-2	4e	3m.	1	0.00000	0.00000	0.15302
O 2	-2	4f	3m.	1	0.33333	0.66667	0.94456
O 3	-2	6h	mm2	1	0.17845	0.35700	0.25000
O 4	-2	12k	.m.	1	0.15785	0.31564	0.05217
O 5	-2	12k	.m.	1	0.50475	1.00957	0.15005

Fe_2O_3

Atom	Ox.	Wyck.	Site	S.O.F.	*x/a*	*y/b*	*z/c*
Fe	3	12c	3	1	0.00000	0.00000	0.35574
O	-2	18e	0.2	1	0.31112	0.00000	0.25000

Table 7. Fitting factors, fractional coordinates and cell parameters for barium hexaferrite and hematite in the sample with Fe:Ba ratio = 13.97 (Sample 15%).

χ^2:	**1.13**		
Phase 1	$BaFe_{12}O_{19}$		
Cell parameters	a = 5.8954 Å c = 23.228 Å	*B.R-factor:*	2.58
Cell ratio	c/a = 3.9401	R_F*-factor:*	2.70
Cell volume	699.17 $Å^3$	Wt. %	87.0

Phase 2	Fe_2O_3		
Cell parameters	a = 5.0365 Å c = 13.751 Å	*B.R-factor:*	3.13
Cell ratio	c/a = 2.7305	R_F*-factor:*	2.60
Cell volume	302.10 $Å^3$	Wt. %	13.0

Atomic Parameters

$BaFe_{12}O_{19}$

Atom	Ox.	Wyck.	Site	S.O.F.	*x/a*	*y/b*	*z/c*
Ba	+2	2d	-6m2	1	0.66667	0.33333	0.25000
Fe 1	+3	2a	-3m.	1	0.00000	0.00000	0.00000
Fe 2	+3	2b	-6m2	1	0.00000	0.00000	0.25000
Fe 3	+3	$4f_1$	3m.	1	0.33333	0.66667	0.02735
Fe 4	+3	$4f_2$	3m.	1	0.33333	0.66667	0.18952
Fe 5	+3	12k	.m.	1	0.16889	0.33762	0.89163
O 1	-2	4e	3m.	1	0.00000	0.00000	0.15173
O 2	-2	4f	3m.	1	0.33333	0.66667	0.94465
O 3	-2	6h	mm2	1	0.18229	0.36474	0.25000
O 4	-2	12k	.m.	1	0.15487	0.30965	0.05177
O 5	-2	12k	.m.	1	0.50404	1.00818	0.14959

Fe_2O_3

Atom	Ox.	Wyck.	Site	S.O.F.	*x/a*	*y/b*	*z/c*
Fe	3	12c	3	1	0.00000	0.00000	0.35570
O	-2	18e	0.2	1	0.30620	0.00000	0.25000

Table 8. Fitting factors, fractional coordinates and cell parameters for barium hexaferrite and hematite in the sample with Fe:Ba ratio = 14.99 (Sample 20%).

χ^2:	**1.15**		
Phase 1	$BaFe_{12}O_{19}$		
Cell parameters	a = 5.8929 Å c = 23.217 Å	*B.R-factor:*	2.65
Cell ratio	c/a = 3.9398	R_F*-factor:*	2.77
Cell volume	698.23 Å^3	Wt. %	81.6

Phase 2	Fe_2O_3		
Cell parameters	a = 5.0352 Å c = 13.747 Å	*B.R-factor:*	3.16
Cell ratio	c/a = 2.7304	R_F*-factor:*	2.65
Cell volume	301.86 Å^3	Wt. %	18.4

Atomic Parameters

$BaFe_{12}O_{19}$

Atom	Ox.	Wyck.	Site	S.O.F.	*x/a*	*y/b*	*z/c*
Ba	+2	2d	-6m2	1	0.66667	0.33333	0.25000
Fe 1	+3	2a	-3m.	1	0.00000	0.00000	0.00000
Fe 2	+3	2b	-6m2	1	0.00000	0.00000	0.25000
Fe 3	+3	$4f_1$	3m.	1	0.33333	0.66667	0.02662
Fe 4	+3	$4f_2$	3m.	1	0.33333	0.66667	0.18958
Fe 5	+3	12k	.m.	1	0.16945	0.33877	0.89191
O 1	-2	4e	3m.	1	0.00000	0.00000	0.15341
O 2	-2	4f	3m.	1	0.33333	0.66667	0.94449
O 3	-2	6h	mm2	1	0.18140	0.36294	0.25000
O 4	-2	12k	.m.	1	0.15733	0.31461	0.05263
O 5	-2	12k	.m.	1	0.50474	1.00956	0.15067

Fe_2O_3

Atom	Ox.	Wyck.	Site	S.O.F.	*x/a*	*y/b*	*z/c*
Fe	3	12c	3	1	0.00000	0.00000	0.35562
O	-2	18e	0.2	1	0.31135	0.00000	0.25000

Table 9. Fitting factors, fractional coordinates and cell parameters for barium hexaferrite and hematite in the sample with Fe:Ba ratio = 16.16 (Sample 25%).

χ^2:	**1.17**		
Phase 1	$BaFe_{12}O_{19}$		
Cell parameters	a = 5.8936 Å c = 23.220 Å	*B.R-factor:*	3.02
Cell ratio	c/a = 3.9399	R_F*-factor:*	3.10
Cell volume	698.49 Å^3	Wt. %	75.1

Phase 2	Fe_2O_3		
Cell parameters	*a* = 5.0356 Å *c* = 13.747 Å	*B.R-factor:*	2.27
Cell ratio	*c/a* = 2.7301	R_F*-factor:*	1.30
Cell volume	301.90 $Å^3$	Wt. %	24.9

Atomic Parameters

$BaFe_{12}O_{19}$

Atom	Ox.	Wyck.	Site	S.O.F.	*x/a*	*y/b*	*z/c*
Ba	+2	2d	-6m2	1	0.66667	0.33333	0.25000
Fe 1	+3	2a	-3m.	1	0.00000	0.00000	0.00000
Fe 2	+3	2b	-6m2	1	0.00000	0.00000	0.25000
Fe 3	+3	$4f_1$	3m.	1	0.33333	0.66667	0.02645
Fe 4	+3	$4f_2$	3m.	1	0.33333	0.66667	0.18976
Fe 5	+3	12k	.m.	1	0.16847	0.33684	0.89173
O 1	-2	4e	3m.	1	0.00000	0.00000	0.15381
O 2	-2	4f	3m.	1	0.33333	0.66667	0.94369
O 3	-2	6h	mm2	1	0.17753	0.35526	0.25000
O 4	-2	12k	.m.	1	0.16050	0.32090	0.05198
O 5	-2	12k	.m.	1	0.50541	1.01093	0.15105

Fe_2O_3

Atom	Ox.	Wyck.	Site	S.O.F.	*x/a*	*y/b*	*z/c*
Fe	3	12c	3	1	0.00000	0.00000	0.35550
O	-2	18e	0.2	1	0.31098	0.00000	0.25000

3.2 Discussion: The process of Rietveld refinement revealed full structural information and the relative proportions of iron oxide (α-Fe_2O_3) and BaM phases in each sample. Fig. 4 shows the diffraction patterns of all samples added with different amounts of α-Fe_2O_3. The figure demonstrated the evolution of the structural peaks corresponding to α-Fe_2O_3 (shaded in yellow) as a function of the weight ratio of hematite added to the original powder (Fe:Ba = 11.5). The integrated intensities I_α and I_{BaM} of all structural peaks of α-Fe_2O_3 and $BaFe_{12}O_{19}$ phases, respectively, and the ratio of these intensities for each sample are summarized in (Table 10).

(Fig. 5) demonstrated a linear variation of the relative integrated intensity of α-Fe_2O_3 with increasing Fe:Ba ratio. The linear fit to the data gave a slope of 5.2088 and an intercept of –62.701. These values were used to determine an approximate value (based on integrated intensity) of the optimal Fe:Ba ratio for fabricating single BaM phase with no additional iron oxide phase. The results indicates that the optimal value of Fe:Ba ratio is 12.039.

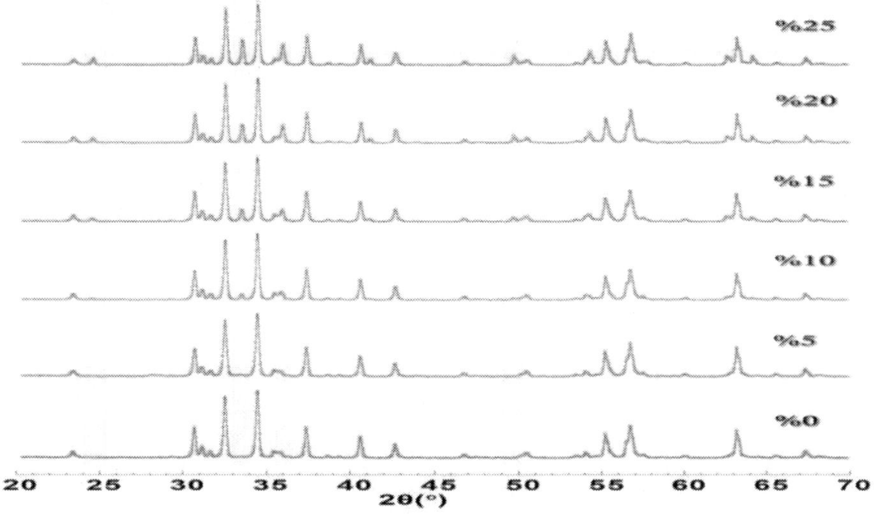

Figure 4. XRD patterns of all samples added with different amounts of α-Fe_2O_3. The main structural peaks of iron oxide phase are shaded in yellow.

Table 10. The integrated intensity of all structural peaks of BaM (I_{BaM}) and of α-Fe_2O_3 (I_α), and the ratio (I_α/I_{BaM}) for the samples with different Fe:Ba ratio. The percentage ratio associated with the Fe:Ba ratio represents the wt. % of α-Fe_2O_3 added to the original powder

Fe:Ba ratio (sample)	**16.158 (25%)**	**14.993 (20%)**	**13.966 (15%)**	**13.053 (10%)**	**12.236 (5%)**	**11.500 (0%)**
I_{BaM}	1234.7	1368.9	1408.9	1727.3	1626.3	1661.8
I_α	265.80	209.20	146.00	80.700	37.100	0.0000
I_α/I_{BaM} (%)	21.530	15.280	10.360	4.6700	2.2800	0.0000

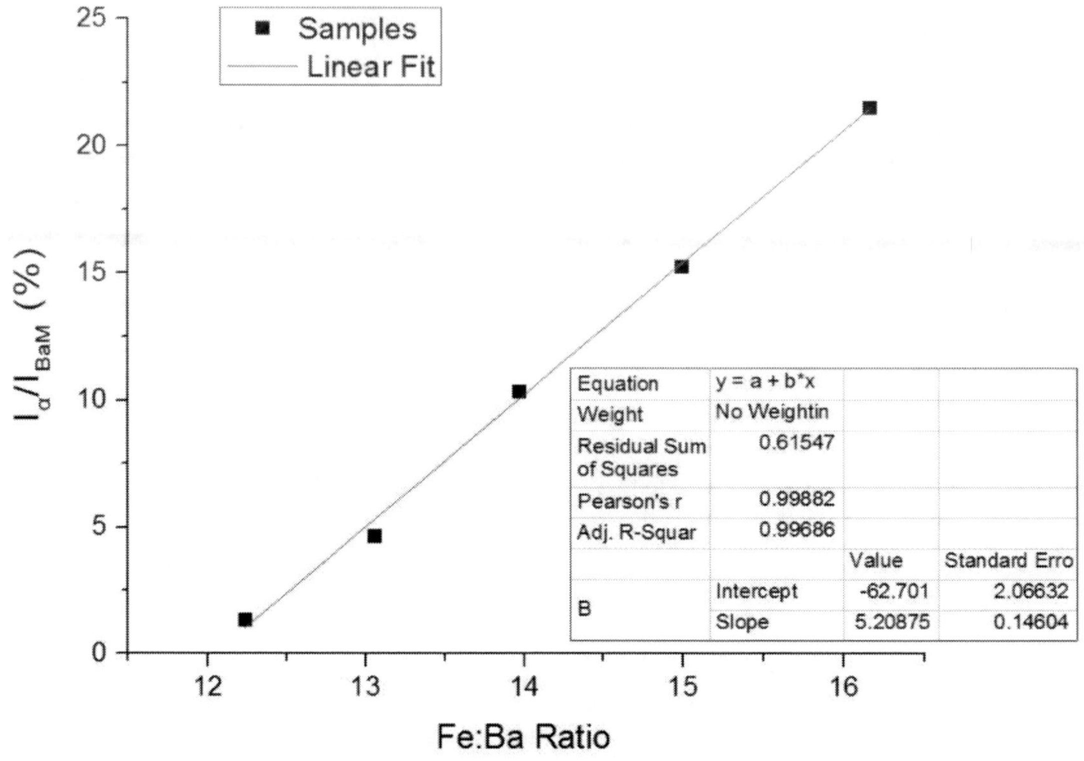

Figure 5. A plot of I_α/I_{BaM} (%) vs Fe:Ba ratio for all samples

The ratio of the secondary iron oxide phase in each sample can also be determined by investigating the integrated intensities of the main peaks of the phases. (Table 11) shows the integrated intensity of the main peak of BaM at 34.1° (I_{MB}) and that of α-Fe_2O_3 located at 33.15° ($I_{M\alpha}$), as well as the ratio of the intensities of these peaks. Fig. 6 shows the relative intensity of the main peak of α-Fe_2O_3 as a function of Fe:Ba ratio. Linear fit to the data gave a slope of 9.0535 and an intercept of –109.565. From these values the optimal Fe:Ba ratio for a single BaM phase was found to be 12.102. This value is close to that obtained from the whole-pattern integrated intensity analysis, and to the stoichiometric composition of 12.0 for M-type hexaferrite. Further, Fig. 6 can be used to determine the weight ratio of α-Fe_2O_3 phase in any hexaferrite sample by evaluating the relative intensity of the main peak and determining the corresponding Fe:Ba ratio, which can be subsequently converted into wt% of the α-Fe_2O_3 phase using the molecular weights of the phases and standard calculations.

Table 11. The integrated intensities of the main peaks of BaM (I_{MB}), α-Fe_2O_3 ($I_{M\alpha}$), and the peak ratios of the two phases for all samples. The percentage ratio associated with the Fe:Ba ratio represents the wt. % of α-Fe_2O_3 added to the original powder

	Fe:Ba ratio	**16.158**	**14.993**	**13.966**	**13.053**	**12.236**	**11.500**
I_{MB} at 2θ of	**34.10**	187.1	208.9	212.00	266.2	248.6	267.9
$I_{M\alpha}$ at 2θ of	**33.15**	69.20	53.50	37.10	21.00	4.000	0.000
$I_{M\alpha}$/ I_{MB} (%)		37.00	25.60	17.50	7.900	1.600	0.000

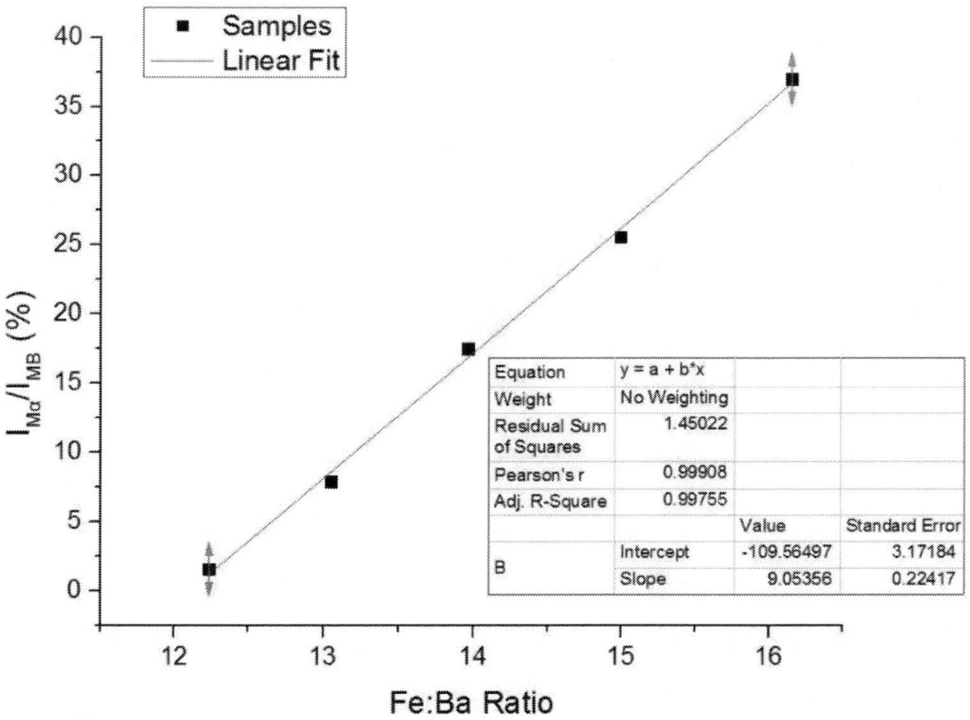

Figure 6. A plot of $I_{M\alpha}/I_{MB}$ (%) vs Fe:Ba ratio for all samples

The weight ratios of the different structural phases in a given sample can be evaluated using Rietveld fitting of the corresponding diffraction pattern. The wt% of α-Fe_2O_3 phase for each sample as determined by the fitting routine is listed in Table 12. These values are generally lower than the wt. % of α-Fe_2O_3 added to the original powder (with Fe:Ba = 11.5), which indicates that the Fe:Ba ratio required for the synthesis of a single BaM phase is higher than 11.5, a result consistent with the optimal ratio determined by the analyses associated with Fig. 5. Fig. 7 shows the variation of the relative main peak intensity of α-Fe_2O_3 with the wt. % of this phase determined from Rietveld refinement of the corresponding diffraction pattern.

Table 12. α-Fe_2O_3 wt. % determined from Rietveld refinement of the diffraction patterns and the corresponding relative main peak integrated intensity.

Sample	**25%**	**20%**	**15%**	**10%**	**5%**	**0%**
α-Fe_2O_3 wt%	24.88	18.37	13.05	6.500	3.530	0.000
$I_{M\alpha}/I_{MB}$ (%)	37.00	25.60	17.50	7.900	1.600	0.000

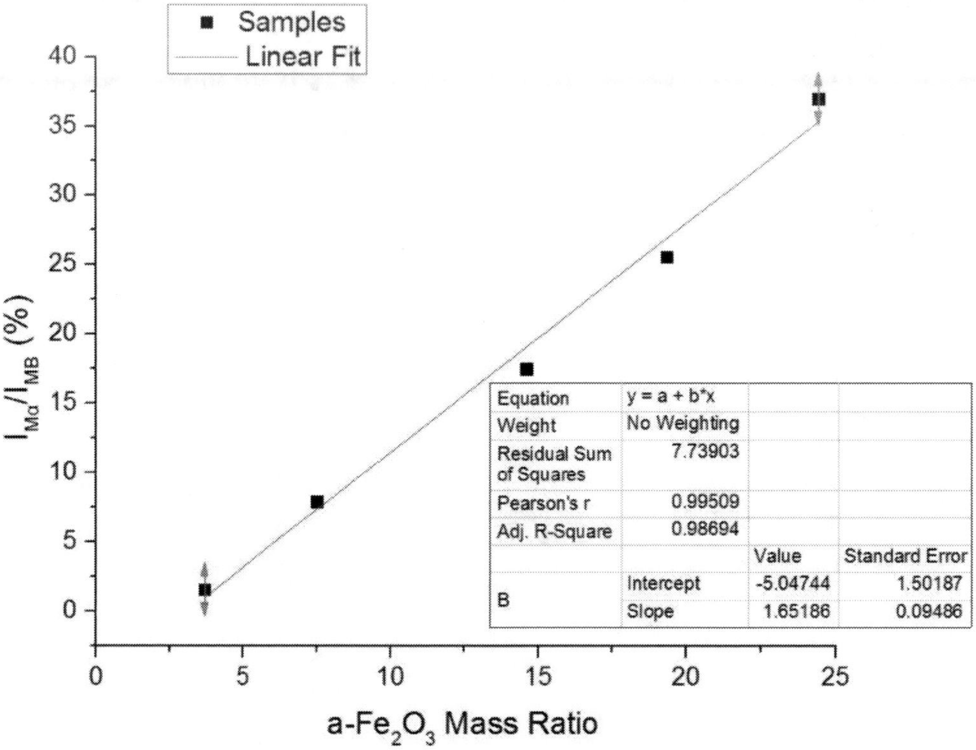

Figure 7. A plot of $I_{M\alpha}/I_{MB}$ (%) as a function of α-Fe_2O_3 wt. % evaluated from the refinement of the patterns of all samples

In the refinement process of the sample added with 5 wt% α-Fe_2O_3, the parameters belonging to the α-Fe_2O_3 phase were left to be generated automatically in order to obtain reliable results with no constraints. This is because the phase peaks are weak in this sample, approaching the background fluctuations, so that the X-Ray absorption contrast cannot play a significant role in this calculation.

The optimum ratio for synthesizing pure sample of barium hexaferrite ($BaFe_{12}O_{19}$) can be obtained from the experimental result by linearly fitting the Fe:Ba ratio of the prepared samples with the α-Fe_2O_3 wt% obtained from Rietveld refinement (Fig. 8). The best straight line representing the data intercepts the Fe:Ba ratio axis at 11.7, which is the refined optimum ratio for a single barium hexaferrite phase. This value is somewhat lower than the theoretical stoichiometric ratio of 12, and lower than the value obtained from analysis of the integrated intensities. These results indicate the sensitivity of the resulting optimal Fe:Ba ratio to the analytical method used for evaluating this ratio. However, the results are close, although Rietveld analysis indicated that the optimal ratio is somewhat lower than the theoretical value.

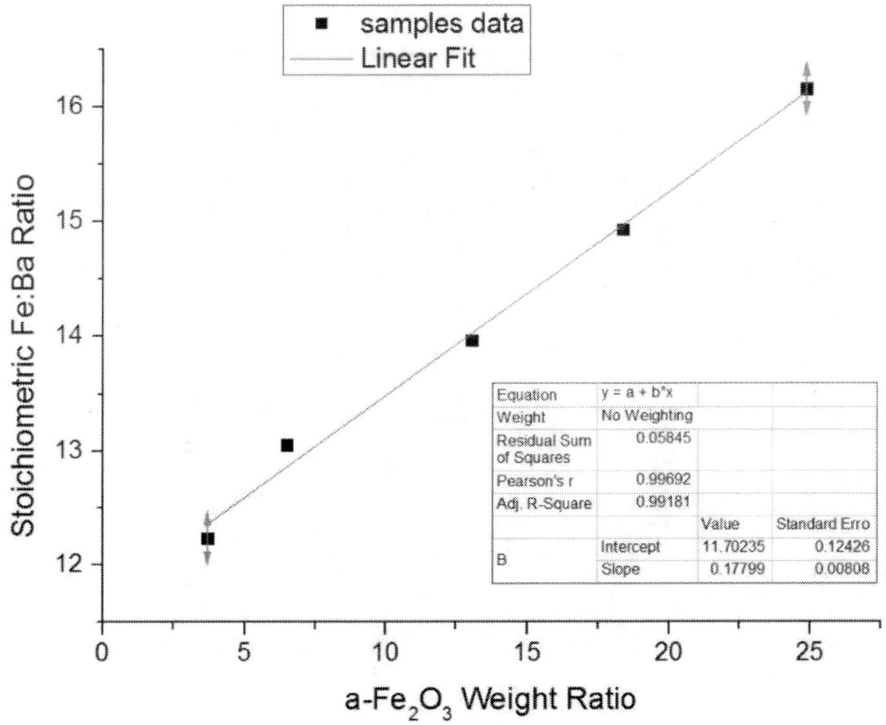

Figure 8. Linear fit of Fe:Ba ratio vs α-Fe_2O_3 wt. % obtained from the refinement of the diffraction patterns.

The refined patterns with rather low Chi-square values gave valuable information on the crystal structure and relative proportions of α-Fe_2O_3 α-Fe_2O_3 secondary phase. Fig. 9 shows the calculated pattern of our hexaferrite sample with Fe:Ba = 11.5, together with standard patterns available in the ICDD library.

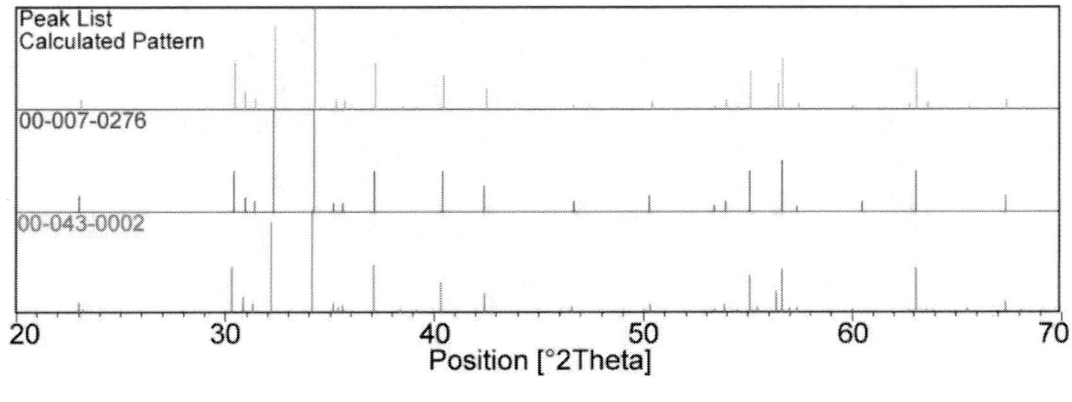

Figure 9. The calculated XRD stick pattern for $BaFe_{12}O_{19}$ phase compared with other standard patterns in ICDD library.

All the data needed to build up the structural model of the unit cell were obtained by the structural refinement of the diffraction patterns. Since no significant differences were observed in the crystallographic structure of the phases in samples prepared with different stoichiometries as can be clearly seen from the results of the refinement, the structural data of the sample with Fe:Ba = 11.5 was used to construct the unit cell. (Fig. 10 and 11) show the unit cell from different angles as can be noted from the direction of the unit cell vectors in each snapshot.

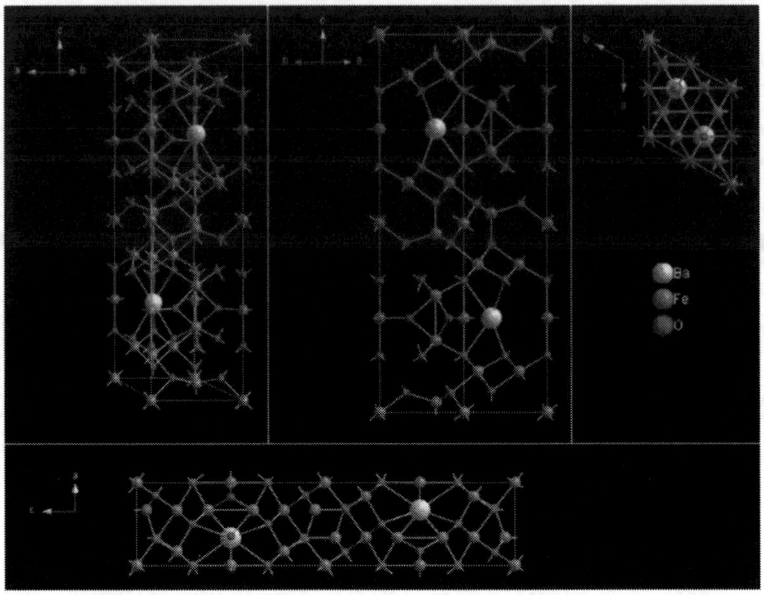

Figure 10. The crystal structure of $BaFe_{12}O_{19}$ unit cell in 3D.

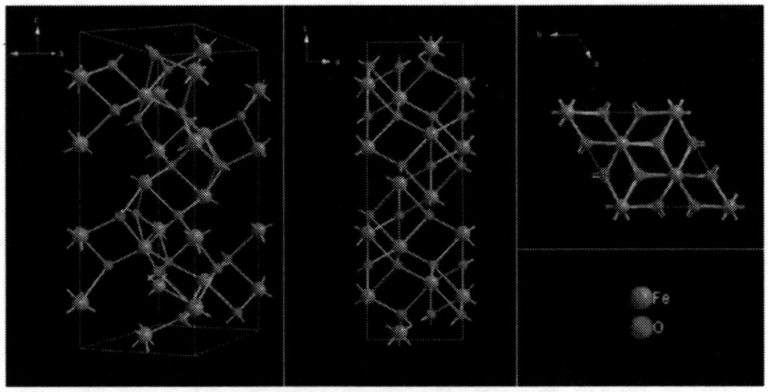

Figure 11. The crystal structure of Fe_2O_3 unit cell in a 3D.

The structural analysis shows that there are no actual structural differences in the two extreme samples (Fe:Ba = 11.5 and 16.16). Table 13 and figures 12-14 show that there are no significant differences between the two samples.

Table 13. The calculated polyhedral volume, average bond length and distortion index for each site polyhedron for the samples of Fe:Ba (11.5 , 16.16).

		2a site	2b site	$4f_1$ site	$4f_2$ site	12k site
Fe:Ba 11.5	Average bond length (Å)	1.9830	1.9357	1.9112	2.0611	2.2027
	Polyhedral volume (Å^3)	10.362	6.5575	3.5760	11.193	10.974
	Distortion index	0.0000	0.0912	0.0007	0.0292	0.0382
Fe:Ba 16.16	Average bond length (Å)	2.0346	1.9183	1.8767	2.0455	2.0127
	Polyhedral volume (Å^3)	11.217	6.3665	3.3912	11.198	11.225
	Distortion index	0.0000	0.0822	0.0120	0.0351	0.0368

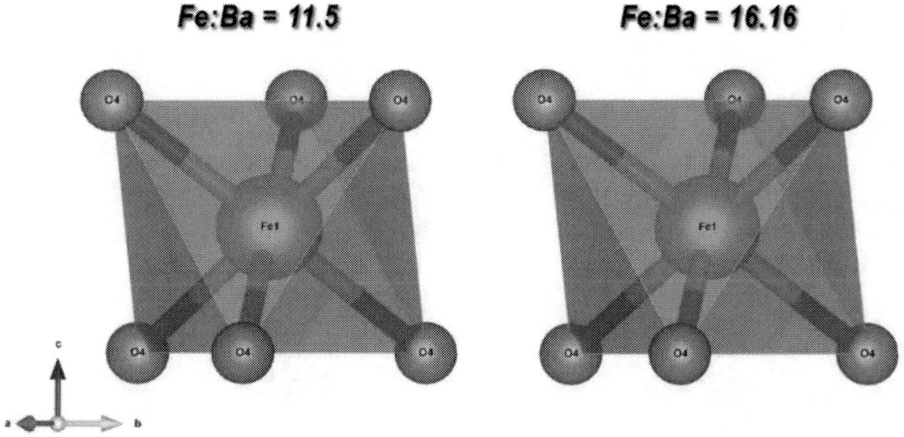

Figure 12. Compression of two octahedron structures in the 2a site, for the samples of Fe:Ba (11.5 and 16.16).

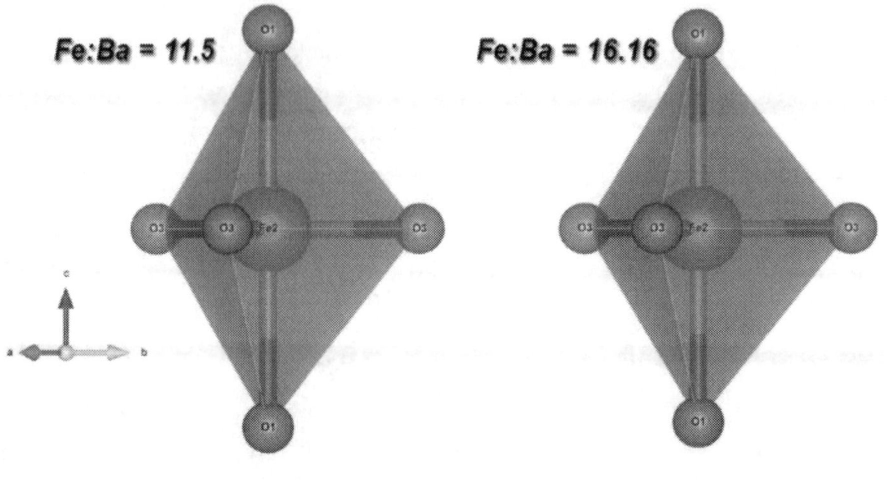

Figure 13. Compression of two Bi-pyramidal structures in the 2b site, for the samples of Fe:Ba (11.5 and 16.16).

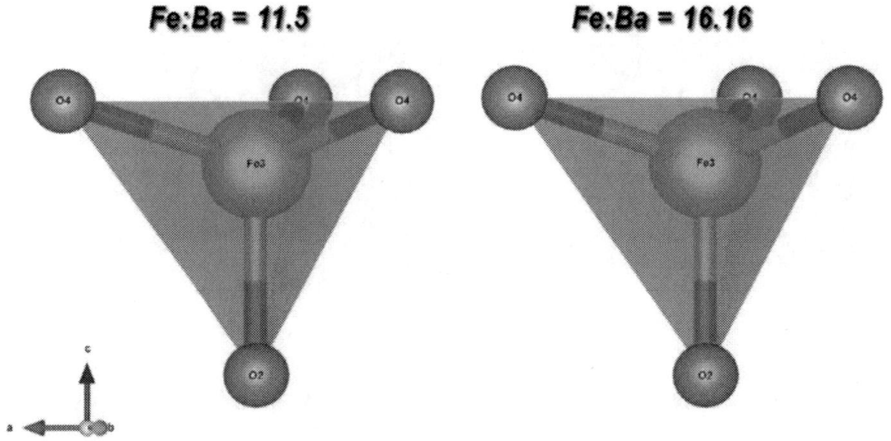

Figure 14. Compression of two tetrahedron structures in the $4f_1$ site, for the samples of Fe:Ba (11.5 and 16.16).

In addition, our refined chemical composition Ba $Fe_{12.18}$ $O_{18.75}$ is in good agreement with the theoretical formula and with the previous reported formula Ba $Fe_{11.80}$ $O_{19.46}$ by Sozeri et al. [36].

Conclusions

The structural analysis of the samples consisting of a mixture of M-type barium hexaferrit (with Fe/Ba ratio of 11.5) and α-Fe_2O_3 iron oxide revealed the presnece of pure M-type phase and α-Fe_2O_3 phase, with no other secondary phases. Also, the structural properties of the M-type barium hexaferrite phase were found to be independent of the α-Fe_2O_3 mass ratio in the mixture. The quantitative analysis using Rietveld refinement for these samples had shown that the optimum iron-to-barium ratio require to prepare a pure M-type barium hexaferrite phase is in the range of 11.7 – 12. This work established a reference scheme for the determination of the mass ratio of iron oxide (α-Fe_2O_3) impurity phase in hexaferrite samples.

Acknowledgement

This work was supported by a generous fund from the Deanship of Scientific Research at The University of Jordan (Grant # 1404). The technical assistance of Y. Abu Salha, The University of Jordan, is acknowledged.

Referances

[1] Ü. Özgür, Y. Alivov, H. Morkoç, Microwave ferrites, part 1: fundamental properties, Journal of Materials Science: Materials in Electronics 20 (9) (2009) 789-834.

[2] R.C. Pullar, Hexagonal ferrites: a review of the synthesis, properties and applications of hexaferrite ceramics, Progress in Materials Science 57 (7) (2012) 1191-1334.

[3] V.G. Harris, A. Geiler, Y. Chen, S.D. Yoon, M. Wu, A. Yang, Z. Chen, P. He, P.V. Parimi, X. Zuo, Recent advances in processing and applications of microwave ferrites, Journal of Magnetism and Magnetic Materials 321 (14) (2009) 2035-2047.

[4] Ü. Özgür, Y. Alivov, H. Morkoç, Microwave ferrites, part 2: passive components and electrical tuning, Journal of Materials Science: Materials in Electronics 20 (10) (2009) 911-952.

[5] J. Smit, H.P.J. Wijn, Ferrites, Wiley, New York, 1959.

[6] B.D. Cullity, C.D. Graham, Introduction to magnetic materials, John Wiley & Sons, 2011.

[7] S. Chikazumi, Physics of Ferromagnetism 2e, Oxford University Press, 2009.

[8] S. Singhal, T. Namgyal, J. Singh, K. Chandra, S. Bansal, A comparative study on the magnetic properties of $MFe_{12}O_{19}$ and $MAlFe_{11}O_{19}$ (M= Sr, Ba and Pb) hexaferrites with different morphologies, Ceramics International 37 (6) (2011) 1833-1837.

[9] M.J. Iqbal, M.N. Ashiq, I.H. Gul, Physical, electrical and dielectric properties of Ca-substituted strontium hexaferrite ($SrFe_{12}O_{19}$) nanoparticles synthesized by co-precipitation method, Journal of Magnetism and Magnetic Materials 322 (13) (2010) 1720-1726.

[10] I. Bsoul, S.H. Mahmood, A.F. Lehlooh, A. Al-Jamel, Structural and magnetic properties of $SrFe_{12-2x}Ti_xRu_xO_{19}$, Journal of Alloys and Compounds 551 (2013) 490-495.

[11] S. Sanghi, A. Agarwal, N. Ahlawat, Structure refinement and dielectric relaxation of M-type Ba, Sr, Ba-Sr, and Ba-Pb hexaferrites, Journal of Applied Physics 112 (1) (2012) 014110.

[12] G. Albanese, B. Watts, F. Leccabue, S.D.a. Castañón, Mössbauer and magnetic studies of $PbFe_{12-x}Cr_xO_{19}$ hexagonal ferrites, Journal of magnetism and magnetic materials 184 (3) (1998) 337-343.

[13] S.B. Narang, I. Hudiara, Microwave dielectric properties of M-type barium, calcium and strontium hexaferrite substituted with Co and Ti, Journal of Ceramic Processing Research 7 (2) (2006) 113-116.

[14] M. Awawdeh, I. Bsoul, S.H. Mahmood, Magnetic properties and Mössbauer spectroscopy on Ga, Al, and Cr substituted hexaferrites, Journal of Alloys and Compounds 585 (2014) 465-473.

[15] I. Bsoul, S.H. Mahmood, Magnetic and structural properties of $BaFe_{12-x}Ga_xO_{19}$ nanoparticles, Journal of Alloys and Compounds 489 (1) (2010) 110-114.

[16] S.H. Mahmood, I. Bsoul, Hopkinson peak and superparamagnetic effects in $BaFe_{12-x}Ga_xO_{19}$ nanoparticles, EPJ Web of Conferences 29 (2012) 00039.

[17] S. El-Sayed, T. Meaz, M. Amer, H. El Shersaby, Magnetic behavior and dielectric properties of aluminum substituted M-type barium hexaferrite, Physica B: Condensed Matter 426 (2013) 137-143.

[18] G. Mendoza-Suarez, L. Rivas-Vazquez, J. Corral-Huacuz, A. Fuentes, J. Escalante-Garcí, Magnetic properties and microstructure of $BaFe_{11.6-2x}Ti_xM_xO_{19}$ (M= Co, Zn, Sn) compounds, Physica B: Condensed Matter 339 (2) (2003) 110-118.

[19] D. Lisjak, M. Drofenik, Synthesis and characterization of A–Sn-substituted (A= Zn, Ni, Co) BaM–hexaferrite powders and ceramics, Journal of the European Ceramic Society 24 (6) (2004) 1841-1845.

[20] S. Choopani, N. Keyhan, A. Ghasemi, A. Sharbati, R.S. Alam, Structural, magnetic and microwave absorption characteristics of $BaCo_xMn_xTi_{2x}Fe_{12-4x}O_{19}$, Materials Chemistry and Physics 113 (2) (2009) 717-720.

[21] I. Bsoul, S.H. Mahmood, A.F. Lehlooh, Structural and magnetic properties of $BaFe_{12-2x}Ti_xRu_xO_{19}$, Journal of Alloys and Compounds 498 (2) (2010) 157-161.

[22] G.H. Dushaq, S.H. Mahmood, I. Bsoul, H.K. Juwhari, B. Lahlouh, M.A. AlDamen, Effects of molybdenum concentration and valence state on the structural and magnetic properties of $BaFe_{11.6}Mo_xZn_{0.4-x}O_{19}$ hexaferrites, Acta Metallurgica Sinica (English Letters) 26 (5) (2013) 509-516.

[23] A.M. Alsmadi, I. Bsoul, S.H. Mahmood, G. Alnawashi, K. Prokeš, K. Siemensmeyer, B. Klemke, H. Nakotte, Magnetic study of M-type doped barium hexaferrite nanocrystalline particles, Journal of Applied Physics 114 (24) (2013) 243910.

[24] W. Zhang, Y. Bai, X. Han, L. Wang, X. Lu, L. Qiao, Magnetic properties of Co–Ti substituted barium hexaferrite, Journal of Alloys and Compounds 546 (2013) 234-238.

[25] Y. Meng, M. He, Q. Zeng, D. Jiao, S. Shukla, R. Ramanujan, Z. Liu, Synthesis of barium ferrite ultrafine powders by a sol–gel combustion method using glycine gels, Journal of Alloys and Compounds 583 (2014) 220-225.

[26] G. Packiaraj, N.R. Panchal, R.B. Jotania, Structural and Dielectric studies of Cu substituted Barium hexaferrite prepared by Sol-gel auto combustion technique, Solid State Phenomena 209 (2014) 102-106.

[27] S. Chawla, R. Mudsainiyan, S. Meena, S. Yusuf, Sol–gel synthesis, structural and magnetic properties of nanoscale M-type barium hexaferrites $BaCo_xZr_xFe_{(12-2x)}O_{19}$, Journal of Magnetism and Magnetic Materials 350 (2014) 23-29.

[28] G.-H. An, T.-Y. Hwang, J. Kim, J. Kim, N. Kang, S. Kim, Y.-M. Choi, Y.-H. Choa, Barium hexaferrite nanoparticles with high magnetic properties by salt-assisted ultrasonic spray pyrolysis, Journal of Alloys and Compounds 583 (2014) 145-150.

[29] I. Bsoul, S. Mahmood, Structural and magnetic properties of $BaFe_{12-x}Al_xO_{19}$ prepared by milling and calcination, Jordan J. Phys. 2 (3) (2009) 171-179.

[30] S.H. Mahmood, A.N. Aloqaily, Y. Maswadeh, Effects of Heat Treatment on the Phase Evolution, Structural, and Magnetic Properties of Mo-Zn Doped M-type Hexaferrites, Solid State Phenomena 232 (2015) 65-92.

[31] S. Mahmood, A. Aloqaily, Y. Maswadeh, A. Awadallah, I. Bsoul, H. Juwhari, Structural and Magnetic Properties of Mo-Zn Substituted ($BaFe_{12-4x}Mo_xZn_{3x}O_{19}$) M-Type Hexaferrites, Material Science Research India 11 (1) (2014) 09-20.

[32] M. Karmakar, B. Mondal, M. Pal, K. Mukherjee, Acetone and ethanol sensing of barium hexaferrite particles: a case study considering the possibilities of non-conventional hexaferrite sensor, Sensors and Actuators B: Chemical 190 (2014) 627-633.

[33] T. Koutzarova, S. Kolev, C. Ghelev, I. Nedkov, B. Vertruen, R. Cloots, C. Henrist, A. Zaleski, Differences in the structural and magnetic properties of nanosized barium hexaferrite powders prepared by single and double microemulsion techniques, Journal of Alloys and Compounds 579 (2013) 174-180.

[34] U. Topal, Improvement of the remanence properties and the weakening of interparticle interactions in $BaFe_{12}O_{19}$ particles by B_2O_3 addition, Physica B: Condensed Matter 407 (12) (2012) 2058-2062.

[35] T. STRUKTURNE, N.I.M.L. PRAHOV, Structural, thermal and magnetic properties of barium-ferrite powders substituted with Mn, Cu or Co and X (X= Sr and Ni) prepared by the sol-gel method, Materiali in tehnologije 46 (3) (2012) 305-310.

[36] H. Sözeri, Z. Durmuş, A. Baykal, E. Uysal, Preparation of high quality, single domain $BaFe_{12}O_{19}$ particles by the citrate sol–gel combustion route with an initial Fe/Ba molar ratio of 4, Materials Science and Engineering: B 177 (12) (2012) 949-955.

[37] X. Gao, Y. Du, X. Liu, P. Xu, X. Han, Synthesis and characterization of Co–Sn substituted barium ferrite particles by a reverse microemulsion technique, Materials Research Bulletin 46 (5) (2011) 643-648.

[38] G. Tan, X. Chen, Structure and multiferroic properties of barium hexaferrite ceramics, Journal of Magnetism and Magnetic Materials 327 (2013) 87-90.

[39] W. Zhang, F. Li, H. Tang, B. Peng, W. Zhang, Morphology control of barium ferrite thin films, Vacuum 86 (7) (2012) 946-948.

[40] A. Grusková, J. Sláma, A.G. Angeles, M. Soka, Measurement of principal microstructure and magnetic properties of doped Ba ferrite particles, J. Elec. Eng 63 (2012) 156.

[41] C.-J. Li, B. Wang, J.-N. Wang, Magnetic and microwave absorbing properties of electrospun $Ba_{(1-x)}La_xFe_{12}O_{19}$ nanofibers, Journal of Magnetism and Magnetic Materials 324 (7) (2012) 1305-1311.

[42] D. Suastiyanti, A. Sudarmaji, B. Soegijono, Influence of Ba/Fe mole ratios on magnetic properties, crystallite size and shifting of X-ray diffraction peaks of nanocrystalline $BaFe_{12}O_{19}$ powder, prepared by sol gel auto combu, in: INTERNATIONAL CONFERENCE ON PHYSICS AND ITS APPLICATIONS:(ICPAP 2011), Vol. 1454, AIP Publishing, 2012, pp. 238-241.

[43] X. Liu, J. Wang, L.-M. Gan, S.-C. Ng, Improving the magnetic properties of hydrothermally synthesized barium ferrite, Journal of magnetism and magnetic materials 195 (2) (1999) 452-459.

[44] S.H. Mahmood, G.H. Dushaq, I. Bsoul, M. Awawdeh, H.K. Juwhari, B.I. Lahlouh, M.A. AlDamen, Magnetic Properties and Hyperfine Interactions in M-Type $BaFe_{12-2x}Mo_xZn_xO_{19}$ Hexaferrites, Journal of Applied Mathematics and Physics 2 (05) (2014) 77-87.

[45] S. Ounnunkad, P. Winotai, Properties of Cr-substituted M-type barium ferrites prepared by nitrate–citrate gel-autocombustion process, Journal of Magnetism and Magnetic Materials 301 (2) (2006) 292-300.

[46] S. Sahoo, K. Agarwal, A. Singh, B. Polke, K. Raha, Characterization of γ-and α-Fe_2O_3 nano powders synthesized by emulsion precipitation-calcination route and rheological behaviour of α-Fe_2O_3, International Journal of Engineering, Science and Technology 2 (8) (2010).

[47] S. Mitra, S. Das, S. Basu, P. Sahu, K. Mandal, Shape-and field-dependent Morin transitions in structured α-Fe_2O_3, Journal of Magnetism and Magnetic Materials 321 (18) (2009) 2925-2931.

[48] C. Tenailleau, M. Allix, J.B. Claridge, M. Hervieu, M.F. Thomas, J.P. Hirst, M.J. Rosseinsky, Modular Construction of Oxide Structure-Compositional Control of Transition Metal Coordination Environments, Journal of the American Chemical Society 130 (24) (2008) 7570-7583.

IOP Conf. Series: Materials Science and Engineering **92** (2015) 012020 doi:10.1088/1757-899X/92/1/012020

Simultaneous determination of vitamins A and D3 in dairy products by liquid chromatography-tandem mass spectrometry (LC-MS/MS)

I S.A.Barakat*, M.K. Hammouri, and I.Habib

Department of chemistry, Jordan University of Science & Technology. Irbid 22110. P.O.Box 3030.

*e-mail:barakat@just.edu.jo

Abstract. A potential method for simultaneous determination of vitamin A and vitamin D3 (25- hydroxyvitamin D3) in fresh milk samples is addressed. The method is based on combination of high performance liquid chromatography and mass spectrometry during the course of analysis. The method applied for determination of vitamins A and D3 on eighteen (18) different fresh milk samples using liquid chromatography along with tandem -mass spectrometry. The work describes the suitability of the proposed method for the simultaneous determination of both vitamins using LC-MS/MS as a specific and quantitative technique. The vitamins of milk were separated by C18 Thermo gold column(100mm x 4.6mm x 5µm) with a flow rate of 1ml/min (using an isocratic mobile phase). The method was validated using duplicate analyses, relative recovery experiment, and comparative analysis with control samples. Liquid- liquid extraction was employed as a pre-concentration step with n-hexane - dichloromethane mixture (90%:10%) as an extraction solvent. The molecular ions (m/z) appeared near 286 and 385nm and for the base peaks were appeared near 255 and 355nm for vitamins A and D3. Good correlation coefficients were obtained, 0.9999 for vitamin D3 and 0.9994 for vitamin A. The limit of detection and the limit of quantification were found to be 0.09ng/ml and 0.54ng/ml for vitamin D3 and 0.32ng/ml and 1.8ng/ml and for vitamin A. The proposed method showed excellent recoveries, about 98 % for both vitamins A and D3.

1.Introduction

A vitamin is an organic compound required by an organism as a vital nutrient in limited amounts [1]. Vitamins generally act as bio-catalysts, reacting with proteins to create metabolic active enzymes that in turn produce thousands of significant chemical reactions throughout the body [2]. However, 13 identified vitamins have been identified as essential to humans. They are classified as fat soluble vitamins such as A, D, E and K and water soluble vitamins such as C and B-group. Vitamin A is a group of nutritionally unsaturated hydrocarbons, which include retinol, retinal, retinoic acid, and several pro-vitamin A carotenoids, among which ß-carotene is the most important [3].
Vitamin D helps the body in the absorption of calcium, needed for the normal development and maintenance of healthy teeth and bones. It also helps also maintain proper blood levels of calcium and phosphorus in the blood[4].
To date, high performance liquid chromatographic ultraviolet/visible (HPLC–UV/Vis) detection for vitamins A, D and carotenes, and fluorescence (HPLC–Fl) detection for vitamins A, E and K have been the methods of choice for the analysis of fat soluble vitamins in milk [3] [5]. Both vitamins were usually assayed individually due to their chemical diversity and varying levels within samples. However, recently an UPLC–UV/Vis method was reported for the simultaneous determination of all trans-retinol, α-tocopherol and β-carotene in milk [6]. High performance liquid chromatography–

tandem [7] mass spectrometry (HPLC–MS/MS) is widely accepted as the benchmark for quantification of a wide range of non-volatile compounds at the parts-per-million (mg/kg) and parts-per-billion (μg/kg) levels [8], this is due to its superior sensitivity and selectivity compared to single quadruple (MS), 3D (Paul) ion trap (MS), and time of flight (Tof) instruments. HPLC–MS/MS has been used for the analyses of vitamins A, D, E, and K in breast 53 milk [10], and other food staffs, [11] [12], infant formula [13] [14] and human blood serum [15] [16].

2. Experimental

A chromatographic system of model API 300 Applied Bio-system consisting of an Agilent stable-bond C18 Thermo gold column with in-built detector was used. The guard and analytical columns were mounted in a thermo stated column compartment set at 30 °C. The peak areas were integrated automatically by computer using Analyst version 1.4 (AB Sciex) software program.
Other apparatus used included a centrifuge, normal evaporator and vortex.

2.1 Apparatus

A chromatographic system of model 3200 ABSciex supplied with an Agilent stable-bond C18 Thermo gold column with in-built MS detector was used. The guard and analytical columns were mounted in a thermo stated column compartment set at 30 °C. The peak areas were integrated automatically by computer using Analyst version 1.4 (AB Sciex) software program.
Other apparatus used included a centrifuge, normal evaporator and vortex.

2.2. Materials

Pure vitamins A, D3 and A- acetate (Sigma- Aldrech Chemicals) with analytical grade methanol, ethanol, hexane, dichloromethane, formic acid and acetonitrile were used.

2.3. Procedure for high-performance liquid chromatography

2.3.1. Chromatographic conditions

Solutions and mobile phases were prepared for the time of use. The mobile phase used was methanol: acetonitrile: de-ionized water: formic acid (68:30:2:0.1 v/v/v/v) with a pH= 3.5. The analytical column used was C18 Thermo gold (100mm x 4.5mm x 5μm). All analysis was performed under isocratic conditions at a flow-rate of 1.0 ml/ min and at room temperature.

2.3.2 Standard Solutions

Three volumetric flasks (100mL each) of individual vitamins A, D3 and A- acetate stock solutions were prepared by dissolving 10mg of each in ethanol. The concentration of each was 1x 105 ng/mL. Serial solutions of vitamin A (50, 100, 200, 300, 400, 500, 750 and 1000 ng/mL) and vitamin D3 (5, 10, 20, 25, 30, 50 and 100 ng/mL) were prepared. A solution of 100ng/mL of vitamin A- acetate as an internal standard was also prepared.

2.3.3 Samples Preparation

1 g of each fresh milk sample was accurately weighed and transferred to a 10-mL plastic tube containing 1mL of internal standard (vitamin A- acetate) and 1 mL of 39% ethanolic sodium hydroxide solution. The mixture was then heated in a water bath at 60oC for 40 minutes. Extraction was performed by adding 4 mL n-hexane-dichloromethane mixture (90%:10%) and manually shacked vigorously for 1 minute to yield two different layers, one organic and one aqueous layer. The hexane-dichloromethane extract was removed, washed twice with 2 ml water. The hexane- dichloromethane extracts was then taken and evaporated until dryness and filtered before acquisition with LC-MS/MS.

IOP Conf. Series: Materials Science and Engineering **92** (2015) 012020 doi:10.1088/1757-899X/92/1/012020

2.3.4. Calibration and linearity

Calibration curves were constructed in the range 5–100 ng/mL for vitamin D3 and 5–950.0 ng/mL for vitamin A to encompass the expected concentrations of measured samples. Curves were obtained by plotting the peak area of the corresponding pure vitamins against their concentrations. Linear calibration curves were generated by linear regression analysis and over the respective standard concentration ranges. The calibration curves equations for the two vitamins are:
$y = 0.00185x$ ($R = 0.9999$) for vitamin A and $y = 0.0028x$ ($R = 0.9994$) for vitamin D3.

2.3.5. Analytical recovery

Absolute recoveries of 5 different concentrations of both vitamin D3 (5–100 ng/ml) and vitamin A (5–950.0 ng/mL) in dairy products were determined by assaying the samples as described above. The peak areas of both vitamins were compared with those obtained from direct injection of the compounds dissolved in the processed blank sample.

2.3.6. Precision and accuracy

The precision and accuracy of the assay was ascertained based on analysis of quality control samples. Dairy product quality control sample concentrations for vitamin D3 and vitamin A were 5–100 ng/ mL and 5.–950 ng/mL, respectively. Five replicate quality control samples at each concentration were analyzed and the mean, standard deviations (SD) and coefficients of variation (C.V.) were calculated by standard statistical methods.

3. Results and discussion

Entire vitamin molecules were ionized by positive ionization mode. Fragmentation of the molecular ions produced fragment ions which were suitable for quantification. A full mass spectra scan protocol was studied using the following: Vitamin D3 parent mass 384.64 m/z, scan range of 0-377.2 m/z, quantification ions m/z 357.2 and m/z 355.2 and vitamin A, parent mass 286.45 m/z, scan range of 0-340 m/z, quantification ions 268 m/z and 255 m/z . Various extraction solvents were used such as hexane and dichloromethane, but hexane-dichloromethane mixture emerged as the best extraction solvent. Formic acid was also used in the mobile phase to enhance resolutions . During reconstitution of the evaporated prepared samples, the residue reconstitution was far better with mobile phase than with methanol due to the lower polarity of the vitamins (A and D3) in the sample. (see figure 1,2 and 3).

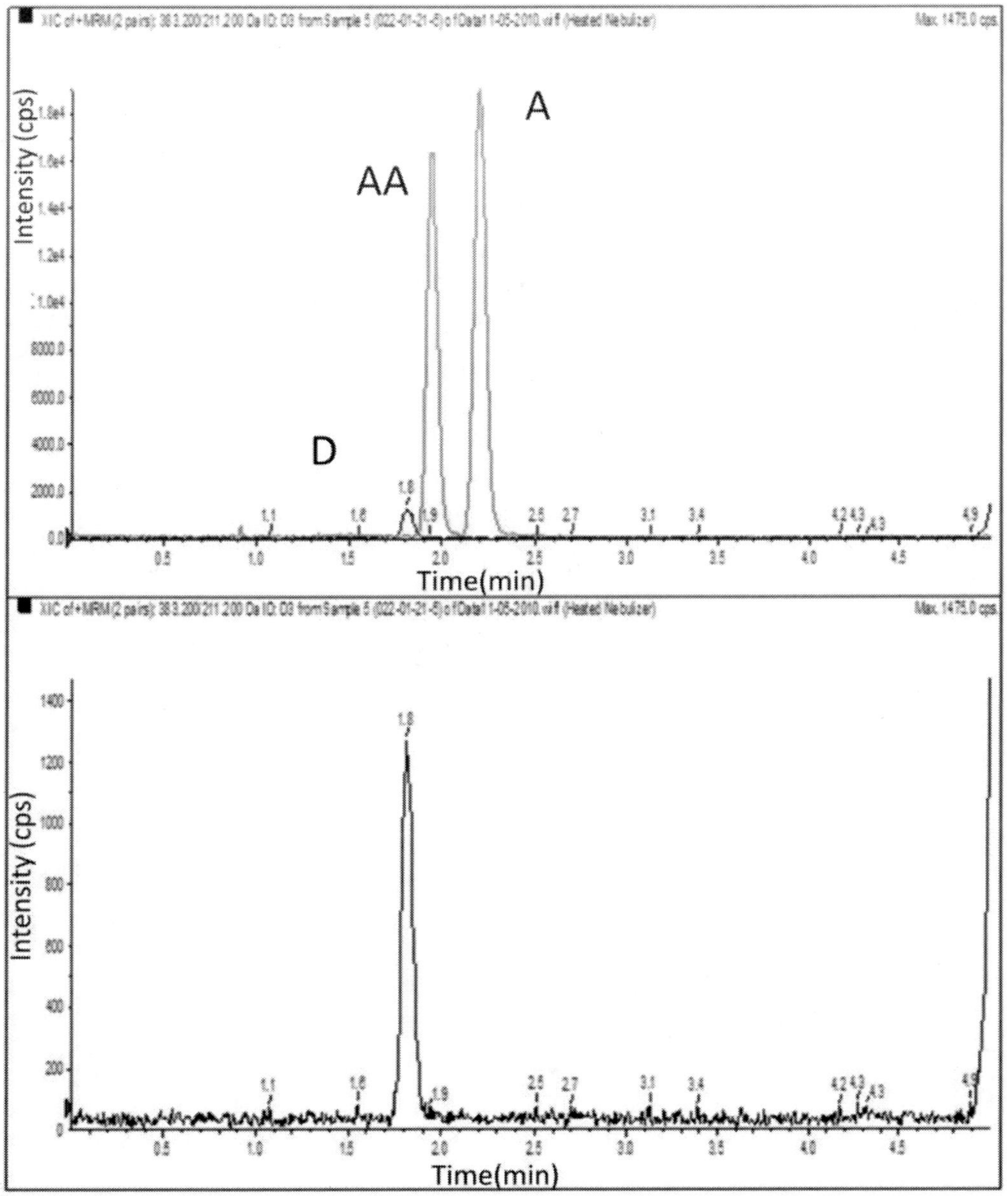

Fig 1. A typical Chromatograms for the simultaneous determination of vitamins A , D3 and internal standard Where retention times of: internal standard = 1.9 min, vitamin A= 2.2min and vitamin D3 1.8min.

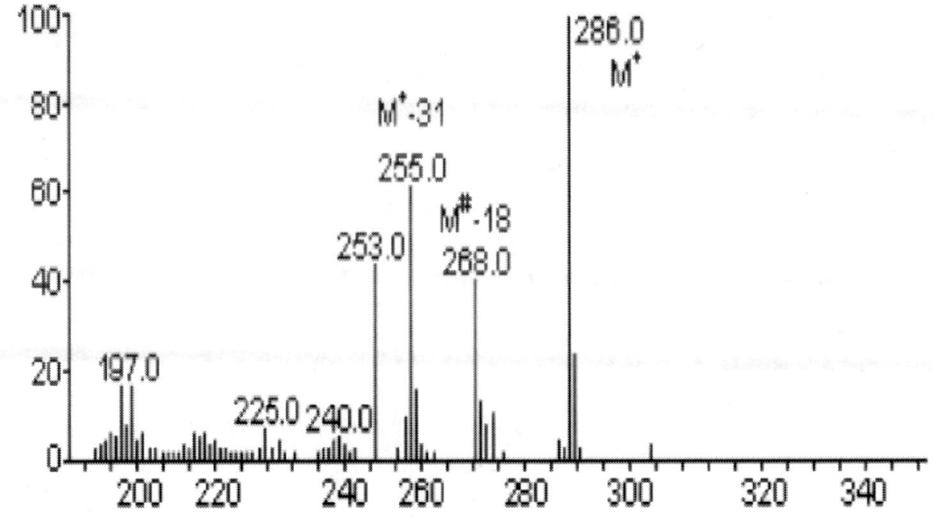

Fig 2. Mass spectra of vitamin A

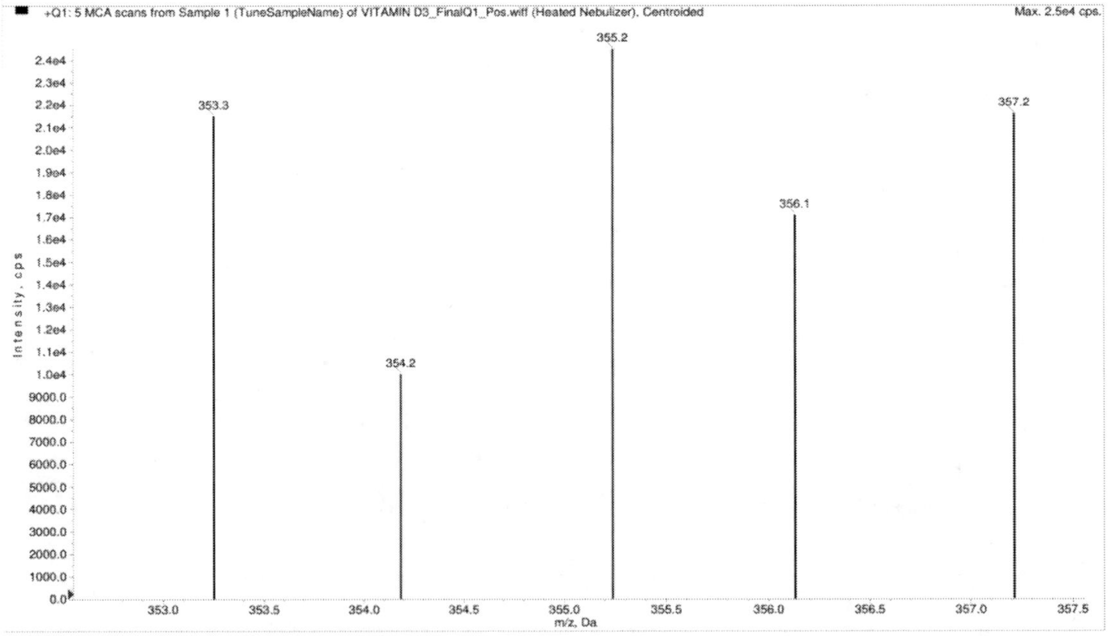

Fig 3. Mass spectra of vitamin D3

Table 1. The average concentrations of Vitamin A (ng/mL) and Vitamin D3 (ng/mL) of each local milk and milk product (18 samples):

S/NO	Type	Code	Vitamin A (ng/ml)	Vitamin D3 (ng/ml)
1	LOCAL	SPL001	46.35	14.60
2	LOCAL	SPL002	6.47	4.91
3	LOCAL	SPL003	39.80	8.88
4	LOCAL	SPL004	12.55	13.65
5	LOCAL	SPL005	30.00	10.00
6	LOCAL	SPL006	7.83	10.36
7	LOCAL	SPL007	7.01	6.76
8	TEEBA FRESH LABANEH	SPL009	9.67	2.69
9	YOGHURT HAMMOUDEH	SPL010	10.00	7.51
10	TEEBA NATURAL YOGHURT	SPL011	16.20	6.78
11	FRESH YOGHURT BALADNA	SPL0112	56.75	5.35
12	JUICE MILK	SPL013	8.87	33.9
13	BANANA MILK HAMMOUDEH	SPL014	1.87	1.76
14	SHANINA BALADNA	SPL015	16.15	8.55
15	TEEBA SHANEENA	SPL016	10.30	9.01
16	ACTIVIA		17.60	1.71
17	ALMARAI MILK		10.78	8.29
18	FRESH MILK HAMMOUDEH		8.89	1.88

The achieved limits of detection (LODs, based on SD of 5 measurements of replicate samples x 3) and the limits of quantitation (LOQs, based on SD of 5 measurements of replicate samples x 10) for both vitamins A and D3 are shown in table2

Table 2. Limits of detection and quantitation

Replicate Samples	Conc. Of vitamin D3 (ng/ml)	Conc. Of vitamin A (ng/ml)
1	1.710	17.600
2	1.730	17.700
3	1.770	17.850
4	1.750	17.900
5	1.790	17.460
Average	1.750	17.700
SD	0.032	0.180
CV (%)	1.83	1.02
LOD (3x SD)	0.096	0.54
LOQ	0.32	1.80
Relative error %	1.32	0.70

The absolute recovery was calculated by comparing the areas under the peaks obtained from standard working solutions with the peak-areas from standard samples. The recoveries of vitamins A and D3 were 98.08 %±1.01 and 98.00 %±1.26 in fresh milk samples, respectively.

Table 3. Recovery of vitamins A and D3 added to fresh milk samples

Amount added (ng/ml)		Amount found (ng/ml)	Recovery %	Amount added (ng/ml)		Amount found (ng/ml)	Recovery %
Vit A	Vit D3	Vit A	Vit A	Vit D3	Vit A	Vit D3	Vit D3
50	14.60	49.7	99.40	25	12.55	24.80	99.20
50	10.00	49.0	98.00	25	9.67	24.80	99.20
50	10.00	48.4	96.80	25	12.55	24.10	96.40
50	13.65	48.8	97.60	25	7.83	24.30	97.20
50	14.60	49.3	98.60	25	9.67	24.50	98.00
Mean %			98.08	Mean %			98.00
SD			0.99	SD			1.23
CV%			1.01	CV %			1.26

The concentration of each sample was measured in duplicate. The range of average calculated concentrations were 1.76 ng/ml – 14.60 ng/ml and 1.88 ng/ml – 56.75 ng/ml for vitamins D3 and A respectively as listed in Table 1. Also the range for the standard concentrations were 5.22 ng/ml – 52.50 ng/ml and 44.00 ng/ml – 922.00 ng/ml for vitamins D3 and A respectively as listed in table 1. This shows that some of the samples have an appreciable amount of vitamins D3 and A, while some have not. This can either be as a result of pasteurization, nature of storage of the milk samples, addition of preservatives , elapsing of expiry date or possibly the exact amount of vitamins A and/or D3 was not added as claimed by the milk producers or that the amount added was not up to the minimum required as recommended by the international standard. This method showed satisfactory recovery of 98.08% for vitamin A and 98% for vitamin D3.

The present study revealed that there is little or no interference , hence no endogenous or exogenous substances were observed during the analysis. The retention times for vitamin D3 was in the range (1.66 - 1.76) min while elution range for vitamin A was (1.88 - 2.46) min.

Conclusion

The research work has potentially provided an accurate, precise, cheap, simple, reproducible and sensitive LC-MS/MS method for the simultaneous determination of vitamins A and D3 in dairy products. Vitamin A- acetate was used as internal standard for better resolution.

The internal standard used was found to be helpful in ensuring better resolution during the analysis.

The method has been successfully applied for the simultaneous determination of Vitamins A & D3 in a set of Jordanian Diary products. From the results obtained we can deduce that the fresh milk samples contain an appreciable amount of vitamins A and D3 compared to long life products.

References

1. Lieberman, S and Bruning, N (1990). The Real Vitamin & Mineral Book. NY: Avery Group, 3, ISBN 0-89529-769-8.

2. U. Satyaranayana & U. Chakrapani (2006). The Vitamins. Biochemistry (Revised Third Ed.,). Chp7, Pp. 116-160.
3. Owen (2008). Fennema's Food Chemistry. CRC Press taylor &Francis Group. pp. 454–455. ISBN 9780849392726.
4. Gerosa and Skoet (2012). "Milk availability – Trends in production and demand and medium-term outlook". FAO, United Nations.
5. Ball, G. F. M. (1998). Fat-soluble vitamin assays in food analysis. A comprehensive review. Essex: Elsevier Science Publishers Ltd.
6. Chauveau-Duriot, B., Doreau, M., Noziere, P., & Graulet, B. (2010). Simultaneous quantification of carotenoids, retinol, and tocopherols in forages, bovine plasma, and milk: Validation of a novel UPLC method. Analytical and Bioanalytical Chemistry, 397(2), 777–790.
7. Zhang, P., Rose, G., & Trenerry, V. C. (2009). Advances in pesticide monitoring: The impact of liquid chromatography–tandem mass spectrometry on the Victorian
8. Willoughby, R., Sheehan, E., & Mitrovich, S. (2002). A global view of LC/MS. Pittsburgh: Global View Publishing.
9. Kamao, M., Tsugawa, N., Suhara, Y., Wada, A., Mori, T., Murata, K., et al. (2007b). Quantification of fat-soluble vitamins in human breast milk by liquid chromatography–tandem mass spectrometry. Journal of Chromatography B:Analytical Technologies in the Biomedical and Life Sciences, 859(2), 192–200.
10. Byrdwell, W. C., DeVries, J., Exler, J., Harnly, J. M., Holden, J. M., Holick, M. F., et al.(2008). Analyzing vitamin D in foods and supplements: Methodologic challenges. American Journal of Clinical Nutrition, 88(2).
11. Huang, M., Laluzerne, P., Winters, D., & Sullivan, D. (2009). Measurement of vitamin D in foods and nutritional supplements by liquid chromatography/tandem mass spectrometry. Journal of AOAC International, 92(5), 1327–1335.
12. Heudi, O., Trisconi, M. J., & Blake, C. J. (2004). Simultaneous quantification of Vitamins A, D3 and E in fortified infant formulae by liquid chromatography–mass spectrometry. Journal of Chromatography A, 1022(1–2), 115–123
13. Lim, Y. O., Kim, B., Ahn, S., & Kim, J. (2011). Improvement of accuracy for the determination of vitamin A in infant formula by isotope dilution-liquid chromatography/tandem mass spectrometry. Food Chemistry, 126(3),1393–1398.
14. Blum, M., Dolnikowski, G., Seyoum, E., Harris, S. S., Booth, S. L., Peterson, J., et al. (2008). Vitamin D3 in fat tissue. Endocrine, 33(1), 90–94.

Effect of insulating layer on the Field Electron Emission Performance of Nano-Apex Metallic Emitters

Ala'a A. AL-Qudah[1], Marwan S. Mousa[1*], A. Fischer[2]

[1] Department of Physics, Mu'tah University, Al-Karak 6170, Jordan
[2] Institute für Physik, Technische Universität Chemnitz, Chemnitz, Germany.

*e-mail: mmousa@mutah.edu.jo & marwansmousa@yahoo.com

Abstract : This paper deals with the process of electron emission from the surface of metals (before and after coating with controlled layers of dielectric materials) into the vacuum due to an intense applied external electric field. This process is usually called cold field electron emission (CFE). The research work reported here includes the current-voltage (I–V) characteristics presented as Fowler-Nordheim (FN) plots and scanning electron micrographs in addition to the spatial emission current distributions (electron emission images). The process of coating the clean tungsten (W) emitters by layers of dielectric epoxylite resin was easy, and the measurements were performed under UHV ~ 10^{-8} mbar. From comparing the results obtained in this work, significant improvement in properties of the emitters after coating are observed.

1. Introduction

Electron emission refers to the release of electrons from the surface of a material into vacuum [1]. Field-induced emission of electrons was explained in 1928, by combining quantum tunneling theory with Fermi-Dirac statistical theory [2]. This was one of the early triumphs of both theories, and a stepping stone towards the modern band theory of solids. The theory of field electron emission from bulk metals was proposed by Ralph H. Fowler and Lothar W. Nordheim [3]. A family of approximate equations, "Fowler–Nordheim-type equations", is named after them. Strictly speaking, these equations apply only to field emission from bulk metals, but the theory is often used – as a rough approximation – to describe field emission from other materials [1]. The material vastly used to carrying out experiments of field electron emission is tungsten [4], due to its superior properties, such as high mechanical strength (hard material due to its weak grain boundaries), high melting point of 3422 ^{0}C, and lowest vapor pressure at temperatures above 1650 ^{0}C, in addition to a remarkable high density and chemical stability.
Other experiments were carried out using the glass coated conductive materials [5], carbon nanotubes [6, 7], and carbon fibers [8, 9]. There are many improvements and developments that has occurred on the Fowler-Nordheim equations [10-16]. The coating of electron emitters by controlled insulating layer technique was used to improve the properties of electron emitters [2, 5, 17].

2. Experimental techniques

The microprint nano-apex metallic emitters used for such measurements were electrolytically etched from tungsten 0.1mm diameter with 99.95% purity wires using a 1M solution of NaOH. Then the tip is ultrasonically cleaned, its profile recorded using SEM, tested at the home built Field electron microscope (FEM) before being coated with epoxylite resin (Clark Electromedical Instruments). The coating procedure involved firstly very slowly dipping a tip into the resin and then accurately removing it to ensure that only a thin film remained on the tip surface. To make sure that this coating forms a regular film located on the tip surface, the tip is then carefully transferred to an oven and subjected to a curing

IOP Conf. Series: Materials Science and Engineering **92** (2015) 012021 doi:10.1088/1757-899X/92/1/012021

cycle of thirty minutes at 100 ^{0}C to drive off the solvents, followed by another thirty minutes at 185 ^{0}C to complete the curing of the resin [4].
The thickness of such a resin film was measured by comparing the radius of the nano tip-apex before and after coating the tip. The radius of tip Nano-apex was measured at 20 KV Scanning Electron Microscope (SEM). A clean W-tip is presented in fig. 1a while in fig. 1b shown after being coated with 0.01945 μm of epoxylite resin.
The experimental facility used to measure the field electron emission characteristics and the emission image data, are carried out in a simple geometry field emission microscope, where the separation between the cathode and the anode was standardized at ~ 10 mm. The emission images were photographed directly from the phosphor screen. All the experiments were performed under pressure ~ 10 $^{-8}$ mbar after baking the ultra high vacuum (UHV) system to ~ 200 ^{0}C for 12 hours.

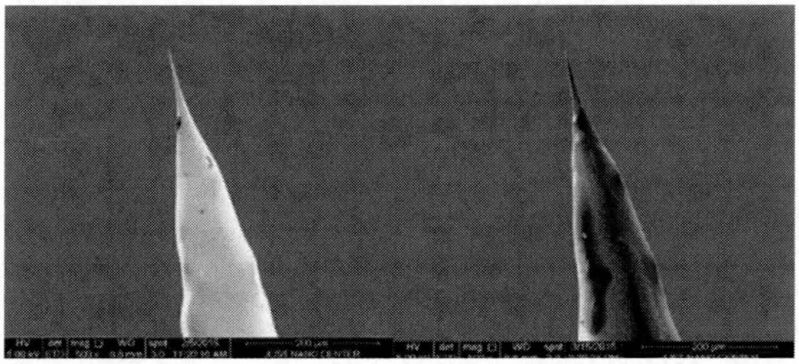

(a) (b)
Fig.1: Micrographs of W-tip (a) before coating and (b) after coating with 19.45 nm of epoxylite resin.

3. Experimental findings
The measurements described below include the I-V characteristics and the emission images of the W-tips before and after coating with 0.01945 μm of epoxylite resin by dipping.

3.1 Emission characteristics
The general form of the I-V characteristic acquired from the composite emitters is presented in fig.2. From the trend shown, it can be seen that as the applied voltage is slowly increased across the virgin composite emitter, a point is reached, *V*sw, where the emission current switches-on unawares from an effective zero to a stable saturated value VSAT. The saturated region 3 to an upper voltage limit *V*max, marked by the onset of current instabilities and possibly followed by a tip explosion. By slowly decreasing the switching voltage to a lower value the emission current starts to be decreasing until a limit *reached*, beyond which the emission current falls smoothly to zero as the applied voltage is decreased to the threshold value *V*th. However, as shown in the figure, this part of the characteristics can be routinely divided into a transition region 2, limited by *V*L, and a low-field region 1 where the I-V data gives a linear Fowler-Nordheim (F-N) plot similar to what has been reported elsewhere[4, 17].

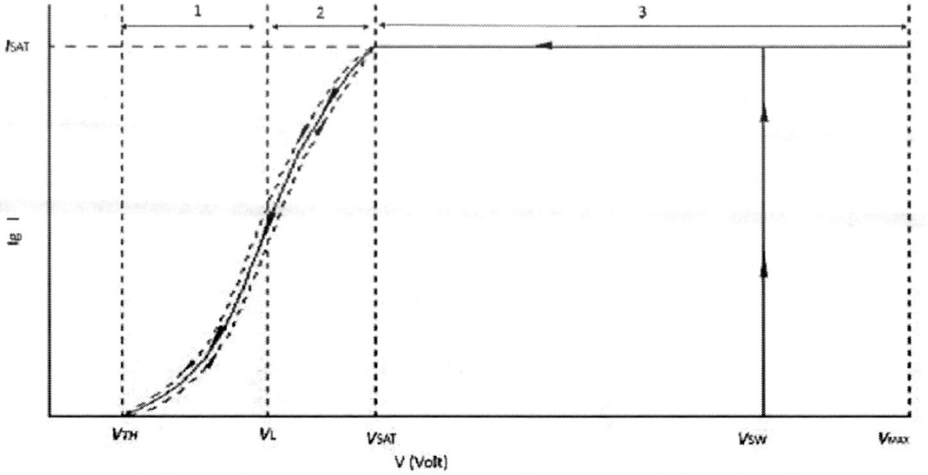

Fig.2: The generalized form of the current-voltage (I-V) characteristic of a W-emitter coated with insulating layer.

The characteristics have been presented using two methods. The first one, which provides very useful information about the emission behavior, is the typical simple I-V plot. This plot demonstrates the dependence of the emission current *I*, on the externally applied voltage *V*, between the emitter and anode [4, 17]. In Fig. 3, curves (A) and (B) represent the comparison of the *I-V* characteristics of the tungsten microprint emitter before and after the coating by 19 nm insulating layer, and curve (C) shows the I-V characteristics after switching-on of the electron emission current. In Fig.3, curve (A), the characteristics of the emitter is shown for before coating the electron emission initiated at applied voltage = 540 V. For the coated emitter, curve (B) shows that the electron emission is initiated at an applied voltage = 100 V. Curve (C) demonstrates the electron emission characteristics that is initiated by a high conventional switch-on the voltage *V*SW of 2800 V, generating a saturation current, V_{SAT} = 1.1 E -5 A. On lowering down the applied voltage, the V_{SAT} remained constant to the saturated voltage value of *V*SAT = 2100 V, and then started to fall smoothly to VL= 500 V. It then continues to decline to zero as the applied voltage is lowered down to a threshold value *V*TH = 100 V. After that, the electron emission is initiated at an applied voltage of 100 V, with emission current = 6.7 nA and reached to 1.1 μA emission current at applied voltage = 500 V. This refers to a significant improvement in the performance of the electron emitter.

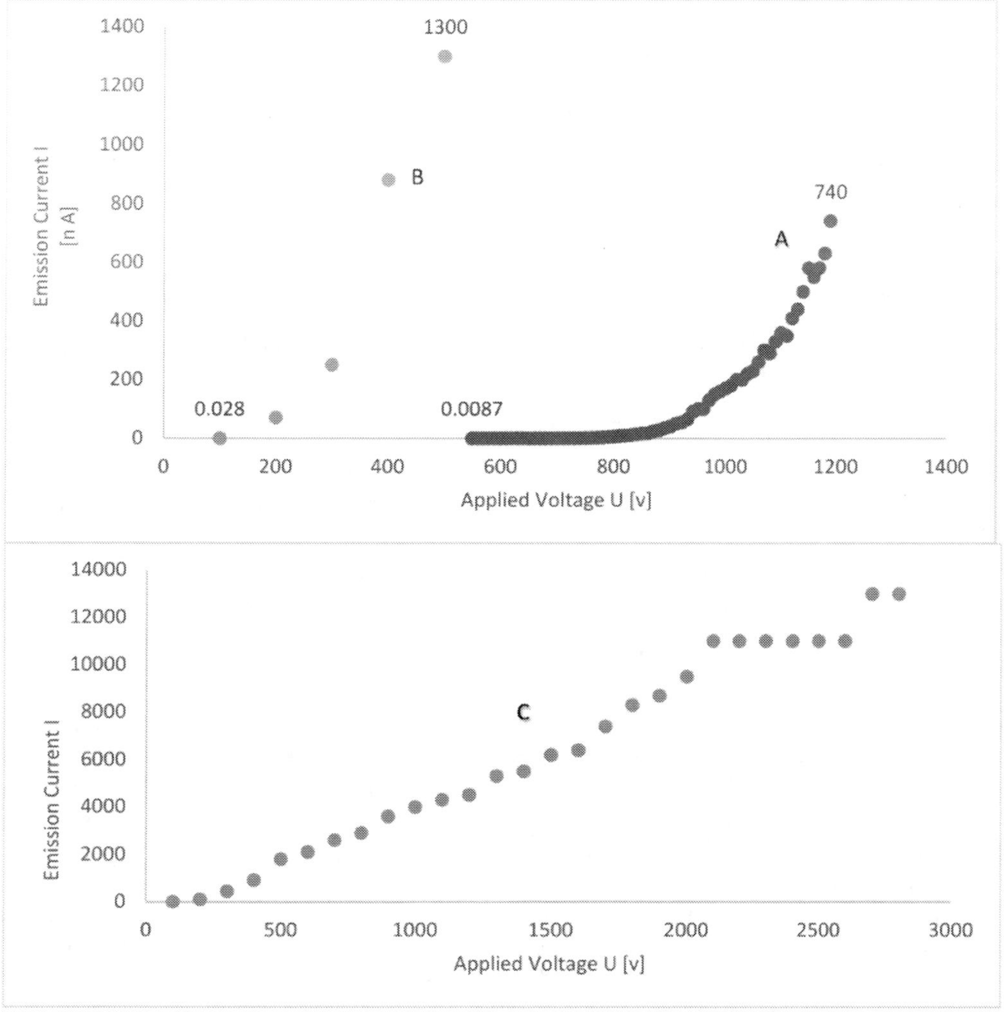

Fig. 3 The *I-V* characteristics of an uncoated tungsten emitter (curve A), after coating with 19 nm layer of epoxylite resin (curve B), and (curve C) presents the I-V characteristics after switching-on of the electron emission.

The second method gave the Fowler-Nordheim (F-N) plots of the clean tungsten emitter (curve A) and that obtained from the composite emitter (curve B) as shown in Fig. 4., in a similar trend to what has been reported elsewhere [4, 17].

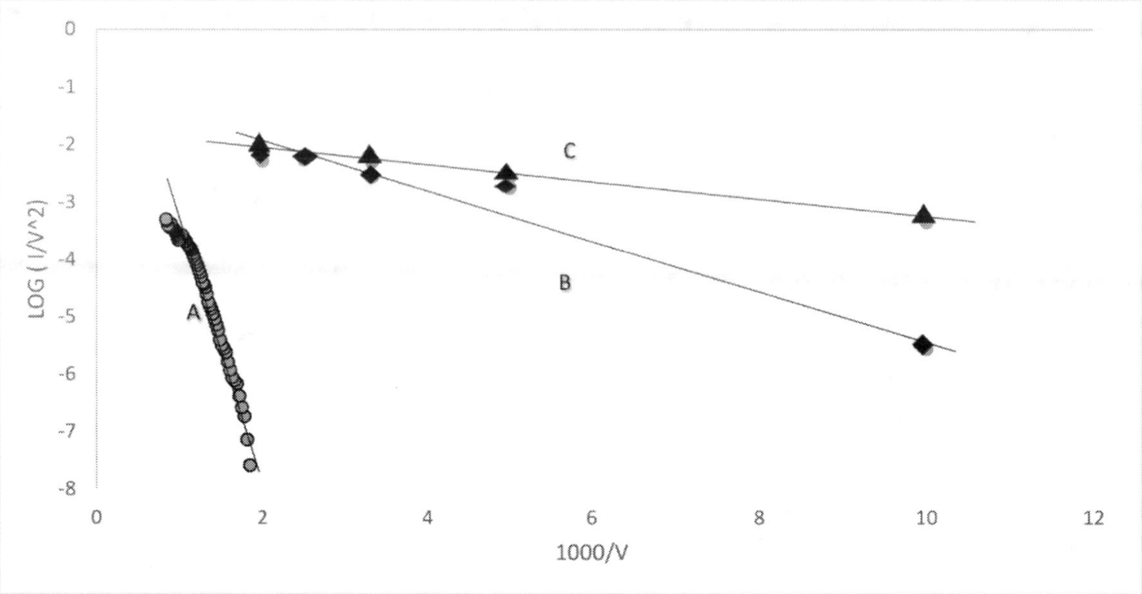

Fig. 4: Fowler-Nordheim plots of the electron emission of clean emitter (curve A), the electron emission of coated emitter (curve B), and (curve C) the electron emission after switch electron emission on of the coated emitter whose *I-V* characteristics are presented in Fig.3.

3.2 Emission images

The electron emission images were photographed directly from the phosphored conductive screen (anode) of the field emission microscope. The emission was concentrated into a single very bright spot that showed no obvious structure within the image. In order to highlight the deference between the emission of the emitter before and after coating, Fig.5 compares the images obtained at a standard emission current of 4.6E-7 A for the same emitter before and after coating. It is shown that, for a given emission current, the composite electron source is much brighter than the clean uncoated metallic source.

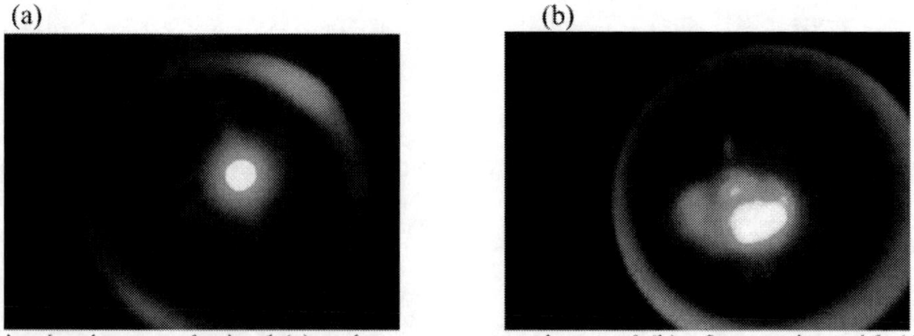

Fig. 5: Projection images obtained (a) a clean tungsten emitter and (b) after coating with 17 nm insulating layer. Both images were recorded with the same distance between the tip and the screen and the same emission current 4.6E-7 A.

IOP Conf. Series: Materials Science and Engineering **92** (2015) 012021 doi:10.1088/1757-899X/92/1/012021

The switch-on phenomenon discussed in the previous section and demonstrated in Fig. 3, was recorded and shown in the sequence of four images in Fig. 6. When the applied voltage reaches a value of 2800 V, the emission current reaches the saturated value ISAT = 1.3E -5 A. In this example, the emission image was concentrated in a single bright spot without any apparent structure within the image.

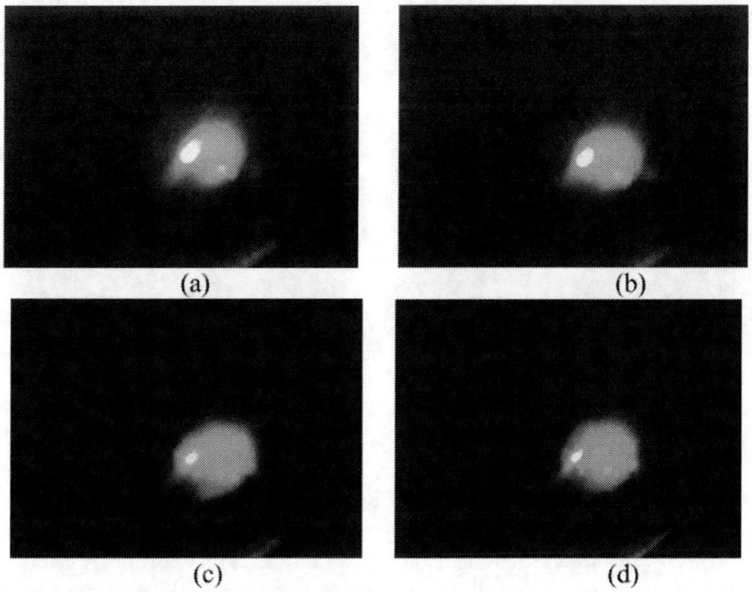

Fig. 6: Sequence of projection images reproduced at actual size showing how the spot size improve with emission saturated current value.

For the tungsten emitter that was coated with an insulating layer of 17 nm thick, the emission appeared to come from several well defined sub-emission centers. This is shown in Fig. 7. The treatment that produced such effects included repeated thermal treatment of the source, and relaxation process by which the tip was left emitting under vacuum overnight. Such effects and the mechanism behind, can be studied and discussed in a future work.

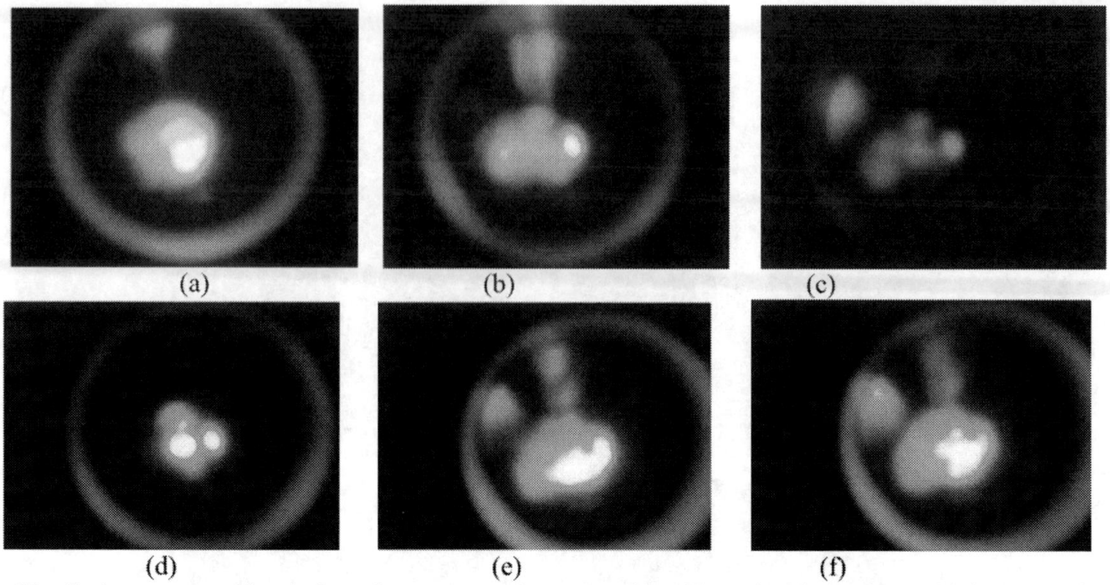

Fig. 7: A sequence of emission of tungsten emitter coated by 17 nm insulating layer thick.

Table 1: the rsm calculated from the scanning electron microscope for the three samples before and after coating. This shows the minimum voltage needed to start the emission before coating, the thickness of insulating layers on the samples, switch-on voltage (Vsw), saturation current (Isat), minimum saturation voltage and threshold voltage (Vth) of the emission after coating. Where: rsm: Radius of tip calculated by using scanning electron microscope. dr: the thickness of insulating layer on the tip

Before coating			After coating					
Tip.No	rsm (nm)	Vlow (volt)	rsm (nm)	dr (nm)	Vsw (volt)	Isat (μA)	Vsat (volt)	Vth (volt)
1	102	540	106.85	4.85	2800	13	2100	100
2	109.3	570	126.3	17	2000	9	1100	100
3	120.2	1410	140.85	20.65	4200	25	3400	200

4. Conclusion

It has been demonstrated that the field emission characteristics of tungsten electron source are drastically changed by coating the emitter with a layer of insulating material. So, such a composite emitter exhibits a switch on phenomenon with a subsequent smooth and reversible current-voltage characteristic whose threshold field lower than that of the clean tungsten electron emitter. The next table shows the emission data obtained from tungsten emitters before and after coating with dielectric epoxylite resin.

References

1. Fowler, R. H., Nordheim, L. W. (1928). Electron emission in intense electric fields. Proceedings of the Royal Society of London, Series A: Mathematical, Physical and Engineering Sciences. 119(781):137–181.
2. Richard G. Forbes. (2008). Physics of generalized Fowler-Nordheim-type equations. Vacuum Science and Technology. B 26, 788.
3. Fowler, R.H.; Dr. L. Nordheim (1928). Electron Emission in Intense Electric Fields . Proceedings of the Royal Society A 119 (781): 173–181.
4. Latham, R. V., Mousa, M. S. (1986). Hot electron emission from composite metal-insulator micropoint cathodes. Applied Physics. 19: 699-713.
5. Mousa, M. S. (1991). Effect of an internally conductive coating on the emission from glass tips. Surface Science. 246: 79-86.
6. AL-Rabadi, A. N., Mousa, M. S. (2012 a). Field emission–based many-valued processing using carbon nanotube controlled switches, Part 1: Fundamentals. Facta Universities, series: Electronics and Energetic. 25: 1-14. DOI: 10.2298/FUEE1201001A.

7. Walt A. de Heer A. Châtelain, D. Ugarte. (1995). A Carbon Nanotube Field-Emission Electron Source. Science.270.5239.1179.

8. Mousa, M.S, Kelly, T. (2003). Stabilization of carbon-fiber cold field-emission cathodes with a dielectric coating. Ultramicroscopy 95: 125-130.
9. Mousa, M. S. (1996). Electron emission from carbon fiber tips. Applied Surface Science 94/95: 129-135.
10. P. H. Cutler, Jun He, N. M. Miskovsky, T. E. Sullivan and B. Weiss. (1993). Theory of electron emission in high fields from atomically sharp emitters: Validity of the Fowler–Nordheim equation. Journal of Vacuum Science and Technology. B 11, 387

11. Forbes, R. G. (1999 a). Field emission; new theory for the derivation of emission area from a Fowler-Nordheim plot. Journal of Vacuum Science and Technology. 17: 526-533.
12. Y.L. Chiou J.P. Gambino. M. Mohammad. (2001). Determination of the Fowler–Nordheim tunneling parameters from the Fowler–Nordheim plot. Solid-State Electronics. Pages 1787–1791.
13. Forbes, R. G., Deane, J. H., Hamid, N., Sim, H. S. (2004). Extraction of emission area from Fowler-Nordheim plots. Journal of Vacuum Science and Technology B. 22: 1222-1226.
14. Edgcombe C.J. (2005). Development of Fowler–Nordheim theory for a spherical field emitter. Phys. Rev. B. 72,045240.
15. Forbes, R. G., Deane, J. H. (2008). Correction for Forbes and Deane, Reformulation of the standard theory of Fowler-Nordheim tunnelling and cold field electron emission. Proceedings of the Royal Society of London, Series A: Mathematical, Physical and Engineering Sciences. 464: 3378.
16. Forbes, R. G., Fischer, A., Mousa, M. S. (2013). Improved approach to Fowler-Nordheim plots analysis. Journal of Vacuum Science and Technology B. 31: 1-8.
17. Mousa, M.S. (1988). Effect of lacomit films on cold-cathode hot-electron emission. Journal de Physique. C9-237.

IOP Conf. Series: Materials Science and Engineering **92** (2015) 012022 doi:10.1088/1757-899X/92/1/012022

Investigating of the Field Emission Performance on Nano-Apex Carbon Fiber and Tungsten Tips

Marwan S. Mousa[1*], Shadi Alnawasreh[1], Mazen A. Madanat*, Anas N. Al-Rabadi[2]

[1]Dept. of Physics, Mu'tah University, Al-Karak 61710, Jordan
[2]Computer Engineering Dept., University of Jordan, Amman, Jordan

*e-mail: marwansmousa@yahoo.com & mmousa@mutah.edu.jo

Abstract: Field electron emission measurements have been performed on carbon-based and tungsten microemitters. Several samples of both types of emitters with different apex radii have been obtained employing electrolytic etching techniques using sodium hydroxide (NaOH) solution with different molarities depending on the material used. A suitable, home-built, field electron microscope (FEM) with 10 mm tip to screen separation distance was used to electrically characterize the electron emitters. Measurements were carried out under ultra high vacuum (UHV) conditions with base pressure of 10^{-9} mbar. The current-voltage characteristics (I-V) presented as Fowler-Nordheim (FN) type plots, and field electron emission images have been recorded. In this work, initial comparison of the field electron emission performance of these micro and nano-emitters has been carried out, with the aim of obtaining a reliable, stable and long life powerful electron source. We compare the apex radii measured from the micrographs obtained from the SEM images to those extracted from the FN-type I-V plots for carbon fibers and tungsten tips.

1. Introduction

A high electric field applied to the surface of a specially prepared pointed metallic surface enables extraction of electrons from the metal by a quantum tunneling mechanism, which is called field electron emission (FE) [1]. On the practical side, the field emission can be used in the display devices by controlling the extracted electrons to hit a desired point on the tin-oxide conductive and phosphor-coated glass transparent screen [1].
Cold field electron emission (CFE) is an electron emission regime in which most electrons escape by tunneling from states below the emitter Fermi level [2]. With CFE, the characteristics of the emitted electrons are controlled by the width and height of the surface potential-energy (PE) barrier through which electrons tunnel [3]. CFE from the bulk is described by approximate equations known as Fowler-Nordheim-type (FN-type) equations [2].
The technical complete (generalized) F-N-type equation for the emission current I can be written in the form [3-5]:

$$I = A_f \, a\varphi^{-1} F^2 \exp[-\nu_F b\varphi^{3/2}/F]$$

where F is the characteristic local barrier field, φ is the local work-function of the base material, a and b are the first and second Fowler-Nordheim constants, and ν_F is the barrier form correction (which accounts for the particular shape of the potential barrier model), and A_f is a parameter known as the formal emission area.

Many kinds of materials such as carbon fibers and tungsten have showed great potential to be used in vacuum electronic devices such as, electron microscopes and other cold cathodes [6].

The attractiveness of particular types of field emitter is due to their favorable emission properties. The operating principle of such emitters is simple. In recent years, the interest in developing new types of field emitters as high brightness electron source for various technologically devices continues [7]. The most commonly used materials for such devices is tungsten [7-9].

In previous studies, Mousa et al. [10] and Baker et al. [11] found that carbon fibers can emit fairly stable currents in relative poor vacuum of lower than 1×10^{-7}mbar. There have been serious studies to employ carbon fibers as electron sources in a variety of applications. This includes scanning tunneling microscopy (STM) [10,12], X-ray sources operated at high emission fluxes [13], and microfocus cathode ray tubes [10,14]. Inefficiency in having carbon fibers employed in electronic equipment is mainly due to stability problems. Forbes [10,15] illustrated this in a comprehensive review of the importance of carbon films and carbon-based materials in science and technological applications.

2. Experimental

Carbon fibers and tungsten emitters from micro and nano structures have been fabricated. For this purpose, tungsten wire with a diameter of 0.1 mm, and carbon-based fibers with a diameter of 7 μm were electrochemically etched in a 2 Molar and 0.1 Molar solutions of sodium hydroxide respectively [16].

After dipping about 2 mm from the sample in the solution, the etching process is started by using a power supply to provide the dc voltage until a certain initial current of about 30 μA for carbon fibers and 15 mA for tungsten is reached. Choosing a suitable etching current produces sharp tips at the liquid surface. After the etching process is completed, the sample is subjected to an ultrasonic cleaning process by using an ultrasonic device [17].

Following the cleaning process, the tip is mounted in a standard, home built, field emission microscope (FEM) with a tip screen distance of 10 mm [18,19]. The anode is formed as a conductive coated phosphor tin-oxide screen in order to allow for the recording of the emission current and the emission images [18,19]. The FEM was evacuated to achieve ultra high vacuum (UHV) condition, using two different systems incorporating a turbo molecular pumped system and an oil diffusion pumped system. The oil diffusion pumped system was connected to a liquid nitrogen (LN_2) trap. Each of the two systems was connected with a rotary pump that reduced the pressure from atmosphere to 10^{-3} mbar. Afterwards, the turbo or the oil diffusion pump starts working and gave a base pressure of 10^{-7} mbar. To achieve the ultra high vacuum pressure (10^{-9} mbar), the system was baked at 180 °C over night [17, 20].

To record the emission behavior in the vacuum, a high tension (EHT) power supply is used to apply negative voltage on the cathode (tip). The voltage is increased slowly until the emission current reaches about one microampere that is measured on a Kiethley 485 auto-ranging pico-ammeter. Then the applied voltage is slowly decreased until the emission current disappears. From the emission behavior of clean and sharp micro and nano emitters, a linear FN-type plot is expected [17,21].

3. Results and Discussion

The clean tungsten and carbon fiber microemitters presented in this work have been prepared with various apex radii to study and compare the properties of the field electron field emission from the respective materials. Selected results, including SEM micrographs of the tips' apex and the FN-type plots of the

IOP Conf. Series: Materials Science and Engineering **92** (2015) 012022 doi:10.1088/1757-899X/92/1/012022

electron field emission I-V characteristics, are presented in Figs 1-4. In this research work, a comparison was carried out between tungsten and carbon fiber apex radii measured from SEM micrographs and apex radii extracted from the FN-type plots as shown in Table 1. The relationship between the two methods will be discussed in the following sections.

3.1 Tungsten

SEM micrographs with high magnification (20 000X) (Figs. 1-4a) gives us the possibility to study the tips' apex shape profiles. The tip apex radius (*r*SEM) has been derived from the graphically best fitted circle to the tip SEM micrograph.
FN-type plots were used to derive the radii of the tips in more physical means as presented in Figs 1-4b. One has to analyze the I-V -type FN plots to obtain the plot slope, which afterwards will be the basis of deriving the second radius (*r*FEM)

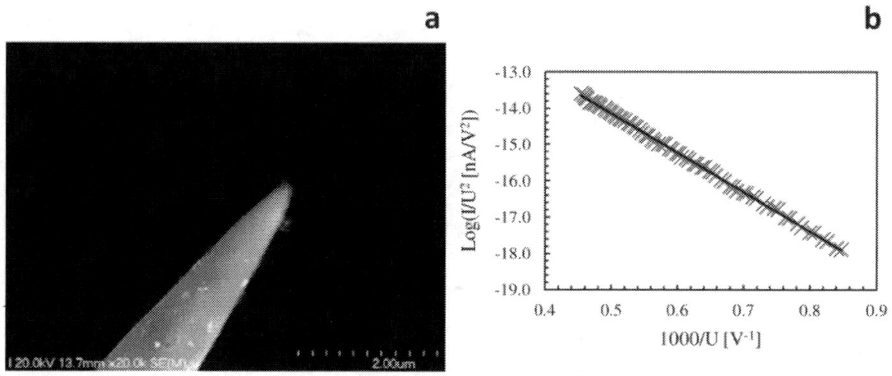

Fig. 1. SEM micrograph (a) of tungsten (W 1) tip at 20 000 times magnification. The FN-type plot (b) of this tip shows a normal behavior of clean tungsten without any fluctuation during the applied voltage change. The slope of the FN-type plot was used to extract the tip radius (r_{FEM}). The extracted values are given in Table 1 below.

Many samples of both tungsten and carbon fibers were tested, with different apex radii. A selection of these samples is presented in this table. As it can be observed from (*r*SEM) and (*r*FEM) for the tungsten type, results showed usually that the (*r*SEM) is less than (*r*FEM) For carbon fibers the results obtained showed opposite trend, where the (*r*SEM) is larger than (*r*FEM).

Table 1

Sample no.	Radius (SEM) r_{SEM} (nm)	Radius (extracted) r_{FEM} (nm)
W 1	139	183
W 2	220	288
CF 1	90	80
CF 2	115	94

The SEM micrograph presented in (Fig. 1a), shows clean and regular tungsten (W 1) tip apex shape. A radius of $r_{SEM,W1}$ = 139 nm has been extracted. Due to the high regularity and symmetric status of the tip apex, there was no problem appeared in fitting the measuring circle to the tip apex.

From the FN-type plot (Fig. 1b), a typical linear behavior of the clean tungsten tip was obtained.

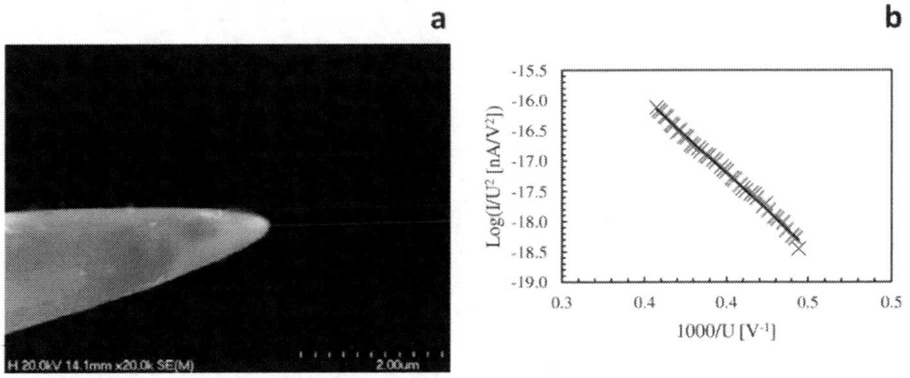

Fig. 2. SEM micrograph (a) of a tungsten (W 2) tip at 20 000 times magnification. The FN-type plot (b) of this tip shows a typical behavior obtained from clean tungsten emitter, without any fluctuation during the voltage change. The slope of the FN-type plot was used to extract the tip radius (r_{SEM}). The extracted values are given in Table 1.

No fluctuation, due to the voltage change, was observed. This meant that the field electron emission was obtained from a certain specific point on the tip apex, which enabled us to perform an accurate analysis. The slope of the plot was derived and analyzed, where the extracted radius is $r_{FEM,W1}$ = 183 nm.

Fig. 2a presents a SEM micrograph for clean tungsten sample (W 2). The regularity and symmetrical shape and curvature of the tip were found. The radius of the tip apex was obtained, by fitting a circle on the SEM micrograph $r_{SEM,W2}$ = 220 nm.

The FN-type plot for the same tip (Fig. 2b) produced a linear straight line which is consistent with the behavior of clean tungsten microemitters. Thus, the slope was obtained and analyzed, from which we extracted the radius of the tip apex $r_{FEM,W2}$ = 288 nm.

IOP Conf. Series: Materials Science and Engineering **92** (2015) 012022 doi:10.1088/1757-899X/92/1/012022

3.2 Carbon Fibers

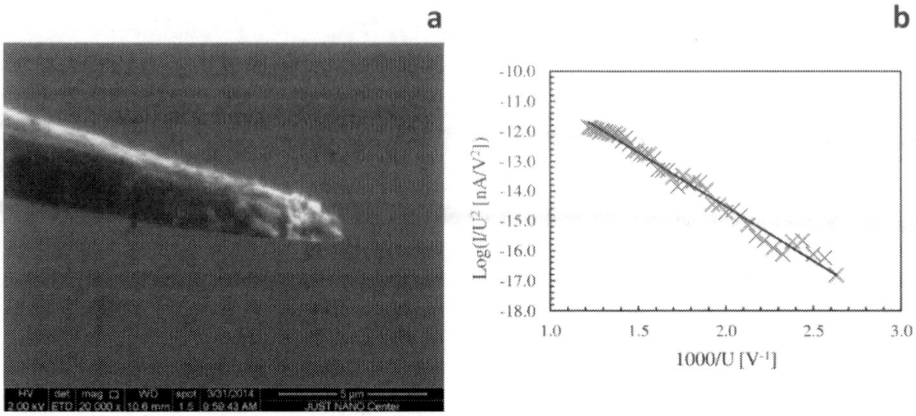

Fig. 3. SEM micrograph (a) of carbon fiber (CF 1) tip at 20 000 times magnification. The FN-type plot (b) of this tip demonstrates a typical behavior with a fluctuation effect which is probably resulted from the voltage change. The slope of the FN-type plot was used to extract the tip radius (r_{FEM}). The extracted values are given in Table 1.

Using carbon fiber as an electron source, we faced more challenges towards the tip preparation and later on analysis of the results. Due to the small size of the fibers, it was hard to fix it into the stainless steel tube used to serve as the holding base for the fiber, as well as producing sharp tips through the etching process.

Fig. 3a shows a SEM micrograph of carbon fiber tip (CF 1). A non-uniform apex shape was obtained following the etching process. This might be as a result of leaving the tip in the etching solution for long time –less than 1 second– after the etch ended. Due to the non-ideal tip shape, we approximated the error Δr_{SEM} as 20 nm. Applying the same techniques as in the case of tungsten, the tip emitting radius was obtained as $r_{\text{SEM,CF1}} = 90$ nm.

The FN-type plot (Fig. 3b) shows a usual behavior of an ideal sharp tip of clean carbon fiber with small fluctuation. The fluctuation results from the applied voltage change and from the possibility of existing sub-emission centers on the emitting area. This could be observed in the emission images (not presented here). The slope was extracted and the radius was derived as $r_{\mathrm{FEM,CF1}}$ =80 nm.

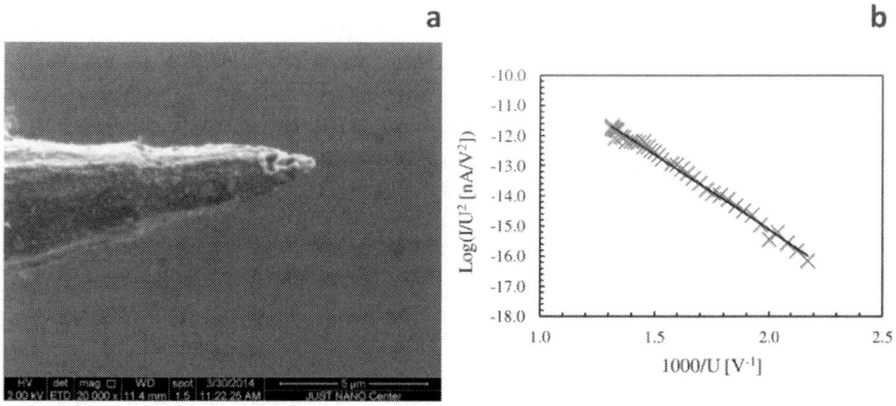

Fig. 4. SEM micrograph (a) of carbon fiber (CF 2) tip at 20 000 times magnification. The FN-type plot (b) of this tip shows a typical behavior with slight fluctuations' effect which probably would be as a result from the voltage change. The slope of the FN-type plot was used to extract the tip radius (r_{FEM}). The extracted values are given in Table 1.

The SEM micrograph presented in (Fig. 4a) for sample (CF 2) from which the tip apex radius of $r_{\mathrm{SEM,CF2}}$ = 115 nm has been obtained. Non-symmetric or uniform tip apex shape was observed form the SEM micrograph which can be considered happening for the same reasons as explained previously for sample (CF 1). One should notice also that for this sample, the curvature of the tip apex has some regularity in shape, which means the behavior of the recorded FN-type plot should have a linear shape.

The emission characteristics presented as a FN-type plot for the sample (CF 2) (cf. Fig. 4b) demonstrated the behavior of a clean carbon fiber. A small fluctuation was observed as a result of changing the applied voltage and might be from the irregularity in the curvature of the tip apex. The slope was extracted, and the radius was calculated $r_{\mathrm{FEM,CF2}}$ = 94 nm. Comparison between both radii; namely (r_{SEM}) and (r_{FEM}) of clean tungsten and carbon fiber tips, is presented in (Table 1 and Fig. 5). On the one hand the tungsten emitter was found to fulfill and support the relation ($r_{\mathrm{FEM}} \simeq 1.35 \times r_{\mathrm{SEM}} - 20$ nm) (see Fig. 5), which was discussed in previous studies [4,22,23]. On the other hand, the carbon fibers were not found to fulfill or follow this relation.

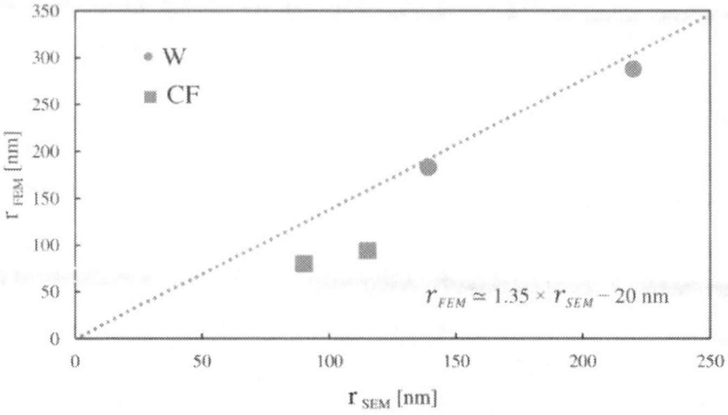

Fig. 5. Comparison between the values of r_{SEM} and r_{FEM} of clean tungsten and carbon fiber tips.

4. Conclusions

Electron micro and nano emitters from tungsten and carbon fibers have been compared in a systematic study. Using electrochemically etching techniques, similar to those employed by Müller in 1937 [24], emitters with different apex radii have been produced. Two completely different methods were used to determine the apex radii. The first was based on SEM micrographs, by graphically best-fitting circles on the SEM images to give r_{SEM} directly. The second was based on FN-type plots, where the slope has been extracted, and the radius of the tips apex was calculated based on the extracted slope (r_{FEM}).

Tungsten based microemitters were found to fulfill the relationship obtained in this work which supports previous arguments [4,22,23]. The relation was used to relate the radii determined by two different methods namely SEM and FEM. Thus, the carbon fiber did not follow the same trend as tungsten. This might be because the tungsten tip apex usually had a more regular hemispherical shape. Thus the radius obtained from the SEM images of a tungsten tip is precise, and more directly comparable to the one extracted from the FEM. With the carbon fiber tips, due to the blurriness of the edges and the non-regular apex shape, it was more challenging to obtain precise apex-radius values from the SEM images. This made it hard to make a full and exact comparison with the radii extracted from the FEM measurements.

Acknowledgements

The authors wish to thank Andreas Fischer from Chemnitz Technical University, Germany, and Richard Forbes from Surrey University, UK, for advice and long-time discussions which enriched this work. Thanks are also extended to Nelia Wanderka from Helmholtz Zentrum Berlin for hosting Mazen Madanat on a student fellowship; thus enabling the use of HZB microscopes, in addition to Marian Jaskula from Jagiellonian Univ. Cracow, Poland who hosted Madanat and facilitated the use of JU microscopes.

References

1- Han Seungwu, Lee. M. H., Ihm Jisoon, Physical Review B, Volume 65, (2002) 085405.
2- Forbes R.G., IOP Publishing, Nanotechnology 23 (2012) 095706
3- Forbes R.G., Fischer A., and Mousa M. S., Journal of Vacuum Science and Technology B 31(2013) 1.
4- Madanat M. A, M.Sc. Thesis, Mu'tah University, Jordan (2014).
5- Alnawasreh S., M.Sc. Thesis, Mu'tah University, Jordan (2014).
6- Zhao Zhigang, Liu Shuhe, Liu Chang, Bai Jin-Bo, and Cheng Hui-Ming, Physica B 396 (2007) 44–48.
7- Mousa M.S. and Al Share M., Ultramicroscopy, 79 (1999) 195-202.
8- Crewe A.V., Isaacson M., Johnson D., Rev. Sci. Instr. 40 (1969) 241.
9- Shirkov E.G., Sov. Phys. Tech. Phys. 14 (1970) 1134.
10- Mousa M.S., Kelly T.F, Ultramicroscopy 95 (2003) 125-130.
11- Baker F.S., Osborn A.R., Williams J., Nature 239 (1972) 96.
12- Lea C., J. Phys. D 6 (1973) 1105.
13- Erickson G.F., Mace P.N., Rev. Sci. Instrum. 53 (7) (1983) 586.
14- Latham R.V., Salim M.S., J. Phys. E 20 (1987) 181.
15- Forbes R.G., Solid State Electron. 45/6 (2001) 779.
16- Mousa M.S., Journal de Physics. 49 (1988) C6 237- C6 242.
17- Alnawasreh S., Mousa M.S., AL-Rabadi A.N, Jordan J. Phys. In press, to appear in Vol. 8, No. 1 (2015).
18- Moran Meza J.A., Lubin C., Thoyer F., Villegas Rosales K.A., Gutarra Espinoza A.A., Martin F. and Cousty J., Carbon, 86 (2015) 363-370.
19- Mousa M.S., Ultramicroscopy 73 (1998) 23-30.
20- Mousa M.S., Surface Science 231 (1990) 149-159.
21- Mousa M.S. and Al Share M., Ultramicroscopy, 79 (1999) 195-202.
22- Madanat M. A., Mousa M.S., Al-Rabadi, A.N. and Fischer. A., Jordan J. Phys. In press, to appear in Vol. 8, No. 1 (2015).
23- Mussa K.O., Mousa M.S. and Fischer A., Ultramicroscopy, 132 (2013) 48.
24- Müller E.W. (1937). Electron microscope observation of field cathodes. Zeitschrift für Physik. 106 (1937) 541-50.

IOP Conf. Series: Materials Science and Engineering **92** (2015) 012023 doi:10.1088/1757-899X/92/1/012023

Construction of Uranyl Selective Electrode Based on Complex of Uranyl Ion with New Ligand Carboxybenzotriazole in PVC Matrix Membrane

M A Abu-Dalo[1*], N A F Al-Rawashdeh[1,*], I R Al-Mheidat[1] and N S Nassory[1]

[1]Department of Applied Chemical Sciences, Jordan University of Science & Technology, P. O. Box 3030, Irbid-22110, JORDAN

E-mail: nathir@just.edu.jo (N A F Al-Rawashdeh); maabudalo@just.edu.jo (M A Abu-Dalo)

Abstract. In the present study uranyl selective electrodes in polyvinyl chloride (PVC) matrix membrane were prepared based on a complex of uranyl ion (UO_2) with carboxybenzotriazole (CBT) as ligand. The effect of the nature of plasticizer in PVC matrix were evaluated using three different plasticizers, these are dibutyl phthalate (DBP), dioctyl phthalate (DOP) and bis(2-ethylhexyl) sebacate (BHS). The results of this study indicated that the best plasticizer could be used is the DBP, which may be attributed to its lowest viscosity value compared to DOP and BHS. The electrodes with DBP as plasticizer exhibits a Nernstian response with a slope of 28.0 mV/ decade, over a wide range of concentration from $3.0x10^{-5}$-$6.0x10^{-2}$ M and a detection limit of $4.0x10^{-6}$ M. It can be used in the pH range of 4.0-10.0 with a response time of less than 10 s for DBP and 25 s for both DOP and BHS. The effects of ions interferences on the electrode response were evaluated. The di- and tri-valent cations were found to interfere less than univalent cations, which was attributed to the high diffusion and the exchange rate between the univalent ions and the uranyl ion solution. The electrodes were characterized by Fourier Transform Infrared Spectroscopy (FTIR) and Scanning Electron microscopy (SEM). The results of the standard addition method were satisfactory with errors less than 7%. The developed electrode was found to be fast, sensitive and reliable indicated its potential use in measuring the uranly ion concentration in the field.

1. Introduction

Uranium contamination can be introduced to the environment naturally from certain types of soils and rocks. Also, it can be introduced from mining, manufacturing activities, coal and other fuel combustion and nuclear power production. Therefore detection of uranyl ion percentages in water is inevitable due to its toxicity. Furthermore, measurement of uranium concentration is having a great importance in nuclear industries due to the application of metal fuel in nuclear reactors.

The presence of uranium in both water and soil environment and the limitations associated with measuring uranium by the reported techniques [1-4] create a need to develop a fast, sensitive and reliable method for uranium determination in various environmental samples. Ion- selective electrodes methods developed due to their unique advantages such as speed, sensitivity and ease of preparation. In literature, various ligands have been used to prepare uranyl ion selective sensors, such as 1-(4-carboxyphenyl)-1H-1, 2,3-triazole-4-carboxylicacid [5], crown ether [6], and bis(2-hydroxy acetophenone) ethylene diimine [7].

Uranyl ion selective electrodes were prepared using crown compounds in presence of different plasticizers by Nassory et al. [8], and have been tested for determination of uranyl ions in environmental water samples Tyagi et al. [9] have succeeded in preparation of uranyl ion selective electrode based on electro active p-tert-butyl-biscalix[4]arene, and have been used for analysis of tap and sea-water samples. The construction and reliability of uranyl membrane electrodes based on vastly available and affordable amino-(trimethyl)-phosphate were investigated by Badr et al. [10], and have been used in nuclear safeguards verification purposes. PVC membrane electrode for uranyl ion based on 2,2'-[1,2-Ethandiyl bis(nitriloethylidene)]bis(1-naphthalene) was prepared by Shamsipur et al. [11], and was successfully used to the recovery of uranyl ion in tap water. Uranyl selective electrode based on a membrane containing a complex of UO_2-bis[di-4-(1,1,3,3-tetramethylbutyl)phenyl phosphate] as an anion-exchanger and tri-tolyl phosphate as a mediator was prepared by Nassory [12], and used for titration of ammonium diuranate. Uranyl ion-imprinted polymers (IIP) resins were prepared by Metilda et al. [13] by dissolving stoichiometric amounts of uranyl nitrate and different ligands. These IIP resins were tested for the recovery of uranium from seawater samples [13].

There are various techniques have been used to determine uranyl ions concentration, such as determination of uranyl in sea water by spectrophotometric method [14,15], determination of uranyl adsorption onto poly(acrylamide-g-ethylenediaminetetraacetic acid) hydrogels in the presence of cadmium and lead by polarographic method [2]. The structural properties of UO_2 complex with macro cyclic ligands were studied by NMR [4].

In this study a uranyl selective electrodes using new ligand carboxybenzotriazole in PVC matrix along with different solvent mediators were prepared. These new ion selective electrodes could be used as a sensor, that provide a simple and sensitive method for uranyl determination in the presence of interfering metal ions in environmental water samples and uranium salts. Furthermore, electrode parameters such as pH, ion interferences, detection limit, response time and life time was evaluated.

2. Experimental

2.1. Apparatus

Potentiometric and pH measurements were carried out at room temperature (25.0 ± 0.1 °C) by a professional bench top TRANS MP3001 potentiometer. Potentiometric measurements were carried out using the fabricated sensor in conjugation with silver/silver chloride (Ag/AgCl) as a reference electrode (Orion, Beverly, MA, USA). Thermo Scientific Orion combination pH glass electrode was used for pH measurements. The potential measurements made with a constant moderate stirring at sensitivity of 0.1 mV. Construction of the electrode body and immobilization of uranyl complex in the PVC matrix membrane were done using a reported method by Craggs et al. [16]. A circular disk (10 mm in diameter) of the PVC membrane was attached to the end of Tygon tube using a concentrated PVC/ THF (tetrahydrofuran) solution as an adhesive. The other end of the Tygon tube was fixed into a glass tube in which silver wire coated with silver chloride was inserted, and then the tube was filled with 0.06 M UO_2 solution for uranyl electrode and 0.06 M CBT solution for CBT electrode. The electrodes were preconditioned by soaking them in 0.06 M of UO_2 or CBT solution for 60 min.

FTIR spectra were recorded using Shimadzu FTIR Affinity-1 spectrometer (Japan), at the range of 400-4000 cm^{-1} with a resolution of 4 cm^{-1}. Scanning electron microscopy (SEM) measurements were viewed using F50 electron microscope (Netherland). Spectrophotometric measurements were recorded using a THERMO ELECTRON UVG-9423 Spectrophotometer (U.S.A).

IOP Conf. Series: Materials Science and Engineering **92** (2015) 012023 doi:10.1088/1757-899X/92/1/012023

2.2 Reagents and standard solutions

All chemicals were of analytical grade and used as received without further purification. Dioctyl phthalate 99% (DOP), Dibutyl phthalate 97% (DBP) and Bis (2-ethylhexyl sebacate) 97% (BHS) were obtained from ACROS Organics. Carboxybenzotriazole (CBT) was obtained from University of Colorado (Boulder, USA). Hydrochloric acid 37% (from Scharlau), sodium hydroxide (from LOBA chemie), tetrahydrofurane (THF) and N,N-Dimethyleformamide 99.5% (DMF) (from GCC, Cyprus), polyvinyl chloride (from Retest company, China), uranyl nitrate hexahydrate 98% (from RIEDEL-DE HAEN AG), tartaric acid 99.5% (from Sigma-Aldrich), ethylene di-amine tetra acetic acid (BDH, England) and arsenazo (III) 98% (from Fluka Chemie Gmbh, Switzerland) were purchased and used in the processes of membranes preparation and evaluation. Analytical grade of nitrate salts of sodium, potassium, copper, barium, calcium cobalt, and lanthanum were used in ions interferences study.

2.3. Uranyl complex formation and membrane preparation

Uranyl-carboxybenzotriazole complex was prepared by dissolving 1.35 mmol CBT (0.22 g) in a mixture of 1 M NaOH (4 mL) and ethanol (4 mL). A 0.675 mmol of $UO_2(NO_3)_2.6H_2O$ was dissolved in 5 mL HNO_3. The two solutions were mixed with stirring until a pale yellow complex precipitate was formed. The precipitated complex was filtered and dried at room temperature for 48 h. Then the dry complex was grounded in a mortar and sieved to particle sizes less than 90 µm. Membranes were prepared by dissolving 0.01, 0.02 or 0.4 g UO_2-CBT complex in a mixture of DMF (0.5 ml) and THF (5.5 mL), followed by adding a 0.4 mL plasticizer and 0.17 g PVC. The mixture was stirred until a clear homogeneous viscous state was obtained. The solution was poured into a glass ring (5 cm in diameter), and the ring was placed on a glass plate and covered with filter paper for 48 h for slow evaporation of the solvent.

2.4. Preparation of Standard Solutions

A stock standard solution of 0.06 M of uranyl nitrate ($UO_2(NO_3)_2$) was prepared freshly and daily with ultrapure water, then a working solutions of different concentrations ranged from $1x10^{-5}$ to $6x10^{-2}$ M were prepared by appropriate dilution with ultrapure water.

A 0.06 M stock solution of the following interfering ions (Na^+, K^+, Ba^{2+}, Ca^{2+}, Cu^{2+} ,Cr^{3+}, and La^{3+}) was prepared using the following nitrate salts of these ions, then a working solutions of different concentrations ranged from $1x10^{-5}$ to $6x10^{-2}$ M were prepared by appropriate dilution with ultrapure water. A 0.1 M HCl solution and 0.1 M NaOH solution were prepared and used for pH adjustments. Tartaric acid (10%), EDTA (2.5%) and Arsenazo (III) (0.25 %) were prepared and used in the spectrophotometric study. All solutions were prepared using ultra-pure water obtained from water purification system (Aqua Max Ultra 370 series, Anyang, Korea) which provides water with resistivity of 18.2 ohm cm (Ω.cm).

3. Results and discussion

3.1. Effect of the complex amount

Several uranyl electrodes were constructed based on a complex of uranyl ion with a new carboxybenzotriazole (CBT) ligand (the structure of CBT is shown in Figure 1).

The FTIR spectra of uranly nitrate, CBT and UO_2-CBT complex are shown in Figure 2. The FTIR spectrum of CBT (Figure 2B) shows a sharp intense stretching vibrational band of N-H at 3450 cm^{-1}, a broad intense stretching vibrational band of O–H at 3200 cm^{-1} and an intense stretching vibrational band of carboxyl group (C=O) in carboxylic acid at a band at 1700 cm^{-1}. The bending vibrational bands of N-H are shown at about 1500-1600 cm^{-1} and the stretching vibrational bands of C-N are shown at about 1300-1450 cm^{-1}. The probable formation of uranyl-CBT complex has been investigated by FTIR (Figure 2). By comparing the major vibrational mode's relative intensities and positions of free CBT spectrum (Figure 2B) with those of the uranyl-CBT complex spectrum (Figure 2C) one can observed that the vibrational band positioned at 1700 cm^{-1} (C=O stretching mode, Figure 2C) is disappeared completely when the complex is formed.

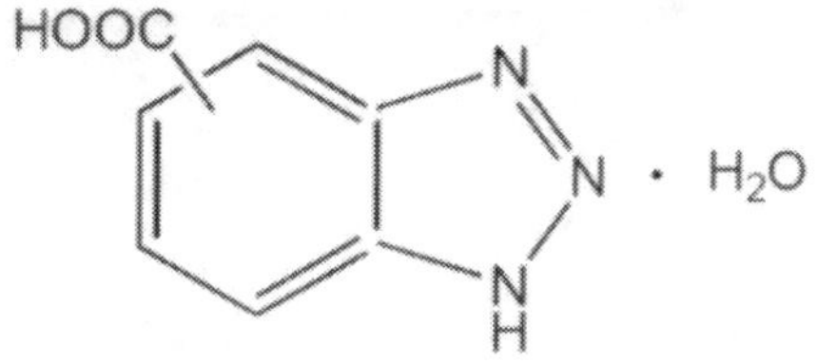

Figure 1. Structure of carboxybenzotriazole used as a ligand

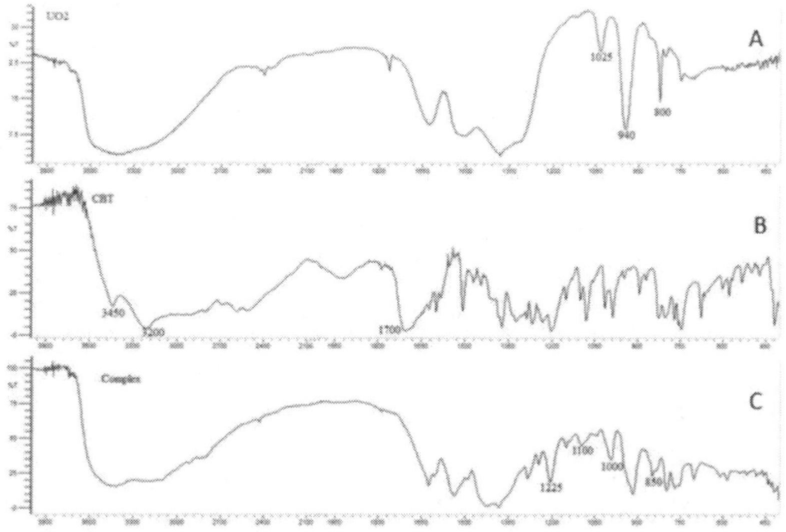

Figure 2. FTIR spectrum of (A) uranyl nitrate (UO_2), (B) carboxybenzotriazole (CBT) and (C) UO_2-CBT complex.

Furthermore, Figure 2C showed a dramatic reduction of intensity and shifted to lower energy of the major bands allocated to U–O vibration at 800 and 940 cm^{-1} and the band at 1025 cm^{-1} was vanished completely. These results demonstrate the formation of the uranyl-CBT complex, by forming a bonding interaction between carboxyl group and the uranyl ion. In spectrum of the complex (Figure 2C), the broad and intense band positioned at 3300-3600 cm^{-1} is allocated to O-H's starching vibrations. The observed broadening of this band indicates the formation of an ice of water over the uranyl ion, through a hydrogen bonding. Figure 2C shows that the N-H stretching vibrational band of CBT in the region of 3300-3600 cm^{-1} is unresolved, however the stretching vibrational band of aromatic C–H located just above 3000 cm^{-1} designating the presence of CBT over the uranyl surface [17]. The intense band located at 1620 cm^{-1} is allocated to OH_2 bending vibration (Figure 2C) [17]. In Figure 2C, the sharp and intense band located at about 1000 cm^{-1} is assigned to O–U–O vibration. The weak vibrational band positioned at 1100 cm^{-1} has been allocated to C–O–M stretching mode [18]. The other observed bands located at 850 cm^{-1}, and 1225 cm^{-1} are allocated to C – C–N bending vibration, and O – C – N stretching vibration, respectively. From all these results of FTIR study, the formation of a uranyl-CBT complex is evident, and most probably this occur through bonding interaction between uranyl ion and the carboxyl group of CBT.

IOP Conf. Series: Materials Science and Engineering **92** (2015) 012023 doi:10.1088/1757-899X/92/1/012023

The morphology of the prepared membrane using the UO_2-CBT complex with and without conditioning has been assessed by scanning electron microscope, and images are shown in Figure 3.

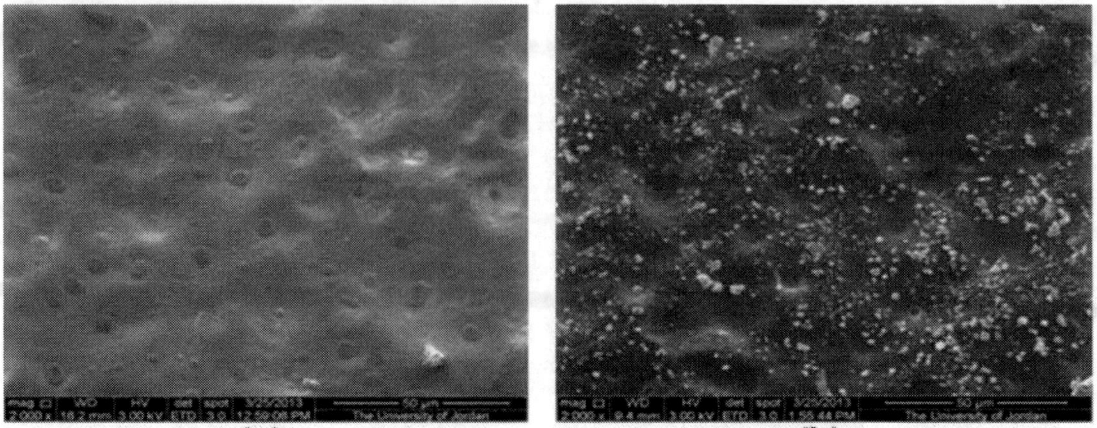

(a) (b)

Figure 3. Scanning electron micrograph of UO_2-CBT complex (a) before conditioning, (b) after conditioning.

Figure 3 shows the scanning electron microscopy (SEM) image of the complex UO_2–CBT before and after conditioning with 0.06 M uranyl nitrate solution. Inspection of Figure 3a (before conditioning) reveals that the particles of the complex are formed in a regular spherical shape with an average of about 50 µm in diameter. In other hand, Figure 3b (after conditioning) shows that the formed particles look like a colloidal particles growing in a solution, this might take place due to a presence of excess of uranyl and nitrate ions that form ionic atmosphere surrounding the complex and create the formation of electric double layers. The image of the membrane that has been observed in this study is very similar to that observed by Coroian et al. [19] for a membrane that prepared using crown ether and used as potentiometric sensor for potassium ion.

The structure of CBT was found to be suitable to form metal complexes due to the presence of active sides [17,18]. Furthermore, the UO_2–CBT complex is extremely soluble in the membrane matrix and has a sufficient lipophilicity. These special properties of UO_2–CBT complex prevent leaching the complex from the membrane to the external solution. The membranes were prepared using different masses of UO_2–CBT complex (0.01 g, 0.02 g and 0.04 g) and di-butyl phthalate (DBP) as a plasticizer. The internal filling solution was 0.06 M UO_2^{2+} for uranyl sensing electrodes and 0.06 M CBT solution for CBT sensing electrodes. The parameters of the prepared electrodes (named with number from 1 to 6) were calculated from the calibration curves and the results were summarized in Table 1.

All potentials were measured at room temperature, and the calibration curve was constructed for each electrode using uranyl standard solutions with a linear response ranged from 1.0 x 10^{-5} to 6.0 x 10^{-2} M. The calibration curves were prepared by plotting the potential in (mV) versus log concentration of uranyl. Electrode parameters including slopes, working concentrations, detection limit, and correlation coefficients were evaluated from the calibration curves. The detection limits for the electrodes calculated according to a reported literature [20] using the relation 18 mV/z (where, z is the charge) and given in Table 1.

The uranyl electrodes number 1, 2 and 3 (Table 1) were constructed and calibrated at different concentration of uranyl ion ranged from 1.0 x10^{-5} to 6.0 x 10^{-2} M. The potential measurements were made from low to high concentration range. The electrodes were kept in 0.06 M uranyl nitrate solution when it was not in use to avoid any damage due to evaporation of the plasticizer or a cracking of the membrane during drying by exposure to the atmosphere.

Table 1. Specific parameters of UO_2 and CBT electrodes using different quantities of the complex.

Electrode No.	Type of electrode	Quantity of the complex	Slope mV/ decade	Linear response range /M	Detection limit /M
1	UO_2	0.01 g UO_2 - CBT	18.6	$2.0x10^{-2}$ – $1x10^{-5}$	$3.0x10^{-6}$
2	UO_2	0.02 g UO_2 - CBT	28.0	$6.0x10^{-2}$ -$3x10^{-5}$	$4.0x10^{-6}$
3	UO_2	0.04 g UO_2 - CBT	21.0	$6.0x10^{-2}$ – $3x10^{-5}$	$7.0x10^{-6}$
4	CBT	0.01 g UO_2 - CBT	35.0	$6.0x10^{-2}$ – $1x10^{-5}$	$2.0x10^{-6}$
5	CBT	0.02 g UO_2 - CBT	29.9	$2.0x10^{-2}$ - $5x10^{-5}$	$2.0x10^{-5}$
6	CBT	0.04 g UO_2 - CBT	22.9	$6.0x10^{-2}$ – $4x10^{-5}$	$2.0x10^{-5}$

The listed results in Table 1 show a slopes of 18.6, 28.0 and 21.0 mV/decade for electrodes number 1, 2, and 3, respectively, indicating that uranyl electrode number 2, which based on 0.02 g complex, gave the best results among the tested electrodes with a slope near to Nernstain slope of 28.0 mV/ decade and having a detection limit of 4.0 x 10^{-6} in a linear response range from 6.0 x 10^{-2} to 3.0 x 10^{-5} M. In addition, the calibration curves of CBT electrode number 5, which based on 0.02 g complex, gave the best results among the tested electrodes with a slope near to Nernstain slope of 29.9 mV/ decade and having a detection limit of 2.0 x 10^{-5} in a working concentration of a linear response range from 2.0 x 10^{-2} to 5.0 x 10^{-5} M. Therefore, for construction of UO_2 and CBT electrodes, the best mixing ratio of the complex, plasticizer, and PVC is found to be 0.02, 0.30, and 0.17 g, respectively. A typical calibration curve for UO_2 electrodes (electrode number2) is shown in Figure 4.

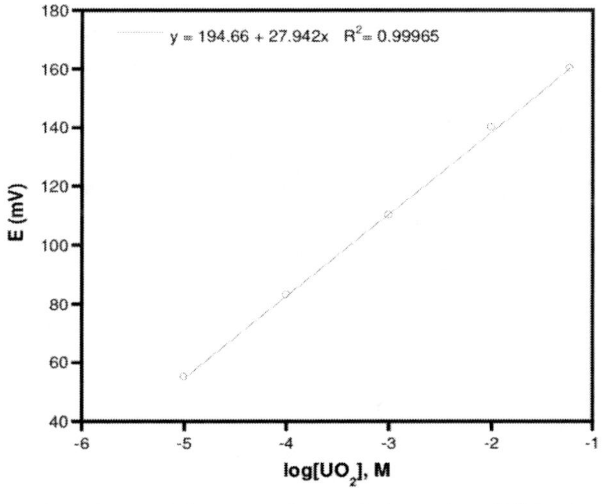

Figure 4. A typical calibration curves of uranyl electrode based on DBP plasticizer. Membrane composition is complex-DBP (electrode number 2). Mass of the complex UO_2-CBT = 0.02 g.

3.2 Effect of the nature of plasticizer

In ion selective electrodes, the proper plasticizer that could be used in preparation of membranes should have desirable nature properties such as, low viscosity, low toxicity, low volatility, compatibility with the polymer, and low solubility in aqueous solution [21]. The nature of plasticizer influences the dielectric constant of the membrane phase, the mobility of the complex and state of ligand. The potentiometric response characteristics of membranes based on 0.02 g UO_2 complexes as an electroactive material and three types of plasticizers, these are dibutyl phthalate (DBP), dioctyl phthalate (DOP) and bis(2-ethylhexyl) sebacate (BHS) in PVC matrix were examined. The effects of the plasticizers were studied with respect to the values of slope, response time, linear working concentration range, life time and detection limit. Electrode parameters of uranyl electrodes based on different plasticizers were summarized in Table 2.

Table 2. Electrode parameters of uranyl electrodes based on different plasticizers.

Membrane composition	Slope (mV/ decade)	Linear response range (M)	Detection limit (M)	Response time (s)	Life time
Complex +DBP	28.0	6.0×10^{-2} - 3×10^{-5}	4.0×10^{-6}	10	> 6 week
Complex +DOP	23.5	4.0×10^{-2} - 1.0×10^{-5}	1.0×10^{-5}	25	-
Complex +BHS	23.8	6.0×10^{-2} - 1.0×10^{-5}	1.0×10^{-5}	25	-

The data shown in Table 2 indicates that using DBP as plasticizer gave the best results among the tested plasticizers. It is believed that low viscosity of DBP (17 cP at 25 °C), which is less than the viscosity of the other two tested plasticizers (viscosity of DOP = 56 cP at 25 °C, and for BHS = 21.14 cP at 25 °C), seems to be a part from this good behavior of DBP, which make DBP the proper one among the tested plasticizers. The low viscosity of DBP provides a free mobility of the complex in the membrane and high compatibility of the complex with the DBP. Therefore, we have used the DBP as a proper plasticizer in other experiments.

3.3 Response and life time

Dynamic response time is very important factor for ion-selective electrodes. In this study the potential response time was recorded by changing the uranyl concentrations. The response time for electrode based on DBP plasticizer (electrode 2) at 10^{-6} M has found to be around 10 s while that for other tested electrodes based on DOP and BHS plasticizers have found to be around 25 s (Table 2), this designated to high viscosity of DOP and BHS relative to that of DBP which hinders the mobility of the complex and takes longer time for equilibrium. The life time for the electrode of uranyl based on DBP plasticizer has found to be around 6 weeks as shown in Figure 5. After 6 weeks, Figure 5 shows that the slopes become constant around 22 mV/ decade this behavior indicates that the electrode can be used furthermore after 6 weeks.

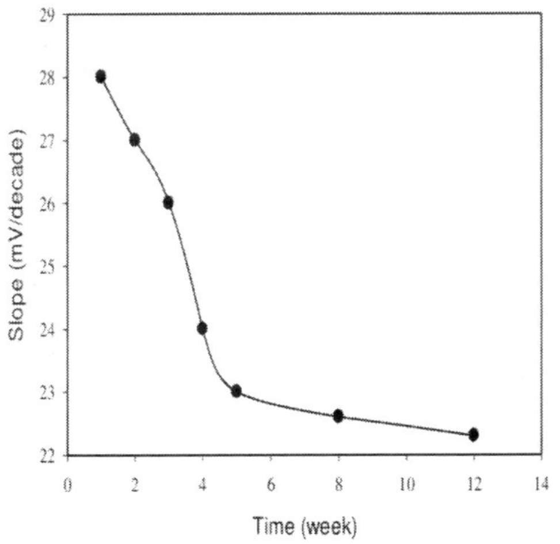

Figure 5. Variation the slope of UO_2 electrode based on DBP plasticizer with time.

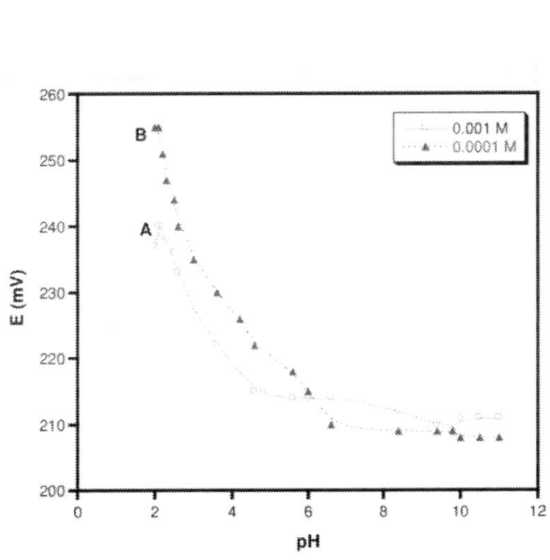

Figure 6. Effect of pH on electrode potential using UO_2 electrode. $[UO_2]$ = 1.0 x 10^{-3} M ((A) and 1.0 x 10^{-4} M (B). Membrane composition is complex-DBP (electrode number 2). Mass of the complex UO_2-CBT = 0.02 g

3.4 Effect of pH of solution

The effect of pH on electrode response was studied for the prepared electrodes. The pH ranges for measuring uranyl nitrate solutions with concentrations of $1.0x10^{-3}$ M and $1.0x10^{-4}$ M were studied at pH range of 2-11. A shift and increase in potential were found at low pH due to hydrogen ions interference (Figure 6). Figure 6 shows better electrode response at higher pH. This result may be attributing to the increase in the negative charge on CBT surface as the pH increases which lead to a stronger interaction between the negatively charge CBT and uranyl ion. The observed small potential drift at higher pH value could be due to the formation of some hydroxyl complexes of UO_2^{2+} (Figure 6). The data that summarized in Table 3 shows that the UO_2 electrode worked properly at pH range from 4 to 11, while the CBT electrode worked properly at pH range of 6-10.

3.5 Selectivity coefficient

The selectivity coefficients have been evaluated using separated solution as recommended by Umezawa et al. [22]. The potential of the electrode is measured using two separate solutions, one containing uranyl ions (A) of activity a_A and charge z_A (in absence of B), the other solution containing the interfering species (B) with charge z_B when ($a_A = a_B$). The selectivity coefficient (K^{pot}) calculated using the method described by Umezawa et al. [22], using Eq. 1.

$$\log K_{A,B}^{pot} = \left[\frac{(E_B - E_A)z_A F}{2.303RT}\right] + \left(1 - \frac{z_A}{z_B}\right)\log a_A \qquad (1)$$

where E_A, E_B are the potential obtained by using solutions A and B (mV), z_A and z_B are the charges of A and B ions, a_A is the activity of A solution, T is the temperature in kelvin, R is the gas constant (8.314 joule/K.mole), and F is the Faraday constant (96500 coulombs). The slopes (in mV/decade) of tested electrodes calculated from the calibration curve (Eq. 1) and were summarized in Table 1 and 2.

IOP Conf. Series: Materials Science and Engineering **92** (2015) 012023 doi:10.1088/1757-899X/92/1/012023

Table 3. Values of the pH range for UO_2 and CBT electrode based on DBP plasticizer.

Electrode	Membrane composition	$[UO_2] = 1x10^{-3}$ M	$[UO_2] = 1x10^{-4}$ M
		pH range	
UO_2	0.02 g UO_2 + CBT	4.2 – 9.8	4.6 – 11.0
CBT	0.02 g UO_2 + CBT	6.7 – 10.0	6.4 – 10.3

Table 4. Values of selectivity coefficients for interfering ions at different concentrations of uranyl ions using electrode based membrane composition of complex-DBP (electrode number 2). Mass of the complex UO_2-CBT = 0.02 g.

Conc./ M	$Cr(NO_3)_3$	$La(NO_3)_3$	$Ba(NO_3)_2$	$Ca(NO_3)_2$	$Cu(NO_3)_2$	$NaNO_3$	KNO_3
$1.0 x10^{-5}$	0.347	1.260	16.700	1.270	0.410	$2.68 x 10^{-5}$	$9.904 x 10^{-4}$
$1.0 x10^{-4}$	0.451	0.580	4.350	0.400	0.440	$1.88 x 10^{-4}$	$2.39 x 10^{-4}$
$1.0 x10^{-3}$	0.577	0.260	1.100	0.140	0.460	$1.17 x 10^{-3}$	$5.76 x 10^{-3}$
$1.0 x10^{-2}$	0.752	0.120	0.290	0.050	0.490	84.02	$1.42 x 10^{-3}$
$6.0 x 10^{-2}$	0.918	0.060	0.110	0.020	0.540	10.74	$4.90 x 10^{-2}$

The results in Table 4 indicate no interferences for divalent and trivalent ions. However, the univalent ions, Na^+ and K^+, was found to interfere with the determination of uranyl designating the presence of a strong interaction between univalent ions and CBT, in which the stability constant of CBT with sodium and potassium is expected to be much higher than that of CBT with uranyl ion. Therefore, it is difficult to measure the uranyl ion in samples containing univalent ions. The interference of trivalent ions, La^{3+} and Cr^{3+}, was found to depend on the electronic structure of the tested ions which determine the type of interaction with CBT. The results of this study show that the selectivity coefficients for the Cr^{3+} ion increases from 0.374 to 0.918 as the concentration increases form $1.0x10^{-5}$ M to $6x10^{-2}$ M, an opposite behavior was found for La^{3+} (Table 4). Figure 7 shows the variation of selectivity coefficient of La^{3+} with log $[La^{3+}]$. Based on the results of this study the selectivity coefficients have found to be dependent on the type (charge) and concentration of interfering ions (Table 4).

3.6 Quantitative analysis

Standard addition method (SAM) was used to determine the concentration of uranyl ions in synthetic samples at three concentrations $1.0 x 10^{-3}$, $1.0 x 10^{-4}$ and $1.0 x 10^{-5}$ M. 50 ml of a synthetic uranyl nitrate solution was used as unknown and the electrode used is based on electrode number 2. The results of standard addition method were summarized in Table 5.

Table 5. Results of standard addition method for three concentrations of synthetic uranyl solutions using UO_2 electrode based on DBP plasticizer (electrode number 2).

Standard deviation	Percent recovery	Measured UO_2 concentration	Calculated UO_2 concentration/ M
± 0.073	107%	$1.07 x 10^{-3}$	$1.0 x 10^{-3}$
±0.070	100.8%	$1.008 x 10^{-4}$	$1.0 x 10^{-4}$
±0.716	83%	$0.83 x 10^{-3}$	$1.0 x 10^{-5}$

Satisfactory results of percent recovery were found at high concentration of uranyl ion (at 10^{-3} M and 10^{-4} M), while the percentage of recovery decreases to 83% at low concentration of uranyl ion (10^{-5} M). This result is consistent with the fact that the concentration of uranly ion at 10^{-5} M is located in the non-linear response range of the calibration curve. A typical plot for standard addition method for uranyl solution at concentration of 10^{-4} M is shown in Figure 8. Uranyl electrode based on DBP plasticizer (electrode 2) was used for uranyl ion determination in water samples taken from different sites. The experiments showed that the uranyl ion concentration cannot be detected in the samples because it is under the detection limit of the tested electrode. Also the water samples were checked for uranyl concentration by using UV-vis spectrophotometer method [23]. For a synthetic solution of 10^{-5} M uranyl nitrate, the percentage of recovery was found to be around 97%. Furthermore, uranyl concentrations in all tested water samples couldn't be detected by spectrophotometric method because it's under the detection limit (3.5×10^{-7} M).

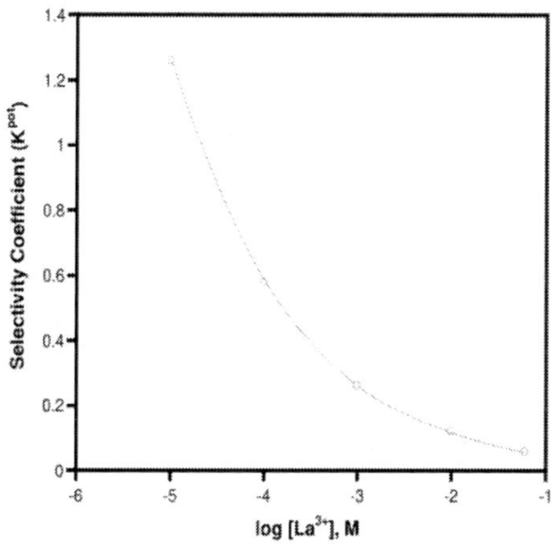

Figure 7. Variation of selectivity coefficient of UO_2 electrode with different concentration of lanthanum nitrate.

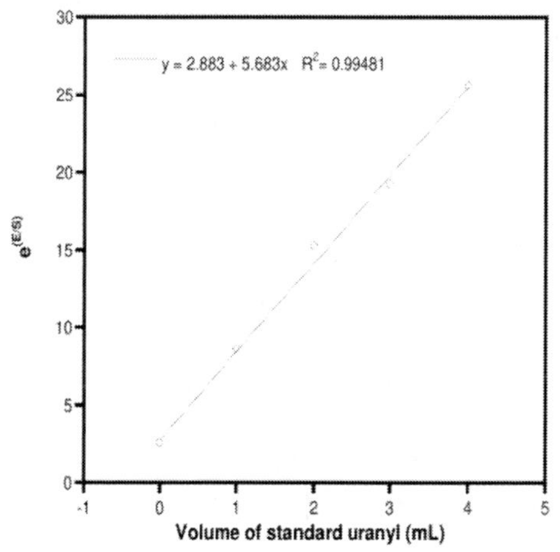

Figure 8. Standard addition method for water sample UO_2 electrode (electrode number 2). Concentration of synthetic solution = 1.0×10^{-4} M. E = potential and S = slope.). Membrane composition is complex-DBP. Mass of the complex UO_2-CBT = 0.02 g.

4. Conclusion

In this study new uranyl selective electrochemical sensors in polyvinyl chloride (PVC) matrix membrane were prepared based on a complex of uranyl ion (UO_2) with carboxybenzotriazole (CBT) as ligand and have been successfully evaluated for uranyl ion determination is synthetic water samples. The best plasticizer could be used is in PVC matrix was found to be dibutyl phthalate (DBP). The electrodes with DBP as plasticizer exhibit a Nernstian response. The developed electrode exhibited short response time and it was found to be sensitive, selective and reliable indicated its potential use in measuring the uranly ion concentration in the field. Also, it has a stable potential over a wide pH range and good stable shelf life for about six week using electrodes with DBP.

Acknowledgment

The authors are grateful to Jordan University of Science and technology (JUST) for the support and to the Scholar Rescue Fund (SRF) Iraq the sponsors of the visiting professor Nabil S. Nassory.

References

[1] Konstantinou M and Pashalidis I 2004 *Mediterranean Marine Sci.* **51(1)** 55
[2] Inam R, Caykara T and Kantogglu O 2003 *Nuclear Instruments and Methods in Physical Research B.* **206** 400
[3] Smith N A, Cerefice G S and Czerwinski K R 2013 *J. Radioanal. Nucl. Chem.* **295** 1533
[4] Amato M E, Ballistreri F P, Pappalardo A, Tomaselli G A and Toscano R M 2010 *Molecules* **15** 1442
[5] Knope K E and Cahill C L 2010 *Crust. Eng. Comm.* **13(1)** 153
[6] Kakhki R M Z and Rounaghi G 2011 *Mater. Sci. Eng.* **31** 1637
[7] Sholrollahi A, Ghaedi M, Montazerozohori M, Khanjari N and Najibzadeh M 2009 *J. Chines Chem. Soc.* **56** 812
[8] Nassory N S and Karim N H 1994 *Turk. J. Chem.* **18** 126
[9] Tyagi S, Agarwal H, Ikram S, Gupta M K and Singh S 2011 *Anal. Bioanal. Electrochem.* **3(4)** 350
[10] Badr I H, Zidan W I and Akl Z F 2012 *Electroanalysis* **24(12)** 2309
[11] Shamsipur M, Saeidi M, Yari A, Yaganeh-Faal A, Mashhadizadeh M H, Azimi G, Naeimi H and Sharghi H 2004 *Bull. Korean Chem. Soc.* **25(5)** 629
[12] Nassory N S 1990 *Fresenius' J. Anal. Chem.* **337(4)** 377
[13] Metilda P, Gladis J M, Venkateswaran G and Prasada Rao T 2007 *Analytica Chimica Acta* **587(2)** 263
[14] Konstantinouand M and Pashalidis I 2004 *Mediterranean Marine Sci.* **51(1)** 55
[15] Smith N A, Cerefice G S and Czerwinski K R 2013 *J. Radioanal. Nucl. Chem.* **295** 1533
[16] Craggs A, Moody G J and Thomas J D R 1974 *J. Chem. Educ.* **51** 541
[17] Ravichandran R, Nanjundan S and Rajendran N 2004 *Appl. Surf. Sci.* **236** 241
[18] Patel P K and Patel P D 2009 *E-Journal of Chemistry* **6(2)** 475
[19] Coroian A, Pica E M, Grecu R, Cosma V, Hopartean E and Hopartean I 2002 *Leonardo J. Sci.* **1** 1
[20] Peter P and Bailey L 1976 *Hayden & Son Ltd.* 20
[21] O'Rourke M, Duffy N, De Marco R and Potter I 2011 *Membranes* **1** 132
[22] Umezawa Y, Umezawa K and Sato H 1995 *Pure Appl. Chem.* **67** 507
[23] Marczenko Z 1976 *Spectrophotometric Determination of Elements* (New York: John Wiley & Sons. Inc.) p 574

The UV and Laser Aging for PMMA/BDK/Azo-dye Polymer Blend Cured by UV Light Beams

A.A. Ahmad [1] **and A.M. Omari** [2]
[1] Department of Physics, Faculty of Science and Arts, Jordan University of Science and Technology, P.O. Box 3030, Irbid-22110, Jordan. [2] Laboratories and Radiological Monitoring Directory, Energy and Minerals Regulatory Commission, P.O Box 1865, Amman-11821, Jordan.

E-mail: sema@just.edu.jo

Abstract. A polymeric-based solution blend composed of Azo-dye methyl red (MR) doped with polymethelmethacrelate (PMMA) solution, in addition, to the BenzylDimethylKetal (BDK) photoinitiator was made with optimum molar ratios and deposited on glass substrate by spin coating technique. The samples were then exposed to UV light beams in order to assist the layers polymerization by the proper exposure process. The photo chemical reaction occurred during the UV light polymerization process induces photo refractive changes which were presented as a function of wavelength or photon energy. Two main strong absorption peaks were observed in the films at around 330 nm (3.75 eV) and 500 nm (2.48 eV) for different curing time periods. This phenomenon enhances the films usage for optical data storage media at these two wavelengths. Since the deposited films were then useful as based layers for Read/Write optical data storage media, they were then tested by UV or laser Read/Write beams independently. The optical properties of the films were investigated while exposed to each beam. Finally, their optical properties were investigated as a function of aging time in order to relate the temporary and/or permanent light-exposure effect on the films compared to their optical properties before the light exposure. The films show a low absorbance at 630 nm (1.97 eV) and high absorbance at 480 nm (2.58 eV). This fact makes it possible to record holographic gratings in the polymeric film upon light exposure. In all cases the optical properties were evaluated by using the very sensitive, non destructive surface testing spectroscopic ellipsometry technique. The films were characterized in the spectral range of 300 to 1000 nm using Lorentz oscillator model with one oscillator centred at 4.15 eV. This study has been supported by the SEM and EDAX results to investigate the effect of the UV and visible beams on their optical properties. The results of this research determined the proper conditions for depositing the proper optical materials for light sensors or Read/Write optical data storage media for the UV or the laser Read/Write heads.

1. Introduction

The PMMA/BDK/Azo-dye polymeric based composite became of a great interest recently due to its potential application in optical data storage media and fast access type memories with read-out efficiency. [1, 2] The efforts of researchers are shed on developing and the improving the storage media capacity and the data storage density for the optical disks. Furthermore, these polymeric systems allow manufacturing of reusable films yielding many thousands of write/read/erase (WRE) cycles. [3] Polymers are used in optical data storage due to their thermal stability and their capability of dissolving. [1, 2, 4-6] Polymers incorporating Azo-dyes chemically or physically have become promising materials for photonic devices due to the reorientation of azobenzene groups through photochemical trans-cis-trans isomerization cycles under linearly and circularly polarized light. [6] Polymerization technique plays crucial role in defining the final physical and optical properties of the films. The way the polymer shrinks together to form ultimately solid film depends on the curing process used for polymerization and time of exposure.[7] The PMMA is the synthetic polymer of methylmethacrylate. This thermoplastic and hard rigid transparent plastic is commonly called acrylic glass. The chemical formula for PMMA is: $[CH_2\text{-}C\ (CH_3)\ CO_2CH_3]n$. The PMMA is often used as an alternative to glass with the difference that, the PMMA is less dense than glass (less than half the density). The PMMA has also higher impact strength, softer and more easily scratched compared to glass. It transmits more light than glass, and does not filter ultraviolet (UV) light. [8] The PMMA is a versatile material and has been used in wide range of fields and applications such as, impact resistant substitute for glass, artistic uses (paints, furniture), producing laserdiscs and DVD's, protective layer in III-V device wafer thinning application. [2, 3, 7, 8] However, the BenzylDimethylKetal (BDK) is used as a photo initiator for the polymerization (photo curing) process through seeding free radical in composite. [1] The main advantage of polymerization started by photo initiators is that; it is temperature independence and easy to be controlled. It can be conducted at very low temperatures and can be stopped simply by removing the light source. [9, 10] The BDK is characterized by its sensitivity to light (changes its colour under light-exposure and temperature). However, it is stable under recommended storage conditions (not to be exposed to light and storage temperature should not exceed 30 ^{o}C. [11] The mechanism involved in polymerization by photo-initiation is that, the BDK undergoes a fragmentation of its structure upon exposure to ultraviolet radiation which initiates the polymerization process. [12] The Azo-dye Methyl-Red (MR) has the name (Azo) comes from azote, the old name of nitrogen. In chemistry, Azo compounds generally have a molecular formula of the form R-N=N-R1, in which R and R1 can be either aromatic or aliphatic. [13] The aromatic azo compounds are usually stable and have different colours, such as red, orange or yellow. Therefore, they are used as dyes which are called azo-dyes. Some azo-dyes, such as methyl red can be used as acid-base indicators, due to their ability to function as weak acids. The colour of methyl red solution depends on the acidity (PH) value. If it is basic, the dye turns yellow. [14] The chemical formula for the methyl red is $C_{15}H_{15}N_3O_3$. For the past few decades, azo-dye compound attracted considerable attention due to their capability of the write/read/erase (WRE) cycles. The basic storage mechanism of azo-dye compound is that the azo-dye molecule exists in two distinct structural forms: (*trans* and *cis*) which leads to the change in the refraction index of the polymer. The material at room temperature is initially in trans state since the *trans*-isomer is thermally stable. After illuminating the material in the *trans* absorption band, light induces photo isomerization transforms from *trans* to *cis* configuration. This change of the molecular structure can be used for memory write and read. In addition, a *cis* form can be returned to a *trans* form by means of thermal relaxation process or illumination. This feature can be used in memory erase. [2] This is shown in more details in figure 1.

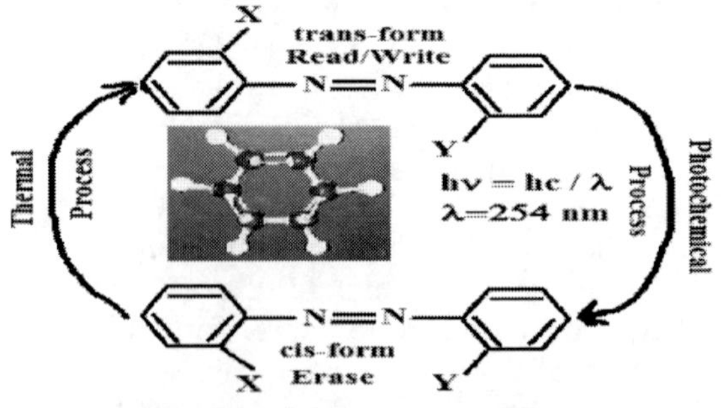

Figure 1. The complete photochemical mechanism involved in the real-time recording and erasing process.

IOP Conf. Series: Materials Science and Engineering **92** (2015) 012024 doi:10.1088/1757-899X/92/1/012024

2. Review and related work

Al-attar and Taqatqa [1] used a solvent composed of an Azo-dye methyl red (MR) doped polymethylmthacrylate (PMMA) polymer with benzyldimthalketal (BDK) as a photo initiator to deposit polymeric-based thin films on substrates which could be used as based layer for optical data-storage media. Upon the exposure to low power UV light (256μW/cm^2) at 254 nm wavelength and using the variable angle spectroscopic ellipsometer (VASE) technique, a large change in the optical constants was found.[1] The related changes reached 90% of their maximum value in five seconds time exposure for the MR/BDK and the MR/BDK/PMMA films. These changes were seen in the refractive index (Δn = - 0.47 at 570nm) and in the absorption coefficient ($\Delta\alpha$ = - 1.4 x 10^{5}cm^{-1} at 470nm) in the MR/BDK film, while the values Δn = 0.03 and $\Delta\alpha$ = - 8 x 10^{4}cm^{-1} were found in the (MR/BDK/PMMA) films. They concluded that these high changes may be attributed to the photo deportation of MR by the radicals formed by the photo decomposition of BDK by UV light. Al-Attar and Telfah [15] used the spectroscopic ellipsometry to study the optical properties for electro chemically prepared thin films of camphor sulfonic acid doped polyaniline/poly(methylmethacrylate) (PAni-CSA/PMMA) blend. The optical constants were modeled using Cauchy or Tauc-Lorentz formulas. They found that the refractive index of the blend system can be adjusted in the wavelength spectral range of (300 – 550nm) by controlling the PAni-CSA percentage contained in the system. The films were transparent and the absorption coefficients vary linearly with the concentration of PAni-CSA ratio. A study by El-Naser and Ali [16] on MR/BDK thin films showed small changes in the refractive index and extinction coefficients before and after exposined to ultraviolet irradiance of 256 μW/cm2 at 254 nm. The MR/BDK/PMMA films were also analyzed and showed relatively large changes in the optical constants. In another study the changes in the optical constants (n and k) due to the effect of environmental parameters (heat and humidity) using the VASE ellipsometer. [17] They found that the optical data storage system (MR/BDK/PMMA) shows a high thermal stability in the temperature range (20 – 70°C) and the film decomposed at temperature above 110°C. The effect of humidity on BDK was less than that on MR. Increasing temperature induces reduction in the optical constants of the BDK. By blending the MR/BDK with the PMMA polymer, the humidity effect on the system was observed to be minimized sufficiently due to the insulating properties of PMMA. Yu JM. *et. al.* [2] studied the changes in refractive index and thickness of spin-coated polymethyl methacrylate (PMMA) thin films upon irradiation by a conventional mercury UV lamp. Significant increase in the refractive index and reduction in the thickness were observed. The thickness reduction of un irradiated PMMA film is consistent with its weight loss which is caused by the escape of the volatile molecules generated during the irradiation process. A slight increase in the refractive index was also observed in the films. Pham *et. al.*[3] characterized azo-dye doped PMMA samples by means of varying different parameters (thickness, concentration of PMMA and azo-dye). Gang *et. al.* [18] employed the technique of all-optical poling to study a system composed of azo-doped PMMA. They proposed a novel method for optical storage based on the mechanism of orientational hole-burning and photoinduced molecular reorientation of azobenzene. They found that the recorded information can be stored for a long time and also can be erased rapidly with a circularly polarized light. The effects of the temperature and the electric field on the electro-optic and photorefractive properties of the PMMA-based polymeric composite have been investigated by Zeng *et. al.*[19] They found that both the electro-optic coefficient and diffraction efficiency increase with increasing temperature to 60 °C. However, the electro-optic coefficient saturates while the diffraction efficiency decreases as the temperature is further increased. Wang *et. al.* [20] reported a new write once optical recording medium (azo-dye in PMMA film). They studied the optical and thermal properties of the azo-dye and its films which are found to have a strong absorption band at 461 nm. The reflectivity was found to be greater than 10% in the visible region which matches well with the wavelength of the argon-ion laser. The reflectivity contrast of the films was found to be more than 30% at writing power (13.3mW) and writing pulse width (600ns) using argon-ion laser operating at 514.5nm. The films were found to have sharp thresholds for optical recording. They concluded that the Azo-dye in the PMMA thin film has a promising potential use for short wavelength optical storage medium. In this work, we investigate the aging factor on the films cured by UV source and faced to either Laser light source or UV light source in order to investigate for how long does it take the films material to be restored or damaged the UV light source exposure. Optical properties for the films have been investigated before light application and during light application and for periods of time after light application. This approach is conducted for the first time. We did not find, up to date, any previous work related to this approach in the literature.

3. Materials and Methods

3.1. Substrates treatment and film preparation

The glass wares and the glass substrates were first cleaned with warm tap – water followed by rinsing with distilled water and finally dipped in acetone. The substrates were bathed for five minutes in ultrasonic path of distilled water and then dried using cold air dryer. One gram of PMMA (Aldrich Co.) with molecular weight of 150000 and 0.3gms of the BDK photoinitiator (Aldrich Co.) and a 0.3gms of Azo-dye MR with PH between 4.2–4.6 (Aldrich Co.) were dissolved separately in 30ml chlorobenzene and the three solutions with the volume fractions of 3 PMMA, 1 MR and 3 BDK values were then filtered using 0.45μm Millipore filter before spin coating on the glass substrates. All films were prepared at room temperature under atmospheric pressure. Thin films were prepared through a volume mixture ratio of 3 PMMA, 1 MR and 3 BDK by using spin coating method on glass substrates. Six to seven drops of the mixed solution were dropped using pasture pipette on top of a glass substrate. The substrate was then rotated by the spinner for 30s at a convenient speed in order to spread the fluid by centrifugal force and to produce almost similar films thickness. The polymerization process was achieved by using ultraviolet source light of 254nm wavelength and ($256\mu W/cm^2$) power intensity. Ten samples were prepared, each at a time, under the same conditions and then cured by the UV source at different periods of time such as: 5, 10, 15, 20, 30, 60, 120, 180, 300 and 540s, respectively. The samples were placed 15cm away from the UV source while curing them and then left to completely polymerize under the normal room air and temperature conditions.

3.2. Ellipsometry characterization

Recent models of ellipsometers embody the use of computerized versions. The variable angle spectroscopic ellipsometer (VASE) manufactured by John A. Woollam Co. Inc. in Lincoln, Nebraska, USA has been used in our measurements and optical analysis approach. The ellipsometer does not measure the optical constants directly; instead, it measures the ellipsometric angles ψ and Δ. From these values, the optical constant were found by solving the inverse problem of ψ and Δ parameter back to the optical constants (ε_1 and ε_2 or n and k) through a proposed mathematical model that gives generated ψ and Δ variables similar to the measured values from the samples.

3.2.1 Measurements

One can use this instrument to measure the beam intensities of reflection and transmission and/or the polarization state (Ellipsometric ψ and Δ as functions of wavelength and angle of incident). Once getting the optical measurements, a model is constructed by which accurate prediction of optical properties can be estimated. This model should contain both known and unknown physical parameters, known parameters such as wavelength of incidence, while the unknown parameters are such as layer thickness and optical constants. Experimental data and generated data, obtained from the mathematical model, are compared to each other by a regression analysis known as fitting process. A normal fit makes similar basic shape and patterns of the experimental ellipsometric ψ and Δ data by varying the model parameters and layers thickness. The optical constants for the assumed mathematical model, are then extracted from the generated ψ and Δ curves that are equivalent to the experimental ψ and Δ curves fitting the optical constants n and k for the model at each incident wavelength. This fit is called; point by point fit which make the experimental and generated values of ψ and Δ values are equal at each point of the experiment. The fit is obtained based on the minimum mean square error (MSE). [21] The angles were experimentally measured based on the ellipsometric equation given by:

$$\rho = tan\ \psi\ e^{-i\Delta} = tan\ \psi\ cos\ \Delta - i\ tan\ \psi\ sin\ \Delta = Rp/Rs \quad (1)$$

where *R*p and *R*s are the complex coefficients of reflectance for light polarized parallel (p) and perpendicular (s) to the plane of incidence, respectively, and ρ is their ratio.[22]

3.2. 2. Structural Model used in ellipsometry characterization

Optical constants of a film vary with both depth and wavelength (optical profile and optical dispersion, respectively). In order to get reliable optical characteristics, an accurate description to the thin film microstructure is required. Our structural model representing the samples is made of glass substrate mathematically modelled with Cauchy dispersions' formula and the film representing the PMMA + BDK + Azo-dye solid cured composite represented by Lorentz oscillator mathematical formula. : [21, 23, 24] An interface layer between the substrate and the film is represented by Bruggeman effective medium approximation (BEMA) mathematical formula containing glass and the film material by certain percentages. [21, 23, 24] The roughness layer was also represented by BEMA formula containing the film material and a percentage of void fractions. The structural model built in the VASE system equipped with the powerful windows based software (WVASE32 ®) is shown if figure 2.

Bruggeman EMA = (PMMA + BDK + Azo-dye) + void % (Roughness layer)
Film = PMMA + BDK + Azo-dye (By Lorentz) (Film material)
Bruggeman EMA = (PMMA + BDK + Azo-dye) + glass % (Interface layer)
Cauchy dispersion Glass substrate modelled

Figure 2. Structural model represented by glass substrate modelled by Cauchy dispersion, interface layer modelled by BEMA which contains glass% and the composite material of the film and the film material represented by and Lorentz oscillator formula.

The complex refractive index is decomposed into real *n* and imaginary *k* parts and is related to complex dielectric function $\varepsilon = \varepsilon_1 + i\varepsilon_2 = N^2 = (n + ik)^2$ where ε_1 is the real part of complex dielectric function and ε_2 is the imaginary part and *N* is the complex index of refraction. [21, 23, 24] The index of refraction n and extinction coefficient k for glass substrate was represented by Cauchy model through a slowly varying functions of wavelength λ, such as: [21, 23, 24]

$$n(\lambda) = A + B/\lambda^2 + C/\lambda^4 \quad (2)$$

$$k(\lambda) = \alpha\ exp[\beta\{\ 1240[(1/\lambda) - (1/\gamma)]\}] \quad (3)$$

where *A*, *B* and *C* are constants, α is the coefficient amplitude, β is the exponent factor, γ is the band edge and all are fitting parameters for *n* and *k*, respectively. (7) The interface thin layer is then modelled in terms of Bruggeman effective medium approximation (BEMA) which contains a mixture of glass and the (PMMA/BDK/Azo-dye) film material.(7) The PMMA/BDK/Azo-dye is modelled by Lorentz oscillator model. In dielectric materials, such the PMMA/BDK/Azo-dye, electrons bonded with nucleus by a force that can be considered as a mass attached to a spring oscillator. [25] The Lorentz oscillator model formulated in the WVASE32 ® software is a multi-oscillator dispersion representing the film layer is described by: [25]

$$\varepsilon\ (E) = \varepsilon_1(\infty) + A/(E_o^2 - (h\upsilon)^2 - iB.\ h\upsilon) \quad (4)$$

Where $\varepsilon\ (E)$ is the dimensionless complex dielectric function that depends on photon energy $E=h\upsilon$, $\varepsilon_1(E)$ is the value of the real part of the dielectric function at very large photon energies. The oscillator is described by three parameters, the amplitude of the oscillator A having the units of $(eV)^2$, B the broadening of the oscillator which has units of (eV) and E_o is the energy location centre of the oscillator measured in units of (eV).

4. Results and Discussion

4.1. Spectroscopic analysis

The VASE software (WVASE32 ®) is enabled to generate ellipsometric ψ and Δ data based on the structural model (figure 2) including the mathematical representation of each layer, sub layer and the substrate. Detailed description for the method of extracting the optical constants for the samples are shown in previous works. [21, 23, 24] All measurements were performed in air at room temperature for three angles of incidence (55°, 65°, 75°) and in the wavelength range of (300 – 1000nm) in steps of 5nm.

4.2. Analysis methodology

The results of this work are presented according to the method they have been performed experimentally and then in the sequence of analysis. First; the set of samples have been prepared and categorized according to the 254 nm wavelength UV curing time periods including the reference sample which was treated with normal air drying (zero time). Second; another set of samples have been analyzed where the aging factor (time response) was the intended to be the varying factor. One sample is selected and exposed to UV light source (λ = 336 nm) another has been exposed to laser light source (λ = 632 nm) for a fixed period of time (during measurement of ψ and Δ around 15min). An experimental data has been performed before light exposure, during light exposure and then a sequence of spectroscopic scans has been performed in order to investigate the effect of the light exposure on the films after the light is turned off. The optical constants are then evaluated as a function of aging time after light exposure is turned off. This aging process is helpful to investigate for how long does it take the film while it still affected by the light exposure.

4.3. Results and analysis

4.3.1 Samples cured with 254nm wavelength UV light source.

All the samples have been exposed to a UV light source of 254 nm wave length for a period of time. Based on the comparison between the experimental and generated ψ and Δ values a regression analysis fit has been carried out to solve for the best film thickness as well as the optical constants (index of refraction and extinction coefficient) at each experimental photon energy or wavelength. All the samples cured with UV source light have been analyzed and the results of the films thickness, interfaces, glass fraction in the interface (EMA) and the surface roughness are presented in table 1 below. The final normal fitting values for Lorentz parameters (equation 4) which gives the best normal fitting results for all the samples are given by: ε1(∞) = 2.25, A = 0.1(eV)2, B = 0.04eV and Eo = 4.15eV.

Table 1. Structural model thickness for the composite film (Lorentz oscillator) and interface (BEMA) as related to the UV curing time.

Curing time (s)	0	5	10	15	20	30	60	120	180	300	540
Interface (nm)	20	7	17	18	3	18	3	19	7	8	0
Film (nm)	159	169	165	163	178	163	180	165	169	170	180
Δn x10^{-2}	0	-3	-2	-1	+3	-7	+11	-9	+5	-3	+5

In order to discuss the optical constants for the films, the samples were divided into three groups based on the curing time: low periods, medium periods and long periods, respectively. The index of refraction (n) is shown in figure 3 with low periods (figure 3 a), medium periods (figure 3 b) and long periods (figure 3 c). It is worth noting that the index of refraction is almost constant in the visible range while deviates in the UV and short visible range due to light-material interactions. A peak shift (towards the UV) in the index of refraction at low UV light exposure (5 seconds) is observed as seen in the figure (3 a). More time of exposure (10 to 15s) restores the index of refraction almost to its original peak. The sample with 20 seconds curing time exhibits a peak shift and an increase toward the visible range rather than figure (3 b). More time exposure (30 seconds) produces dramatic UV peak shifting and peak height too. Sixty seconds of time exposure exhibits the highest peak shift and an increase toward the UV range. 540s of time exposure (figure 3 c) makes almost similar effect as the 20s time exposure. It makes a visible shifting and peak enlargement. The sample cured for 120s time exposure restores the index of refraction with rather smaller peak. The sample with 180s of cuing time shows almost similar effect as the sixty seconds period of time exposure. It makes a UV peak shifting but with smaller peak value. The change in the peak value of the index of refraction (Δn) as a function of UV curing time is shown in table 1. Our results are in consistence with the results obtained by Al-attar and Taqatqa [1] where Δn was found to be 0.03 at 5s curing time. However, the maximum change found in our experiment is + 0.11 at 3 hours of UV light exposure. Similar discussion on the extinction coefficient applies to the absorption coefficient due to their defined relation. The absorption coefficient (α) is shown in figure 4 and calculated form the extinction coefficient (k) based on the formula; $\alpha(\lambda)=4\pi k(\lambda)/\lambda$ and represented in terms of photon energy rather than wavelength. They are divide similarly into low periods (figure 4 a), medium periods (figure 4 b) and long periods (figure 4 c) of curing time, respectively. Since our composite material behaves like semiconductors, the absorption coefficient curves indicates of at what photon energy the absorption takes place dramatically and hence, indicates the value of the band gap. A strong absorption band of the film is located in the range of 2.4-3.8 eV as seen in figure 4 a. The film of 5s UV light curing shows an absorption peak at 3.75 eV (330 nm) and at 2.48 eV (500 nm), this indicates that the polymeric material is highly influenced by the light of such photon energy (wavelength) for READ/WRITE memory response. However, the UV curing process for various periods of time causes a blue shift to the absorption (extension coefficient) peaks. The film processed with 540s UV curing time gives the highest absorption peak at similar energy as the 5s UV curing time (See figure 4 a). This means that the material memory has been restored and even enhanced at 540s curing time period (See figure 4 c). However, the 2.48eV absorption peak obtained with 5s UV curing time has been restored without peak value enhancement with 60s UV curing time as seen in figure 4 b. The change in absorption coefficient ($\Delta\alpha$) at 5s curing time is $-2.0 \times 10^4 cm^{-1}$ which is less than what was found by Al-attar and Taqatqa [1] where $\Delta\alpha = -8.0 \times 10^4 cm^{-1}$. However, the change in α is $+7.4 \times 10^4 cm^{-1}$ for the sample cured at 540s. This result has no comparison in the literature.

4.3.2 Optical band gap transmission

The absorption coefficient dependence on the incident photon energy is described by the well-known Tauc plot relation $(\alpha h\upsilon)^2 = B(h\upsilon - E_g)$, where B is a constant represents the slope of linear part above the band gap (E_g) for the relation.[21, 23, 24] Drawing the relationship between $(\alpha h\upsilon)^2$ vs. ($h\upsilon$) gives a linear relation above the band edge. The optical band gap energy (E_g) is extracted by extrapolating the linear part of the curve to the incident photon energy. This approach has been applied for all the samples and the relationship between E_g and the UV curing time is shown in figure 5. Increasing the UV time of exposure to the films causes a sharp increase in the band gap as long as the exposure time is up to two hours (120min). However, a further increase in the UV curing time decreases the band gap but with slower trend.

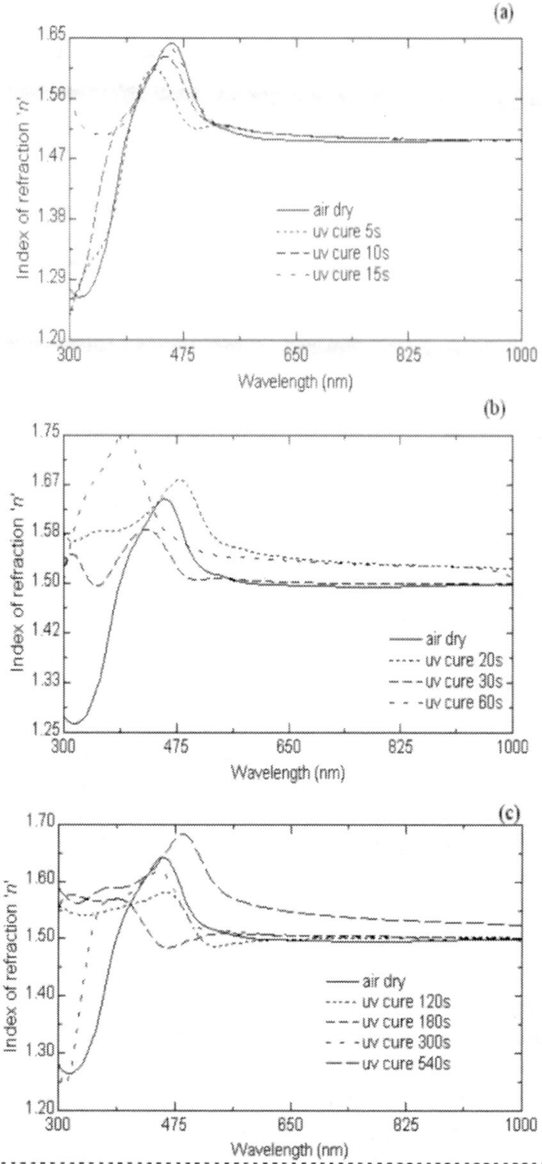

Figure 3. Index of Refraction for samples cured by UV with (a) short periods, (b) medium periods and (c) long periods of time.

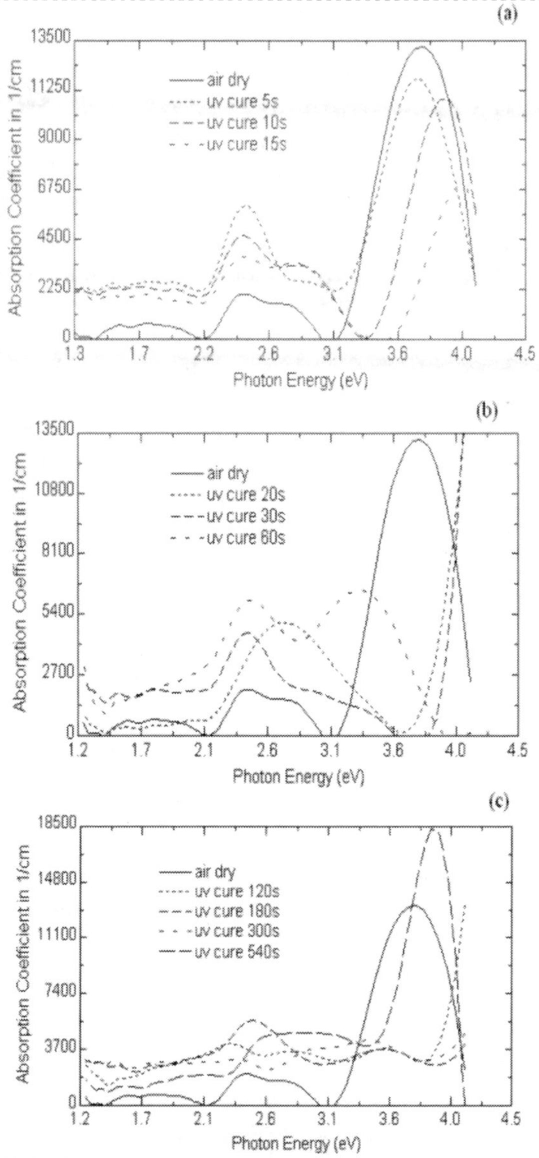

Figure 4. The absorption coefficients (α) for samples cured by UV with (a) short periods, (b) medium periods and (c) long periods of time.

The room temperature Optical transmissions for the films have been taken using the ellipsometer at normal angle of incidence. Figures 6 shows the transmission with respect to incident photon energy for the samples cured at short periods of time and compared with that for glass. The transmission for the samples cured for medium and long periods of time (not shown) were performed and showed almost similar trends as those for the short period of exposure (figure 6). All the films have transmittance over 85%. The dip part of the transmission spectrum is due to the absorption occurs which is not seen for glass curve. The steepness of the transmission at the dip feature is maximum for 5s time exposure while it decreases for higher time exposure which is consistent with the results for E_g obtained from Tauc plot.

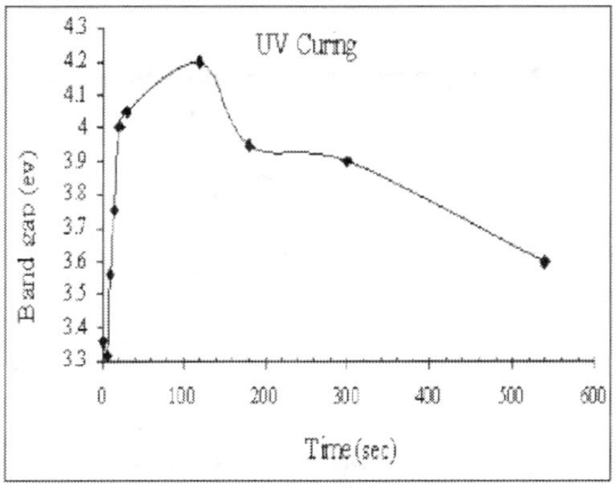

Figure 5. The band gap energy for the films cured for different periods of time.

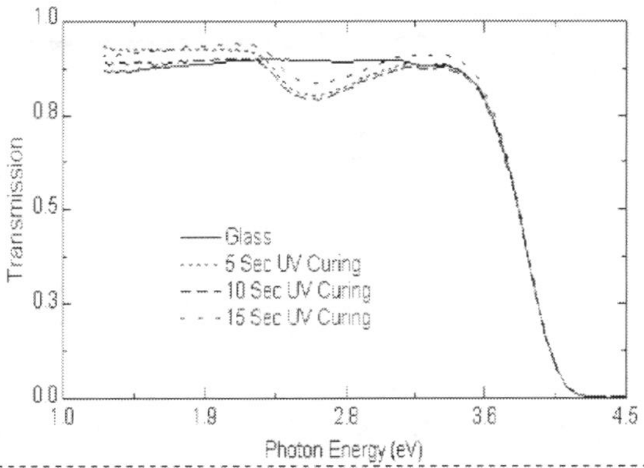

Figure 6. Transmission comparison for samples cured by UV for short period of time.

4.3.3. Film Morphology and structural changes due to aging by laser and UV sources

The microstructural changes are presented and discussed using the Scanning electron microscopy (SEM) images while the chemical changes are presented in Electron dispersive analysis of X-rays (EDAX) analysis. Figure 7 shows the SEM images for the samples as deposited (figure 7 a), cured for 5s (figure 7 b), cured for 20s (figure 7 c) and cured for 540s (9min) (figure 7d), respectively. Small nanoparticle features are shown in figure 7 a, when the film is air dried normally at room temperature. These are expected due to the room light which initiate few defrag mentions in the BDK photo initiator throughout the PMMA/BDK/Azo-dye(BDK) composite in the film. However, as the film is cured for 5s, more nanoparicle sites are observed showing an enhancement of UV light-material interactions with particle size is increased. When the film is cured by UV light for 20s, more particle sites are observed with even larger sizes (see figure 7 c). Long period of UV light curing diffuses the particles features and smoothen the surface of the film. The EDAX analysis shown in figure 8 represents the elemental ratios between carbon and oxygen as a function of curing time. Figure 8 b shows the C/O ratio to be 6.52/1 for the as-deposited film (i.e. air normal air dry curing) with low composite arrangements as seen by the area under the curve. This indicates that the BDK photo initiator is in cis form under room temperature and no light initiation. However, when the film is subjected to the UV light for 5s, the C/O ratio decreases to 5.37/1 which indicates that carbon is released for the oxygen bond under light action. However, the area under the curve is enlarged highly indicating that huge number of BDK defragmentations is taking place in the microstructure. This indicates that the material memory has been transformed form cis-state to tran- state as illustrated in figure 1. This phenomenon is also shown in figure 8 c, where the C/O ratio is decreased to 5.23/1 while cured for 20s with UV light. The memory transition from cis-state to trans- state represents the WRITE/READ action in the film. When the curing time is increased to 540s (9min), the ratio of C/O is returned back almost to original ratio (6.4/1) seen for the film treated with normal air dry curing. Also one can see that the area under the curve of the cumulative composites decrease dramatically to almost vanishes. This phenomenon implies that the memory of the material under goes (returned back) to its cis-state from the trans-state produced with short curing period of time (5s and 20s). This returned memory implies that the memory is ERASED under the action of thermal heat or light.

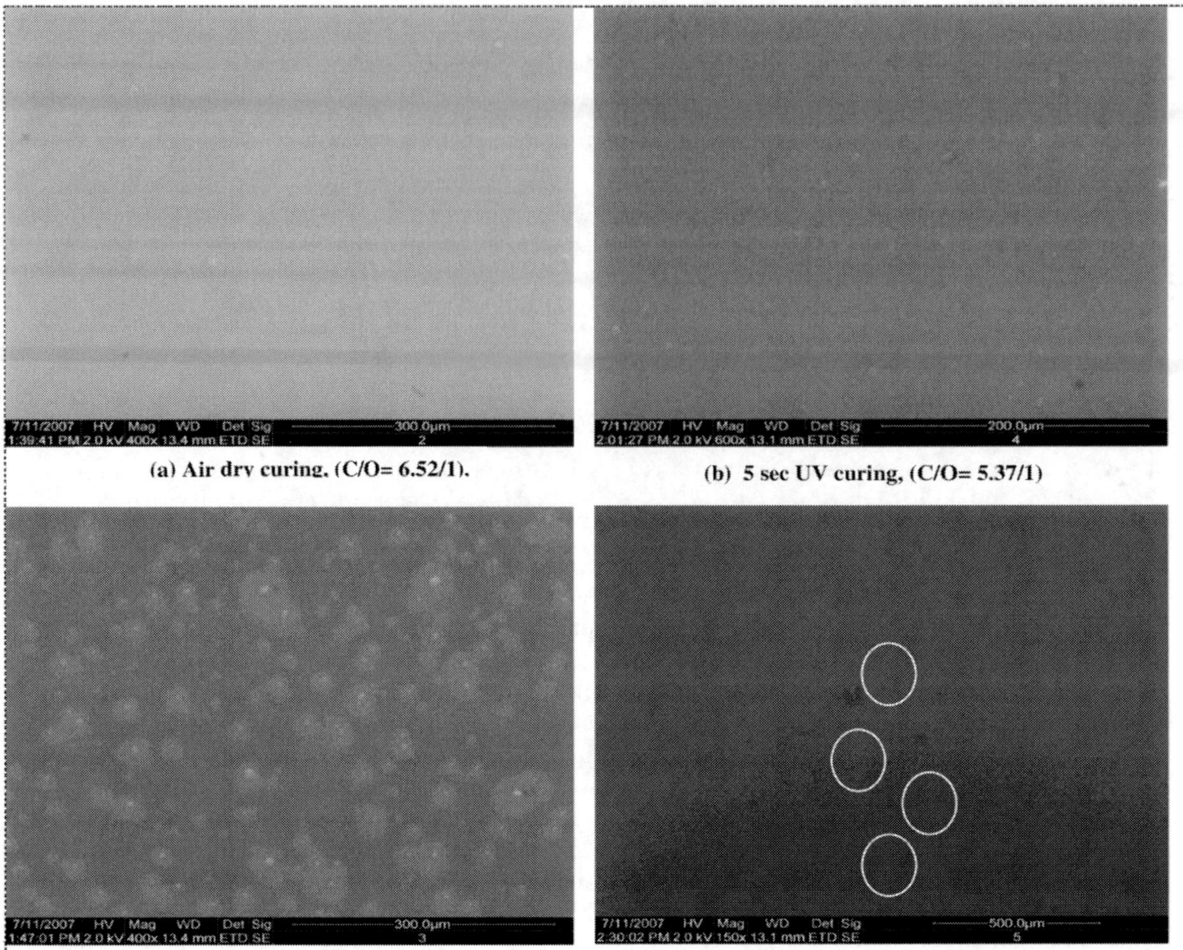

(a) Air dry curing, (C/O= 6.52/1). **(b) 5 sec UV curing, (C/O= 5.37/1)**

(c) 20 sec UV curing, (C/O= 5.23/1) **(d) 9 min UV curing, (C/O= 6.4/1)**

Figure 7. SEM images for samples: as deposited (a), cured for 5s (b), cured for 20 s (c) and cured for 540 s (9min) (d).

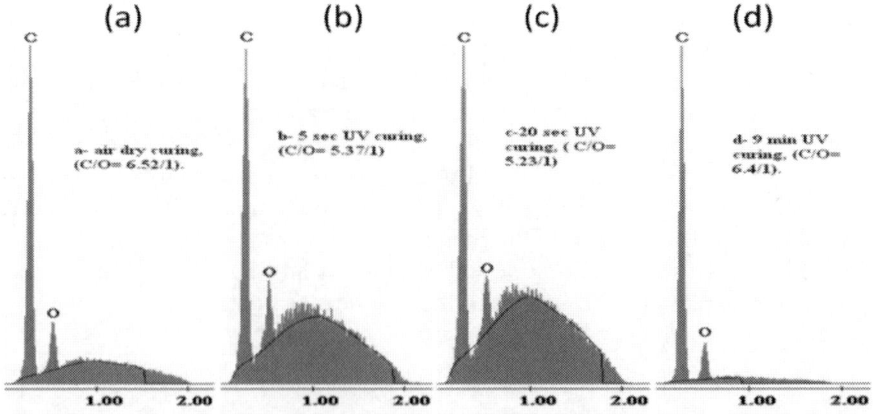

Figure 8. The EDAX analysis for the composites of the PMMA/BDK/Azo-dye solid cured films (a) with air dry, (b) 5s UV, (c) 20s UV and (d) 540s or 9min UV curing periods.

4.4. Analysis for the sample aging by UV and laser

Similar procedure conducted for obtaining the optical properties for the samples cured by UV light is repeated in finding the optical properties for the sample under the aging study by using either laser or UV lights. Our focus is shed towards the effect of light exposure on the absorption coefficient of the sample before, during and after exposure. In this technique, a laser beam of light (He–Ne,632nm wavelength) with 30mW/cm^2 power intensity is applied one-meter distance from the sample perpendicular to the sample surface. The ψ and Δ values were taken before, during and after laser exposure for periods of time (20, 40, 60 and 90min), respectively. Similar measurements and analysis have performed before, during applying 336nm wavelength UV source and after UV light exposure by one or 160min, respectively. The absorption coefficient curves found for the laser treated sample is shown in figure 9, while those obtained by the UV treated sample are shown in figure 10, respectively. The absorption coefficient (α) for the as deposited sample is shown in figure 9 (solid red colour). The absorption coefficient for the film while it is exposed the laser beam is shown in Open Square. It looks that α did not change over all the spectral range, except in the UV region (>3.3eV) where an increase in the absorption is observed. After turning off the laser beam by 20min, the absorption coefficient is restored completely over the whole spectrum, except the peak is slightly increased. However, after 40mn, huge change in the absorption spectrum is observed all over the spectral range. The peak decreases in favour of forming a small glitch of absorption at about 1.4eV while increases sharply above 3.7eV which observed after 40min from switching the laser source off. However, 90min later, the sharp increment in the absorption coefficient vanishes and its value is restored to the value of the as deposited sample (4.1eV). The absorption coefficient for the as deposited film (before exposure) is also seen in solid red colour in figure 10. The absorption coefficient for the sample while it is under UV light exposure as seen in solid line with open squares in figure 10 shows dramatic decrease all over the spectrum, but formed a considerable absorption peak at around 3.85eV which is expected to be due to the defragmentation of the seeds of the photo initiator (BDK) in the solid composite of the film. One minute after the UV source is switched off, the curve for α is decreased further with even the photo initiated peak at 3.85eV (330nm) dies sharply. After 160min, α decreases all over and the photo induced peak vanishes completely in favour of inducing sharp absorption peak at around 1.4eV (886nm). We conclude from this that although the film has not been affected by the laser bean during exposure, UV beam did make a dramatic change in the UV range centred at 3.75eV (330nm).

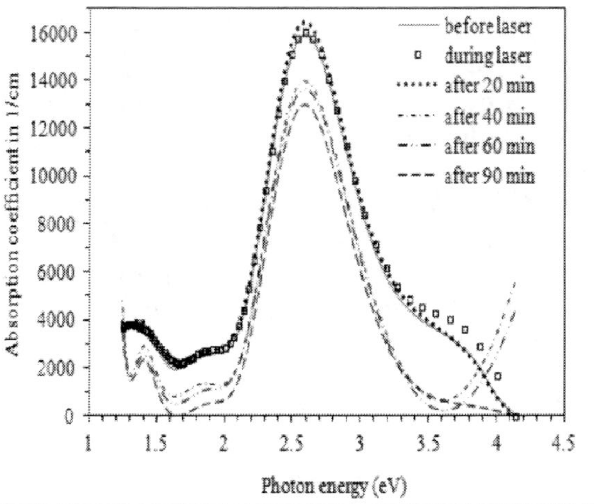

Figure9. Absorption coefficient comparison for laser aging before, during and after exposure..

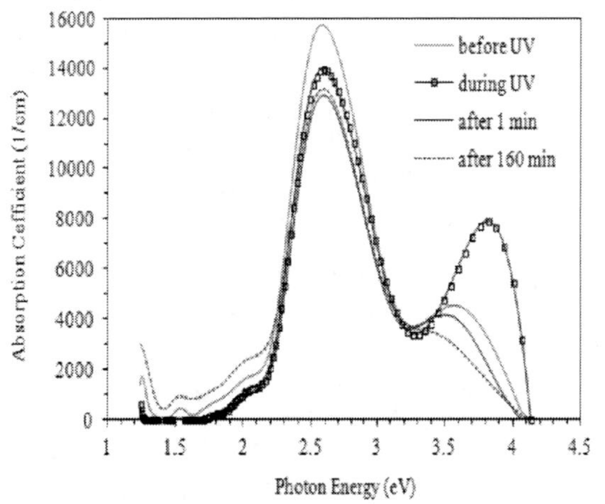

Figure 10. Absorption coefficient comparison for UV aging before, during and after exposure.

5. Conclusion

The optical properties (refractive index, extinction coefficient, absorption coefficient, energy band gap, and transmission) of the azo-dye (methyl red) and the photosensitive benzyldimethylketal (BDK) blend with polymethylmethacrylate (PMMA) have been studied using Variable Angle Spectroscopic ellipsometry (VASE), in the spectral range 300-1000 nm. The Lorentz model is reasonable accurate in representing the (PMMA/ Azo-dye (MR)/ BDK) with one oscillator centered at 4.15 eV.The curing process for polymerization affects the physical and optical properties of the polymer. The thin films material prepared by spin coating method has a strong and a broad absorption band and the transmittance was more than 85% in the visible region. Both azo-dye and BDK materials are known as photo sensitive substances and respond to UV light in two different mechanisms; the first one is photo polymerization using BDK as a photo initiator. The second mechanism is the isomerization process (cis- trans) transition. Changes in the optical properties were obtained in the (PMMA/BDK/MR) film during exposure to UV and Laser which indicates that this media is may be used as an optical data storage media for READ/WRITE using UV light. The characterization data of the films include surface morphology of the films made by scanning electron microscopy (SEM), which shows improvement of nanoparicles initiated by short time of UV curing. The composition analysis for the films performed through the electron dispersive analysis of X-ray (EDAX) shows that they contain mainly carbon and oxygen and the ratio varies with respect to the curing time. This conclusion confirms the cis-trans and trans-cis transformation that occurs due to UV light curing. We conclude that although the film has not been affected by laser beam aging during exposure, the UV beam did make a dramatic change in the UV range centred at 3.75eV (330nm) which makes it so sensitive to such light energy. Changes in the optical properties were obtained in the (PMMA/BDK/MR) film during exposure to UV and Laser which indicates that this media may be used as an optical data storage media or UV and laser light sensors.

Acknowledgments
Authors are wishing to acknowledge the assistant provided by Jordan University of Science and Technology in laboratory works. The authors acknowledge Mr. Ahmad El-Jedi, the Biology Lab. Assistant for his kind assistant.

References

[1] Al Attar H A and Taqatqa O 2003 *Journal of Optics A: Pure and Applied Optics* **5** S487
[2] Yu J, Tao X, Tam H, et al. 2005 *Applied surface science* **252** 1283
[3] Pham V P, Manivannan G and Lessard R A 1995 *Thin solid films* **270** 295
[4] Meng Z, Yang W, Chen P, et al. 2004 *Journal of colloid and interface science* **408** 1
[5] Lee M-J, Jung D-H and Han Y-K 2006 *Molecular Crystals and Liquid Crystals* **444** 41
[6] Jin M, Yang Q X, Lu R, et al. 2003 *Optical Materials* **24** 445
[7] Lindberg A, Dijken J, Horstedt P 2005 Clinical Oral Investigations **9** 77
[8] Murphy E 2007 *Topics in Photographic preservation* **13** 160
[9] Webb R L, Langford S C, Dickinson J T, et al. 1998 *Applied surface science* **127** 815
[10] Meng Z, Yang W, Chen P, et al. 2013 *Journal of colloid and interface science* **408** 1
[11] Israel P D D C W, BAnsch D, BAicker D, Butter C, Chun J, Deisenhofer I, Eckardt L, Geller J C, Hanke T and Klingenheben T *Empfehlungen Firma St. Jude Medical Herzschrittmachertherapie+ Elektrophysiologie* **23** 107
[12] Ginsburg J C 2004 *The Hous. L. Rev.* **41** 263
[13] Duarte F, Maldonado-HÃdar F J, PAorez-Cadenas A F and Madeira L M 2009 *Applied Catalysis B: Environmental* **85** 139
[14] Albert J D and Comiskey B "Curing binder; deformation with mechanical force; suspending, or electrophoretic, fluid; electro-osmotic displays": Google Patents 2000.
[15] Al-Attar H A and Telfah A D 2004 *Optics communications* **229** 263

[16] El-Nasser H M and Ali O D *Iranian Polymer Journal* **19** 57-63
[17] Yesodha S K, Pillai C K S and Tsutsumi N 2004 *Progress in Polymer Science* **29** 45
[18] Xu G, Yang Q G, Si J, et al. 1999 *Optics communications* **159** 88
[19] Zeng F, Wu S, Li F, et al. 2001 *European polymer journal* **37** 459
[20] Wang G, Hou L and Gan F 1999 *physica status solidi A* **174** 269
[21] Ahmad A, Saqaan S, Lahlouh B, et al. 2009 *Physica B: Condensed Matter* **404** 1
[22] Ahmad A A, Alsaad A M, Albiss B A, Al-Akhras M A, El-Nasser H M and Qattan I A 2015 *Physica B: Condensed Matter* **470-471** 21
[23] Ahmad A and Alsaad A 2006 *The European Physical Journal B-Condensed Matter and Complex Systems* **52** 41
[24] Ahmad A A, Alsaad A M, Albiss B A, et al. 2015 *Thin Solid Phsics (submitted)*
[25] Griffiths D J and Reed C *Introduction to electrodynamics*: Prentice Hall Upper Saddle River, NJ, 1999

IOP Conf. Series: Materials Science and Engineering **92** (2015) 012025 doi:10.1088/1757-899X/92/1/012025

Comparison of the inhibitory capacity of two groups of pure natural extract on the crystallization of two types of material compound urinary stones *in vitro* study

Mohamed. Beghalia[1*], Said Ghalem[2], Hocine. Allali[3]

[1]*University Centre of Tissemsilt Algeria*

[2]*Laboratory Environment, Natural Substances Plant and Food Technology (ESNTA), University Center of Relizane Algeria.*

[3]*LASNABIO. Department of Chemistry, Faculty of Sciences, University Aboubekr Belkaïd BP 119, Tlemcen, Algeria.*

[*]e-mail: beghaliamohamed@ymail.com

Abstract: Urolithiasis is defined as the result of an abnormal precipitation within the urinary tract. This precipitation is most often from the normal constituents of the urine. This is a fairly common condition in the population. She is happy and recurrent etiology is often unknown if hypothetical. In Algeria, as in many countries, a large number of patients use herbal medicines in the treatment of their diseases including urolithiasis. Thus the aim of this study is the most widely used to evaluate the effectiveness of aqueous extracts of medicinal plants, in the treatment of calcium urolithiasis oxalo-and magnesium-amoniaco in vitro. The study also examines the effect of these extracts on the states of crystallization (nucleation, crystal growth, crystal aggregation), followed by photography on polarized light microscope.In this regard, we are devoted to studying the crystallization steps from oxalo-calcium and phospho-calcic prepared as artificial urine and supersaturated aqueous solutions, maintained at 37 ° C to remain close to biological conditions. Extracts of the first group of herbs: *Ammodaucus leucotrichus, Ajuga iva, Globularia alypum, Atriplex halimus* are studied on the crystallization calcium oxalate, we cite the Ammodaucus leucotrichus which acts on the stages of nucleation, growth and the aggregation with a total inhibition. The second group of extracts plants tested on calcium phosphate crystallization : *Acacia raddiana, Citrullus colocynthis, Rhus tripartita, Pistacia lentiscu, Warionia saharae,* are able to significantly reduce phosphate crystallization in vitro. It is easily proved by FTIR and optical microscope. In conclusion the results of our work allows us to confirm the use of these plants as an aqueous decoction, in the field of urolithiasis. These activities may help to strengthen the body in depressed situations.

1.Introduction

Nephrolithiasis is a complex disease that results from a combination of various factors related to both urine composition and kidney morphoanatomy. Calcium oxalate is the predominant component of renal stones (70%). Two types of calcium oxalate stones have been distinguished, composed of either calcium oxalate dihydrate (COD) or calcium oxalate monohydrate (COM). COD crystals are thermodynamically unstable and develop only under kinetically favorable conditions, such as a high degree of supersaturation (high calcium concentration >170 mg/L and/or hyperoxaluria), low concentrations of crystallization inhibitors (phytate, citrate), and urodynamically appropriate conditions (e.g., urinary stagnation). Due to thermodynamic instability, COD crystals slowly transform to stable COM crystals mainly in contact with urine [1] .All over the world especially in developing countries, approximately 80% of population continues to use traditional medicine in primary medical problems. In the past decade, therefore, research has been focused on scientific evaluation of traditional drugs of plant origin. There is an urgent need to systematically evaluate the plants used in traditional medicine. Such research could lead to new drug discovery or advance use of indigenous herbal medicines for treatment. This revival of interest in plant derived drugs is mainly due to the current widespread belief that green medicine is safe and more dependable than the costly synthetic drugs many of which have adverse side effects . [2] Urolithiasis is a complex process that occurs from series of several physicochemical event including super-saturation, nucleation, growth, aggregation and retention within the kidneys. Calcium- containing stones, especially calcium oxalate monohydrate, calcium oxalatedihydrate and basic calcium phosphate are the most commonly occurring ones to an extent of75-90% followed by magnesium ammonium phosphate (Struvite) to an extent of 10- 15%, uric acid 3-10% and cystine 0.5- 1%1 Patients suffering from diseases like hyperparathyroid-ism, renal tubular acidosis, cystinuria, hypercalciuria, hyperoxaluria, crohn's disease etc. are more prone to stone formation. [3] The pathogenesis of calcium oxalate stone formation is involved nucleation, crystal growth, crystal aggregation, and crystal retention in multistep process. [4] It is widely accepted that many patients become recurrent stone formers. Therefore, the prevention of recurrence (up to 60%) is crucial. Unfortunately, despite considerable progress noted in the field of medical therapy, there is no satisfactory drug to treat urolithiasis.Calcium oxalate is known in three crystalline forms:

- Calcium oxalate monohydrate (whewellite, $CaC_2O_4.H_2O$)) crystallizes in the monoclinic system.

- Calcium oxalate dihydrate The (weddellite, $CaC_2O_4.2H_2O$) crystallizes in the tetragonal system

- The trihydrate calcium oxalate ($CaC_2O_4.3H_2O$) crystallizes in the triclinic system. This form is exceptional in urine and rarely encounter. Although CaOx is supersaturated in urine, the formation of renal calculi in healthy people is difficult because of all kinds of inhibitors in urine, including citrate, magnesium, osteopontin, and tyrosine hydroxylase [5] . Chemically speaking, CaOx calculi formation is closely related to with the following factors: high concentration of calcium and oxalate in urine, nucleation, growth and aggregation of CaOx crystals, and adhesion of calcium oxalate monohydrate (COM) to renal tubular epithelial cells. [6] CaOx calculi formation is thought to follow two main routes, (slow) formation of subepithelial plaques at papillary tips and (rapid) formation of intratubular plugs [7]. Struvite is the mineralogical name of Ammonium Magnesium Phosphate Hexahydrate [NH4MgPO4 _6H2O] crystals.The general chemical formula for a struvite type compound is X^+Y2^+PO4 _ nH2O, where n=6–8. Potassium Magnesium Phosphate Hexahydrate [KMgPO4 _ 6H2O], also known asstruvite-K, and sodium magnesium phosphate heptahydrate [NaMgPO4 _ 7H2O],or struvite-Na^+, are new struvite type compounds.Struvite Na is the sodium analog to struvite in spite of the excess of water molecule.In struvite-Na ,the Na^+ cations replace the $NH4^+$ cations. Struvite has attracted considerable attention due to its common occurrence in a wide variety of environ- ments.Struvite occurs as one of the components of the urinary calculi in humans as well as in animals [8]. Struvite and carbonate apatite are main components of the so-called infectious urinary stones , which are the result of the activity of microorganisms producing urease, mainly frome proteus species, the aggregation of precipitating particles and bacteria is suspected to be one of the primary causes for urinary stone formation.[9] Urine is normally supersaturated with most stone forming salt components, as well as contains chemicals that prevent or inhibit crystal development in urinary tract. However, the presence of certain molecules raise the level of supersaturation of salts needed to initiate crystal nucleation or reduce the rate of crystal growth or aggregation and prevents stone formation calcium oxalate stones represent up to 80% of analyzed stones.[10]. The techniques consist of exploration, in general, to study the appearance of crystals, their importance and their size in a solution which is added the inhibiting substance. The mode of action of inhibitors also often remains hypothetical, but many of them can act on various stages of crystallization (germination, growth, aggregation). Some act chemically competition, others by changing the ionic strength, others occupant, the surface of the crystal lattice selective sites. It is not necessary to cover the whole of the crystal surface, and some are active not binding on a very small percentage of surface [11]. In Algeria, as in many countries, a large number of patients use medicinal plants to treat their illnesses including urolithiasis. The purpose of this study is to evaluate the efficacy of aqueous extracts of medicinal plants commonly used in Algeria, in the treatment of calcium-oxalate urolithiasis and amoniaco-magnesium in vitro. The study also covers the action of these extracts on the statements of crystallization (nucleation, crystal growth, crystal aggregation) followed by photograph polarized light microscope. In this context, we are devoted to studying the crystallization steps from calcium oxalate-aqueous solutions and phospho-calcic and supersaturated, maintained at 37 ° C to remain close to biological conditions.

2. Materials and Methods

2.1 Synthetic urin*e*

2.1.1. Crystal calcium oxalate

We chose the classical model for the study of oxalate crystallization because of its simplicity and reproducibility.*Anti-calcium oxalate crystallisation activities of Algerian plants* 12 Synthetic urine supersaturated with calcium oxide was prepared according to a previously described method [12] at a constant temperature of 37°C in capped vessels. Chemicals of reagent-grade purity were dissolved in deionized and redistilled water. The artificial urine was prepared immediately before use by mixing in a T-type mixing chamber. For determination of the effects of plant extracts on crystal formation, preparation of the synthetic urine was performed in their presence at various different concentrations.

Simulation of the sedimentary crystal formation

Mixture agitation was maintained to prevent sedimentation. The crystal size development was monitored by polarized microscopy at different time intervals. Sample drops were examined at every five minutes by polarizing optical microscopy. Crystals were identified using a microscope of the Zeiss type with 40 x magnifying lens, equipped with a WINDER M 476079 camera. Percentage of inhibition of crystallisation (I%) was calculated as previously described [13] and based on the formula, I% = [(TSI- TAI) / TSI] x 100, in which TSI and TAI represent number crystals in absence and presence of inhibitors (plant extracts), respectively. Nucleation, growth and aggregation of crystals were visually assessed under the microscope.

2.1.2. Crystal phosphate calcium

Artificial urine is the classical model for the study of phosphate crystallization because of its simplicity and satisfactory reproducibility. This model includes the study of crystallization without inhibitor and with it, in order to assess the inhibiting capacity of any chemical species used. Two solutions of following composition were mixed:
A : 11.02 g/l $Na_2SO_4.10H_2O$, 1.46 g/l $MgSO_4.7H_2O$, 4.64 g/l NH_4Cl, 12.13 g/l KCl et 0.24 g/l Ca^{2+}

B : 2.65 g/l $NaH_2PO_4.2H_2O$, 18.82 g/l $Na_2HPO_4.12H_2O$, 13.05 g/l NaCl, 1 g/l $Na_3C_6H_5O_7.2H_2O$ et 0.05 g/l $C_2O_4^{2-}$. The solution in $C_2O_4^{2-}$ is prepared from oxalic acid 0.05 g.

The precipitation of the solid phase of phosphates from artificial urine at different initial pH value (pH = 8) was the object of our investigation. Artificial urine is prepared by mixing and stirring two equal volumes of 50 ml of solutions A and B at constant temperature (37°C) in capped vessels to give final artificial urine . The pH of solution B was adjusted to required value by adding either HCl or NaOH as appropriate. Mixture agitation was maintained to prevent sedimentation. The crystal size development was monitored by polarized microscopy at different time intervals by proceeding as follows: Sample drops were examined every five minutes by polarising optical microscopy. Crystals were identified with x 40 magnifying lens.

2.2 Preparation of medicinal plants

2.2.1. For calcium oxalate inhibition:

Extracts plants of wild Algerian plants were studied (Table 1). Different parts of the plants were harvested from natural resources in 2006 and 2007, mainly when plants were at flowering stage. Voucher specimens were deposited in the laboratory of Plant Physiology, Oran University [14]. In some cases, a species was sampled at different times. Infusions were prepared daily just before handling by suspending a weighed amount of dry plant material in boiling tap water at room temperature. The suspension was stored at room temperature for 15 min and then filtered through filter paper.

Table 1 a: Plant materials harvested from the wild in west and south of Algeria

Plant species (family)	Part used
mixed aerial parts	*Ajuga iva (Lamiaceae)*
fruits	*Ammodaucus leucotrichus (Apiaceae)*
leaves	*Atriplex halimus (Chenopodiaceae)*
flowers & roots	*Globularia alypum (Globulariaceae)*

2.2.2. For phosphate calcium inhibition:

Table 1 b: Plant materials harvested from the wild in west and south of Algeria

Plan species (family)	Part used
Acacia Radiana	bark
Citrullus Colocynthis	sheet
Pistacia lentiscus	sheet
warionia saharae	sheet
Rhus tripartite	sheet

3. Results

3.1. Experimental details

3.1.1 After 5 minutes

Table 2. Evolution of the size of oxalate calcium crystals in the presence of extracts plants after 5 minutes

	% Inhib [25%]	% Inhib [50%]	% Inhib [75%]	% Inhib [100%]
Globularia (*bark*)	21.20	76.57	89.00	93.58
Globularia (Sheet)	81.41	89.52	89.79	93.71
Ammodaucus	80.20	88.21	92.27	95.41
Atriplex halimus	56.80	66.49	83.76	92.80
Ajuga iva	84.03	89.92	92.80	89.79

3.1.2 After 30 minutes

Table 3. Evolution of the size of oxalate calcium crystals in the presence of extracts plants after 30 minutes

	% Inhib [25%]	% Inhib [50%]	% Inhib [75%]	% Inhib [100%]
Globularia (*bark*)	75.53	90.45	89.73	96.09
Globularia (Sheet)	88.78	95.58	96.77	95.22
Ammodaucus	94.98	96.87	97.25	97.85
Atriplex halimus	63.12	67.18	91.64	89.26
Ajuga iva	89.97	95.22	97.01	96.06

3.2. Inhibition of calcium phosphate :

3.2.1 After 5 minutes

Table 4. Evolution of the size of struvite crystals in the presence of extracts plants after 5 minutes

Temps	**Evolution of the size of struvite crystals and I% inhibition**										**Plants**
	SI	1ml	I%	5ml	I%	10ml	I%	15ml	I%	20ml	I%	
5	30	16	46.7	16	46.7	28	6.7	28	6.7	16	46.7	*Acacia Radiana*
5	30	16	46.7	12	60	16	46.9	16	46.7	12	60	Citrullus Colocynt
5	30	24	20	16	46.7	16	46.7	24	20	16	46.7	Pistacia lentiscus
5	30	20	33.4	20	33.4	G	-	20	33.4	20	33.4	warionia saharae
5	30	24	20	20	33.4	28	6.7	24	20	32		Rhus tripartita

3.2.2. After 30 minutes:

Table 5. Evolution of the size of struvite crystals in the presence of extracts plants after 30 minutes

Temps	Evolution of the size of struvite crystals and I% inhibition											Plants
	SI	1ml	I%	5ml	I%	10ml	I%	15ml	I%	20ml	I%	
30	45	28	37.8	24	46.7	40	11	32	28.9	12	73	Acacia Radiana
30	45	32	28.9	28	37.8	40	11	32	28.9	32	28.9	Citrullus Colocynthis
30	45	28	37.8	20	55.5	20	55.5	28	33.8	20	55.5	Pistacialentis cus
30	45	28	37.8	20	55.5	28	37.8	20	55.5	32	28.9	Warionia saharae
30	45	28	37.8	24	46.7	32	28.9	20	55.5	20	55.5	Rhus tripartita

4. Discussion

The work provides an update on therapy approaches needed depending on the damage type of the urinary tract and the diversity of medicinal plants answered in nature. Many of these plants are active human specialties. Most plant extracts used in our work has proved effective in tests on calcium oxalate inhibition majority component in urinary stones. In this study extracts: *Ammodaucus leucotrichus, Ajuga iva, Atriplex halimus, Globularia alypum* (roots), *Globularia alypum* (flowers), were presented to inhibit the formation of crystals of calcium oxalate monohydrate (Table 2 and 3). From extracts of plants tested, demonstrated potent inhibition higher and therefore the greatest potential Antilithic 97.85 % after 30 minutes *Ammodaucus leucotrichus* is known in Algeria as "Kammun Sofi-es" or "Hairy Cumin" also showed strong inhibition calcium oxalate crystallization. The Marrubium deserti and Ammodaucus leucotrichus L. an Algerian endemic species, has several applications in traditional medicine for example as a remedy for asthma and diabetes, and was found to have antibacterial properties. In this work, an antioxidant and antimicrobial activities was performed on phenolic extracts of Marrubium deserti, Ammodaucus leucotrichus plants.[15]. Calcium interferes with the crystallization of struvite by blocking active growth sites and competiting for phosphate to form calcium phosphate. As in the case of urinary calculi associated with metabolic disorders, the formation of infectious urolithiasis is a result of many physical and chemical processes. Environment of urine is just as important as bacterial factors in the development of the urolithiasis development, which emphasizes the importance of proper diet in the therapy. This diet should not only involve the reduction of Ca2+, Mg2+ and phosphate. Previously, it was shown that the intensity of struvite crystallization is positively correlated with an increase in the pH level.[16]. Behind this second experiment the idea was to know the role of plant extract in dissolving the already formed stones nucleus in renal system. For this artificial calcium oxalate crystal were prepare in the laboratory by standard method.[17] The first step in the formation of urinary stones is the nucleation of urinary minerals from supersaturated urine. The formed nucleus (generally less than 10 nm) grows and/or aggregates to a pathological size (several tens of microns). These crystallites are then retained in the urinary tract or fixed by urinary tract organization, forming urinary stones (millimeters to several centimeters) [18]. *Ammodaucus leucotrichus* was found to potently inhibit nucleation, growth and aggregation.The results for the inhibitory effect of extracts of plants on calcium phosphate crystallization as given in Tables 4 and 5 are identical to the work Chauhan and all have shown that the decrease in size of struvite crystals in the presence of an extract diffusa plant Boerhaavia 13.33 and 30.37% inhibition for 05 and 1% concentration of extract [19]. Our results also with all the plants we studied expresses this inhibition on the sizes of the crystals, for example: the effect of the bark of the *acacia radiana* 60% from its value without inhibitor ,. *Pistacia lentiscu* a decrease of 46% from its value without inhibitor. Kidney stones are hard, solid particles that form in the urinary tract. In many cases, the stones are very small and can pass out of the body without any problems. However, if a stone (even a small one) blocks the flow of urine, excruciating pain may result, and prompt medical treatment may be needed.. [20]. Medicinal plants have played as significant role in various ancient traditional system of medication. Even today, plants provide a cheap source of drugs for majority of world"s population. Several pharmacological investigations on the medicinal plants used in traditional antiurolithiatic therapy have revealed their therapeutic potential in the *in-vitro or in-vivo* models.[21].The same procedure was followed as an example: the Rustipentila plant, has an inhibitory effect only on the size of the crystals, and aggregates. The addition of 10 mL of plant volume provides an inhibition of 28.9% compared to its value without inhibitor. Moreover the minimum crystal size after 4 hours at 15 mL reaches 20 microns or a percentage inhibition of 55.5%. The same parameters (size and number of crystals, the crystallization time) were also studied. Medicinal plants are able to significantly reduce the phosphate crystallization in vitro, can be clearly seen by optical microscope that aqueous extracts of: *Acacia raddiana*, *Citrullus colocynthis*, *Rhus tripartita*, *Pistacia lentiscu*, *Warionia saharae*, produce a significant amount of growth inhibition phosphate crystals. This in vitro study provides useful information for the in vivo studies.

Conclusion

The result of our work does not allow us to confirm the use of these plants in the field of urolithiasis. However, all our extracts gave an inhibitory activity with the aqueous decoction .These activities may help to strengthen the body in depressed situations. In addition, a detailed study of the toxicity would be necessary to achieve determine doses. The plant extracts are generally used in the rough. This is why we utilize high doses for our tests. To overcome this, it is best to isolate the active principles of different plant extracts and present them in an acceptable dosage form. Awareness is still needed on the appropriate use of traditional medicine and the importance of the environment. Through this work, we hope to bring our modest contribution to the promotion of traditional medicine to reach place at the disposal of the population based drugs effective medicinal plants and accessible

References

[1] C. Conti, L. Brambilla, C. Colombo et al., "Stability and transformation mechanism of weddellite nanocrystals studied by X-ray diffraction and infrared spectroscopy," Physical Chemistry Chemical Physics, vol. 12, no. 43, pp. 14560–14566, 2010.
[2] Jain M, Bhandari A, Bhandari Ankansha, Patel P. Isolation, characterization and in-vitro Antiurolithiatic activity of cerpegin alkaloid from Ceropegia bulbosa var.lushii root. Int J Drug Dev & Res 2012; 4(4): 154-160.
[3] Atodariya U, Barad R, Upadhyay S and Upadhyay U. Anti-Urolithiatic Activity of Dolichos Biflorus Seeds. Journal of Pharmacognosy and Phytochemistry. 2013;2(2):209-213.
[4] Kanu Priya Aggarwal, Shifa Narula, Monica Kakkar, and Chanderdeep Tandon. Nephrolithiasis: molecular mechanism of renal stone formation and the critical role played by modulators. BioMed Research International 2013; Article ID 292953, 21 pages, 2013. doi:10.1155/2013/2929

[5] C.-C. Liu, S.-P. Huang, L.-Y. Tsai et al., "The impact of osteopontin promoter polymorphisms on the risk of calcium urolithiasis," Clinica Chimica Acta, vol. 411, no. 9-10, pp. 739–743, 2010..
[6] H. Peng, J.-M. Ouyang, X.-Q. Yao, and R.-E. Yang, "Interaction between sub-micron COD crystals and African green monkey renal epithelial cell," The International Journal of Nanomedicine, vol. 7, no. 8, pp. 4727–4737, 2012.
[7] J. C. Williams and J. A. McAter, "Retention and growth of urinary stones: insights from imaging,"Journal of Nephrology, vol. 26, pp. 25–31, 2013
[8] C.K. Chauhan,M.J.Joshi,In vitro crystallization,characterization and growth- inhibition study of urinary type struvite crystals, J.Cryst.Growth 362(2013) 330–337.
[9] Jolanta Prywer , Rafal R. Sadowski , and Agnieszka Torzewska ,Aggregation of Struvite, Carbonate Apatite, and Proteus mirabilis as a Key Factor of Infectious Urinary Stone Formation Cryst. Growth Des., 2015, 15 (3), pp 1446–1451
[10] Singh SK. Disease Monograph-5: Urolithiasis; Central Council for Research in Homoeopathy Publication, Deluex Printery, 2011; 4
[11] Trolliet P., François B., Gilly R., Bondon P. Identification d'une lithiase urinaire par l'analyse des calculs en spectrophotométrie infrarouge. Press. Med., 1986, 15:20.
[12] Costa-Bauzá A, Perelló J, Isern B, Sanchis P, Grases F. Factors affecting calcium oxalate dihydrate fragmented calculi regrowth. BMC Urol. 2006; 6:16.
[13] Hennequin C, Lalanne V, Daudon M, Lacour B, Drueke T. A new approach to studying inhibitors of calcium oxalatecrystal growth. Urol Res 1993; 21(2):101-108.
[14] Beghalia M1,2, Ghalem S1, Allali H1, Belouatek A2 and Marouf A3 Malaysian Journal of Biochemistry and Molecular Biology (2008) 16(1), 11-15 11
[15] Sifi Ibrahim, Benaddou Fatima Zohra, Yousfi , Mohamed Antioxidant and Antimicrobial Activities of Phenolic Extracts of Endemic Plants Marrubium deserti and Ammodaucus leucotrichus from Algeria,World Academy of Science, Engineering and Technology Bioengineering and Life Sciences Vol:2, No:1, 2015 International Science Index Vol:2, No:1, 2015 waset.org/abstracts/12661
[16] Ariyanto E, Sen TK, Ang HM (2014) The influence of various physico-chemical process parameters on kinetics and growth mechanism of struvite crystallisation. Adv Powder Tech 25: 682–694.
[17] Satish H, Raman D, Kashama D, Shivananda BG, Shridhar KA. In vitro anti-lithatic activity study of Tribulus terrestris fruits and Boerhaavia diffusa roots. Scholars Research library Der Pharmacia Lettre 2010; 2(3):12-20.
[18] J.-M. Ouyang, Z.-Y. Xia, G.-N. Zhang, and H.-Q. Chen, "Nanocrystallites in urine and their relationship with the formation of kidney stones," Reviews in Inorganic Chemistry, vol. 32, no. 2–4, pp. 101–110, 2012.
[19] C.K. Chauhan, M.J. Joshi and A.D.B. Vaidya. Growth Inhibition of Struvite Crystals in the Presence of Herbal Extract *Boerhaavia diffusa* Linn . American Journal of Infectious Diseases 5 (3): 177-186, 2009
[20] Kumkum Agarwal, Ranjana Varma. Ocimum gratissimum L.: A medicinal plant with promising antiurolithiatic activity. Int J Pharm Sci Drug Res, 2014; 6(1): 78-81
[21] Chandaka Madhu L, Harish K, Narsimha Reddy S, Sagar D, Soumya G, Pavan Kumar. Antiurolithic activity of aqueous extract on roots and seeds of Plectranthus tomentora on ethylene glycol induced kidney stones in male albino rats. Asian J Res Pharm Sci, 2012; 2(4): 129-133.

Institute of Physics
Dirac House, Temple Back
Bristol BS1 6BE UK

ISSN: 1757-8981
ISBN 978-1-5108-1463-9

9 781510 814639

CURRAN ASSOCIATES INC.
proceedings.com